Atomic Layer Deposition for Semiconductors

Cheol Seong Hwang
Editor

Atomic Layer Deposition for Semiconductors

Springer

Editor
Cheol Seong Hwang
Department of Materials Science
and Engineering and Inter-university
Semiconductor Research Center
Seoul National University
Seoul
Korea

ISBN 978-1-4614-8053-2 ISBN 978-1-4614-8054-9 (eBook)
DOI 10.1007/978-1-4614-8054-9
Springer New York Heidelberg Dordrecht London

Library of Congress Control Number: 2013951501

Printed on acid-free paper

Springer is part of Springer Science+Business Media (www.springer.com)

Foreword

Atomic Layer Deposition (ALD) is not a new technique, since the first experiments were carried out over 40 years ago. Industrial use began in 1983 with the production of thin film electroluminescent displays. In these displays, the dielectric-luminescent layer-dielectric thin films stack is made in a continuous process. Interestingly, this stack is 1–1.5 micron thick Other industrial applications of ALD remained limited for 20 years, but during the past 10–15 years microelectronics has been the major driver for ALD technology.

In the late 1990s, it became obvious that the continuation of Moore's law would require introduction of new materials into microelectronics. Additionally, new deposition methods were needed in IC technology since materials had to be deposited with atomic level accuracy as very thin films uniformly over the increasing wafer sizes and conformally over the increasingly demanding three-dimensional (3D) device structures. New interest was paid to ALD and enormous research activities went into the development of processes to manufacture high-*k* dielectric materials, metals, and materials for barrier layers.

High-*k* dielectric materials for both gate oxides in metal–oxide–semiconductor field effect transistors (MOSFET) and capacitor dielectrics in Dynamic Random Access Memories (DRAM) have been the most important topics in ALD research during last 10–15 years. Especially, DRAMs require complex 3D capacitor structures and no other thin film technology than ALD can be employed in their production. ALD has been used in these high-*k* applications already for some time. Furthermore, at the moment the dielectrics used in DRAMs are ZrO_2–Al_2O_3–ZrO_2 nanolaminates the preparation of which on a 3D structure is even more challenging. The DRAMs contain electrodes which also need to be conformal. ALD TiN is the material of choice for this application.

About 6 years ago, it was announced that the first microprocessors using MOSFETs containing high-*k* dielectrics were coming to the market. These revolutionary new devices contained ALD-made hafnium-based oxides as dielectrics and metals as gate electrodes. The current MOSFETs are still planar devices and as such not so much dependent on ALD as the 3D DRAMs. On the other hand, the replacement gate approach and the future 3D transistor structures will also make MOSFET processing a 3D task and thereby involve ALD.

A third semiconductor application area where ALD has been extensively studied is interconnects. Apparently, the first real application was tungsten seed layers for tungsten chemical vapor deposition (CVD) contact plug fill using a process based on reduction of WF_6 with either diborane or silanes. For copper interconnects ALD has been explored for years but its implementation has been delayed. Copper appears to be a challenging material to find good ALD chemistry and the problem has not been solved yet. Development of ALD processes for platinum group metals has been successful and they may be employed in semiconductor applications as electrodes or barriers. There are also many other application areas in microelectronics for which ALD have been explored and where it could be used. These include non-volatile memories such as ferroelectric RAM (FeRAM), flash, and phase change memories.

This book *Atomic Layer Deposition for Modern Semiconductor Devices* focuses on its topic ALD in semiconductor industry, but presents also fundamentals of ALD technology and ALD reactors. This book consists of chapters written by prominent scientists and engineers in the field. It is timely since there are enough research and results on the use of ALD in microelectronics for reviewing. This book nicely covers the existing application areas of ALD in microelectronics and presents the emerging areas such as phase change memories. This book is useful for scientists, students, and engineers to achieve an overview of the present situation, as well as the future and challenges in ALD for microelectronic materials and device fabrication.

Helsinki, Finland, April 30, 2012 Markku Leskelä

Preface

How will our world and society look in 100 years? This may be a question that no one can answer clearly. Perhaps, it is even difficult to say how our world and society will look in 10 years. There can be, however, many plausible answers to the question about our future even at this moment. “It will be a world that uses better information technology than now” without a doubt. The basic hardware for the information technology will be certainly take the form of a computer, no matter which level they are at; from supercomputer to small hand-held electronic appliances. Better computers, however, do not simply mean calculation machines with higher speed processors and higher density memories. They must be environmentally benign and sustainable. In addition, the main stream semiconductor technology based on Si-chips will most likely not be radically changed within 10–20 years although there are several potential technology contenders such as carbon-based electronics. On the other hand, it is forecast that pursuit of the present technology trend in computer and semiconductor chips will necessitate too much energy consumption in 10 years. Therefore, computers with better performance but low energy consumption are necessary. Lighter weight and longer battery life time are the key market requirements for portable appliances. These impose a significant challenge to all scientists, engineers, and designers in the information technology field and semiconductor industry.

Two basic strategies could be adopted to accomplish such a demanding goal; one is following Moore’s law, and the other is surpassing it (or pursuing goals beyond-Moore’s law). The enormous improvement in computing technology over the last ~50 years has been accomplished according to Moore’s law. This has been achieved by appropriate collaborations among and competitions between people working in the fields of design technology (such as recent multi-core architecture of microprocessors), material innovations, and chip fabrication technology. As the dimensions of modern semiconductor chips have shrunk to ~20 nm, further miniaturization became extremely challenging due to the complexities of related processes and resultant costs. Therefore, alternatives to blind miniaturization are indispensable; those would include new functional materials, new device structures, and new thin film processes including deposition, etching, and cleaning.

There have been innovative approaches differing from the conventional computing technologies based on the binary digits which have been so successfully implemented on complementary metal oxide semiconductor field effect transistors.

The so-called quantum computing is one of these new directions although it faces critical challenges due to the uncertainty principle. These approaches are usually called more-Moore or beyond-Moore technology. No matter which research trend becomes the main stream in 10–20 years in the field of microelectronics or nanoelectronics, new functional materials processed by an innovative process technology for new innovative devices must be pursued.

Confining the scope of discussion to the field of semiconductor processing, Atomic Layer Deposition (ALD) naturally comes to our attention due to its unique properties in fabricating various nanometer scale thin films. ALD was originally designed for depositing uniform passivation layers over a very large area specifically for display devices in late 1970s and early 1980s when other thin film deposition techniques could not offer this capability. However, it has not been highlighted until the end of the twentieth century; it only became very popular at the dawn of the twenty-first century. This is probably due to the adoption of this technique in highly integrated semiconductor memory devices. ALD started being applied to Dynamic Random Access Memories (DRAM) early in this century, while earlier DRAMs were mostly constructed by chemical vapor deposition. Nowadays, many other types of memory chips also employ ALD processes for various purposes. Another milestone in microelectronics in ALD history was the mass production of microprocessors which was initiated with the production of Penryn in 2007 by Intel, where the replacement gate process with high-*k*/metal gate stack was adopted. Since then ALD has evolved into a standard process in semiconductor industry.

According to these new trends, a book entitled "Atomic Layer Deposition of Nanostructured Materials" edited by N. Pinna and M. Knez was published by VCH-Wiley in 2011. The editor of this monograph contributed a chapter, entitled "Atomic Layer Deposition for Microelectronic Applications" to that book. The book was intended to cover the general aspects of ALD and contained almost all possible applications in nano-research related fields as the title implies. This is a timely and useful guide book for ALD, but may not be detailed enough for the microelectronics or nanoelectronics field despite the huge potential importance of this technique. Inspired by this idea, an initial suggestion on this monograph was made by a senior editor of the publisher Springer, Dr. Kenneth Howell, in the middle of 2010 when I was serving as the general chair of 10th international conference on the ALD held in Seoul, which was a special conference series of American Vacuum Society. After several discussions with colleagues in the field, this monograph took shape. I appreciate all the contributions to this great work, thank all the contributors, and hope that it can be a useful guideline for all engineers and designers in this field as well as for all application-oriented physicists and chemists.

kwanak-ku, seoul Cheol Seong Hwang

Contents

Part I
Introduction

Chapter 1
Introduction

Cheol Seong Hwang and Cha Young Yoo

The enormous improvement in information technology, accomplished mostly by the better computers, has been one of the most important impetuses that have driven the development of modern civilization over the last ~50 years. Better computers are based on better semiconductor chips for logic processors and solid state memories, such as dynamic random access memory (DRAM), as well as disk-type storage devices. The paradigm for the chip fabrication can be well represented by the famous Moore's law, which has been accomplished through strict scaling technology in both memory and logic chips. Although the scaling has been achieved mainly by the improved photolithography and dry etching processes, advanced thin film growth processes can also greatly contribute to the development of the devices.

As the scaling of semiconductor devices proceeds, the demands for a thin film deposition with low thermal budget, higher accuracy in thickness control, and better conformality over three-dimensional (3D) structures are increased. This is mainly due to the highly scaled metal–oxide–semiconductor field effect transistors (MOSFET), high-k gate dielectric/metal gate technology, and capacitors for DRAMs. The extremely small gate length (<20 nm) of most advanced MOSFETs generally leads to the instability in the threshold voltage control due to various reasons (called short channel effect). Among the several methods to overcome this issue, adopting the low thermal budget for chip fabrication is one of the options which generally results in the better diffusion profile control and shallow junction formation. The atomic layer deposition (ALD) often adopts a lower process temperature compared to the chemical vapor deposition (CVD), making it a more desirable method for the low thermal budget. Another key technology for

C. S. Hwang (✉)
Department of Materials Science and Engineering and Inter-university Semiconductor Research Center, Seoul National University, Seoul 151-744, Korea
e-mail: cheolsh@snu.ac.kr

C. Y. Yoo
Semiconductor R&D center, Samsung Electronics Co. Ltd, Yongin, Korea

C. S. Hwang (ed.), *Atomic Layer Deposition for Semiconductors*,
DOI: 10.1007/978-1-4614-8054-9_1,

addressing the issues related with the MOSFET is the high-k gate dielectric/metal gate, which greatly decreases the gate leakage current as well as the threshold voltage instability. The thickness level of high-k gate dielectrics in modern logic chips is only 2–3 nm, meaning that the thickness variation by 0.2–0.3 nm already corresponds to 10 % of the total thickness. It must be noted that 0.2–0.3 nm corresponds to only one mono-layer thickness of most oxides, indicating that self-limiting or self-terminating thin film deposition techniques should be used to fulfill such a tough requirement. In addition, the extremely shrunk dimensions of modern semiconductor devices naturally require physically very thin films. A typical example is the stack capacitor of DRAMs, which requires at least three layers (two capacitor dielectric and one top electrode layers) between two storage nodes meaning that the thickness of each layer must be <7 nm for devices with the 20 nm design rule (more precisely 40 nm pitch device). Regardless of the electrical performances, the materials cannot be placed into the narrow gap if they are thicker than $\sim$7 nm. This consequently imposes significant challenges to many materials. For example, the effective dielectric constant of a very high-k material (e.g., (Ba,Sr)TiO_3) degrades very rapidly with decreasing film thickness [1], which could make the adoption of such a process-wise demanding material less advantageous considering the less severe thickness-dependent properties of the relatively lower (bulk) k material, such as TiO_2 [2–4]. Such intrinsic behavior of thin dielectric film is almost independent of the film growth and the interface controlling methods. This means that the material choice for such extremely scaled devices cannot be determined solely based on the intrinsic or bulk properties. Furthermore, the extreme 3D geometry also requires a thin film deposition process with better step coverage than CVD. The very thin films (<10 nm) usually have different structural and electrical properties compared with bulk materials, particularly for the case with the vapor-phase grown materials. This is a natural consequence of the growth process considering that the very thin layers are deposited on a certain type of substrate which is usually different from the depositing material itself. Once the substrate is coated with the growing materials, the film growth enters into the steady state and more bulk-like films can be obtained. For many cases, the initial growth on the heterogeneous substrates is retarded by the inherent nucleation barriers which usually render an island-like growth and non-uniform film. These factors severely affect the material properties of the very thin films. This is, however, one type of extrinsic effects that could be improved by an appropriate growth control process.

Therefore, ALD appears to be a very timely and indispensable process for microelectronics, even though it was originally expected to achieve film deposition with uniform thickness ($\sim$ μm) over a very large area for other applications [5]. In addition to its unprecedented conformality over the 3D structures when precursor molecules are well saturated on the surface, ALD is suitable for the extremely thin film growth due to the following two reasons. The first is its self-saturating or self-regulating nature, making the thickness control at atomic level feasible. Secondly, in principle, once the substrate surface is covered with the first mono-layer of the growing film, the growth should be in the steady-state, suggesting that the film recovers the bulk-like properties from thicknesses as thin as $< \sim 1$ nm. However, these two are valid only for several exceptional cases, and in most cases they deviate from these ideal behaviors, making all the engineering efforts in ALD field indispensable.

ALD usually suffers from the low growth rate (thickness increase per process time) except for several exceptional cases with an extremely high growth rates [6]. Typical ALD materials, such as HfO_2, benefit from an improved growth rate by the recent development of novel precursors based on the heteroleptic ligand structures [7]. However, it is noteworthy that the generally low growth rate becomes less problematic in microelectronics because the required film thickness decreases gradually as mentioned above. This subsequently makes the adoption of ALD into the future microelectronics and nanoelectronics highly desirable.

During the past $\sim$10 years, these general features of ALD progressively became known to the thin film and microelectronics society and also attracted much of their interest. Although the ALD industry is not fully matured yet, it has already been applied to sizable material and deposition tool industries as well as huge semiconductor chip fabrication lines. Due to the layer-by-layer growth mechanism of ALD, it fits very well to the scientific interests on the surface reaction so that many surface chemists and material scientists have devoted a considerable amount of effort to this field. ALD actually serves as a viable methodology to understand the gas–solid interactions in somewhat complicated material systems. Several in situ, such as Fourier transformed infrared spectroscopy or spectroscopic ellipsometry, and ex-situ, such as X-ray photoelectron spectroscopy, and surface sensitive investigation techniques have been employed.

However, a considerable gap in understanding and application of ALD appeared between surface scientists and semiconductor process/device engineers. This monograph is hence intended to serve as a link between the two groups. The diverse structures and operational principles of semiconductor devices might be difficult to understand for surface scientists, while the engineers and designers in the solid-state device field may have problems in understanding the chemistry-oriented ALD reactions and materials properties. This monograph, therefore, is composed of chapters describing the device structure, operation principle, and material processing issues related with ALD in various types of memory and logic devices. Due to the extreme diversity of the ALD processes in various semiconductor devices, the book was written on a rather arbitrary classification of semiconductor devices into mass-production memory/emerging memory and logic devices. Mass-production level memories include the DRAM and Flash memories, whereas the phase change random access memory (PcRAM) and ferroelectric RAM (FeRAM) belong to emerging memories. There are also other memories such as static RAM (SRAM), magnetic RAM (MRAM, or spin transfer torque RAM (STTRAM)), and resistance switching RAM (ReRAM). However, SRAM is comprised of six transistors forming the cross-coupled latches, which is the same as logic devices from the fabrication point of view. The key component of MRAM and STTRAM is magnetic tunnel junction (MTJ), where an extremely thin dielectric layer ($\sim$1–2 nm), typically MgO or Al_2O_3, is sandwiched between two magnetic conductors. Depending on the relative magnetization directions of the two magnetic layers, the electrical conduction which occur via the tunneling mechanism through the thin dielectric layer modulates, which is the basic principle of the MRAM or STTRAM. Due to the extremely high sensitivity of the tunneling current on the dielectric layer thickness, very tight control (<0.1 nm) over the dielectric layer thickness throughout the large area wafer is crucial. ALD can be a viable solution for the tough target, but presently plasma oxidation of sputter-deposited metal (Mg or Al) layer is used as the industry standard. Several trials of the ALD process were adapted to this area but little noteworthy success has been reported. ReRAM is an interesting area where numerous materials show electrically induced non-volatile resistance switching based on various mechanisms [8]. However, there are several problems in ReRAM from the ALD point of view. One of them is the non-determined resistance switching material. The device structure is not well defined yet either. Therefore, SRAM, MRAM (STTRAM) and ReRAM are excluded from this monograph.

Certainly, there are very diverse logic devices ranging from the state-of-the-art multi-core main processors to fairly simple microprocessors. However, all those logic chips are combinations of high performance transistors without special capacitors or junction devices involved. Therefore, from the ALD point of view, the logic can be treated as a device composed of the front-end and back-end of the line (the sequence of fabrication procedure is called 'line' here) steps. This monograph has two separate chapters on the front-end and back-end of the line processes for Si chips, accordingly. On the other hand, great attention has been paid recently in the adoption of new substrates with high-mobility, such as Ge, compound semiconductors (GaAs and InP-based materials), or even carbon-based materials. These substrates have very distinctive physical and chemical interactions with the growing high-k films. Therefore, this must be treated separately but was not included in this monograph.

In addition to the above mentioned main bodies, there are three more important chapters. Two of them are on the fundamental sides of ALD—one chapter for the precursors and the other for the simulations. The final chapter is for ALD hardware/tools. These three chapters can be regarded as the general background for all types of ALD but are more focused on the microelectronic applications.

In this application field, the ALD layers work not only as the functional materials in the final products but also as the sacrificial layers that enhance several process capabilities. ALD layers discussed above are the functional materials. As a process capability enhancing factor, the double patterning process is a typical example of sacrificial ALD, which largely improves the photolithographic capability without astronomical cost. This kind of application is also discussed in detail in the main text. These two applications require very different film properties and process conditions. The active applications need supreme quality films that are generally accompanied by high process temperatures and low growth rates, whereas the sacrificial applications require low temperatures but rather thick ALD films. These conditions are generally incompatible with each other.

In this introductory chapter, the general aspects of memory and logic devices are shortly covered too, which we believe particularly helpful for more process- and/or chemistry-oriented readers. More detailed operational principles of each device are described in each chapter. Presently, digital binary data stored in memory devices are in the form of either charge (DRAM, Flash, FeRAM, and SRAM) or resistance (PcRAM, MRAM, and ReRAM). Logic operations, which process the stored date according to the Boolean logic functions, proceed by combinatorial operations of complementary MOSFETs (CMOSFET). The most common logic elements are NAND, NOR, and NOT gates. Shannon already revealed that all the necessary Boolean logic operations can be accomplished using the (complicated) combinations of the three basic logic gates [9].

First, let us consider the principles of memory. Since a large amount of digital data must be stored, the memory cells are arranged in a type of array structures. As widely recognized, the most efficient array structure is the so-called "matrix" type, where the two crossing lines, bit- and word-lines, define a memory cell. The 3D cell structure is constructed by extending the two-dimensional memory arrays into the third dimension either by layer stacking or vertical integration.

On a very fundamental level, two types of arrays, passive and active are shown in Fig. 1.1 a and b, respectively. In the passive array, the simultaneous activations of both bit- and word-lines are required to "write" a data bit into a certain cell. In this writing process, a certain threshold level of stimulus intensity must be overcome, which must be greater than the signal intensity of the bit- or word-line. This means that only the cell where both lines accessed becomes active, which makes a "random access" possible. The random access here means the "write" and "read" of a certain memory cell without involving "write" and "read" of other memory cells. Needless to mention, the random access is a more efficient way to operate the memory compared to non-random access methods as in the NAND flash. In this case, a memory cell adopts an orthodox chain-like geometry making random access very complicated. One example of the passive array type memory cell can be found from the NOR type flash memory. Flash memories, both NAND and NOR type, use double gate MOSFETs where the floating gate serves as the charge storage and the control gate drives charge transports between the channel region

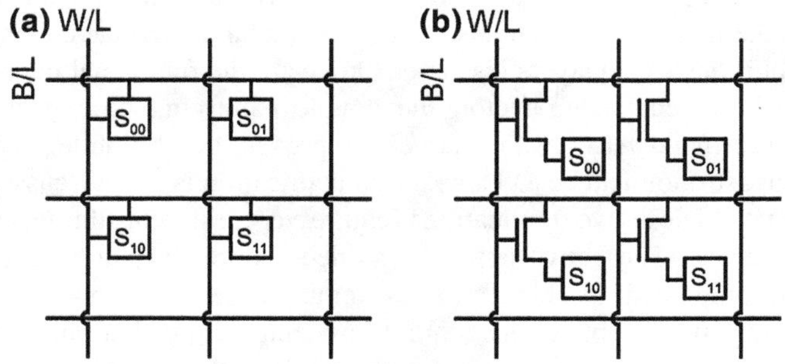

Fig. 1.1 Schematic diagrams of (**a**) the passive matrix array and (**b**) the active matrix array

and floating gate by applying a high voltage between semiconductor channel and itself. When a high control gate voltage ($\sim$10 V) is applied to a certain word-line, all cells connected to this word-line are just about to inject electrons from the inverted channel to the floating gate via the tunneling oxide. Then, the bit-line connected to the target cell is activated, and a current starts to flow between the source and drain in the target cell because the transistor would be already turned on due to the sufficiently high control gate voltage. The flowing carriers heat up the channel region by scattering and the carriers are thermally excited (hot carriers) to be injected into the floating gate particularly near the drain edge with the help of the high control gate voltage. Therefore, only the target cell can be programmed leaving other cells with common word-line unaffected because the current between source and drain does not flow, and, thus, the thermal effect would not be induced. Cells with the common bit-line would not be influenced because there is no control gate voltage and no source-drain current flow due to the off state of the MOSFET. Another type of passive memory cell can be found from the selection diode integrated PcRAM cell. This is explained again in conjunction with the active matrix array below.

On the other hand, the active matrix array utilizes an electrically operating switch between the bit-line and each of storage cell. This array has been extensively used in DRAM where the switch and storage cell are cell MOSFET (mostly n-type) and capacitor, respectively. The n-type MOSFET switches are driven by the word-line voltage. In this case, all the cells connected to a common bit-line are ready to be written when the bit-line is activated. When the word-line connected to the target cell is activated, the MOSFET is turned on and charges flow into the storage capacitor from the bit-line and the memory cell is thus written. Then, the word-line voltage decreased and the written data is kept within the cell. It can be easily understood that other cells sharing the same bit-line are not written because their switches were remained off (the word-line voltage remained low). In addition, other cells sharing the activated word-line with the target cell are not written either because the bit-line connected is not activated. This is a typical operation of DRAM. Overall, the active type memory cell is conceptually simple and easier to design compared to the passive type memory cell, which relies upon certain non-linearity and threshold effect of the transistor and other functional material. However, with an extreme shrinkage of cell size, imperfect operation of MOSFETs as a switch (i. e., incomplete turning on when the gate (word-line) bias was turned on and incomplete turning off when the gate bias was turned off) makes the passive type memory gain renewed interest. While the logic MOSFETs are designed to gain high on-current for higher operation speed, which is generally achieved by the low threshold voltage ($\sim$0.3 V), memory cell MOSFETs have a higher threshold voltage ($\sim$0.7–0.8 V) to keep the stored charge stably. The low and high threshold voltages inevitably suffer from high off-state leakage current and low on-state current problems, respectively. Despite huge amount of engineering work, this is a sort of unsolved inherent problem of present MOSFET, the operation of which is based on the thermal distribution of electrons. These are charge-based memory because the digital data is stored as a form of charge in cell capacitor of DRAM and in floating gate of Flash memory (or charge trap layer in charge trap flash cell). Storing and saving charges in memory cells generally require two contradicting functions of MOSFET, i.e., to be completely on during writing and to be completely off during saving, which cannot be satisfied simultaneously even with practical criteria in highly scaled memory cells. Therefore, alternatives can be found in resistance-based memories.

Figure 1.2 a and b shows (overly) simplified internal read out scheme of a charge-based memory (DRAM) and a resistance-based memory cell (typically PcRAM) when the stored data is read out. For the case of DRAM, the potential (V_{bit}) of bit-line, which is connected to the sense amplifier, is set to be half of the operation voltage ($V_{dd}/2$) initially. Then, the word-line voltage (V_{word}) is pulled up to turn on the MOSFET of the reading cell. V_{bit} either increases or decreases depending on the charge state of the cell capacitor, which triggers the sense amplifier, which is a destructive readout. Right after the destructive read process, the original data is recovered by the self-refreshing function of the read operation. If the stored charge is overly small resulting from insufficient capacitance of the cell capacitor, small write current, and/or large leakage current of the capacitor dielectric and MOSFET, the sense amplifier may not detect the variation of V_{bit}. It can be easily anticipated that this problem becomes increasingly serious as the

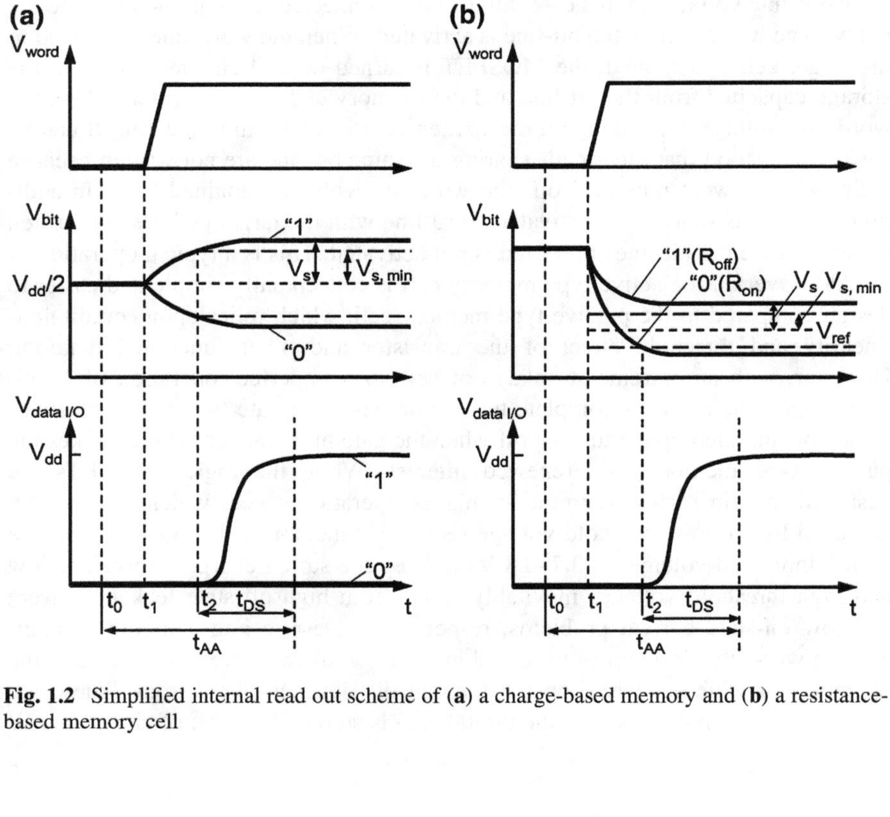

Fig. 1.2 Simplified internal read out scheme of (**a**) a charge-based memory and (**b**) a resistance-based memory cell

device scaling proceeds. On the other hand, the situation is different in the resistance-based memory cell. In this case, V_{bit} was also initially set to a certain value. Then, the cell MOSFET turns on, and V_{bit} decreases depending on the resistance of the resistor (MTJ or phase-change/resistance-change material). The sense amplifier detects the difference and therefore the digital data stored as either high or low voltage, can be read out. It can be understood that the necessary functionality of cell MOSFET in this case is much less strict because there is no need to keep the stored charge. This is therefore a fundamental advantage of these memories compared to the charge-based memory. However, the electrically switchable resistors, which are as reliable, low power consuming, fast operating, and repeatable as DRAM capacitor, are not available yet. MTJ is probably close to these targets but both the extreme difficulty in fabricating such a thin tunneling dielectric layer and the overly high operational current inhibit its widespread use.

There is another problem related with the cell MOSFET. The resistance switching in promising materials including $Ge_xSb_yTe_z$ alloys for PcRAM and metal oxides for ReRAM, requires very high current to change its resistance state. For example, the PcRAM requires a short (<50 ns) peak current (order of 0.1 mA) to reset (i.e., switch from on-state (crystalline state) to off-state (amorphous state)) the phase change cell, which must flow through the cell MOSFET. However, MOSFET generally has a maximum allowable current of <0.1 mA when the gate width is <100 nm. Therefore, scaling of PcRAM with the cell MOSFET has been hindered by the limited current supply, meaning that the active array configuration may not fit well to the PcRAM. Recent PcRAM, therefore, adopts the passive array type cross-bar structure, where the bit- and word-lines cross in a matrix-like shape and the phase-change materials are intervened between the two lines. The voltages of bit- and word-lines can be controlled to induce the desired phase change only at the junction of interest where both lines are activated. Since there is no MOSFET used in this structure, the problem of "limited current" is hence eliminated. However, a so-called read disturbance issue arises with this type of passive array. When an off-state cell is surrounded by on-state cells, the off-state cell can be misread as on-state since the surrounding on-cells provide a fluent current path. A common solution to this problem is placing diodes in series with each of the memory cell [10]. In the present PcRAM, p-n junction type Si-based diodes are employed instead of MOSFETs and each diode is connected with each memory cell [11]. The cross-sectional microscopy image of PcRAM may look like an active array but it is actually a passive array. Old paradigm comes back and revives in a new emerging memory!

As for logic devices, basic functions of the NAND gate are explained and characteristic features of CMOSFET circuit are briefly mentioned below. Other gates can be understood in the same manner. Figure 1.3 shows a simplified diagram of a NAND gate which is composed of two n-MOSFETs and two p-MOSFETs. The truth table of NAND logic is also included in Fig. 1.3. In electric circuit, the "1" and "0" logic states correspond to the high and low levels of voltages, respectively. Although the detailed operations of n- and p-MOSFETs are rather complicated and non-linear in response to the external voltage input, they

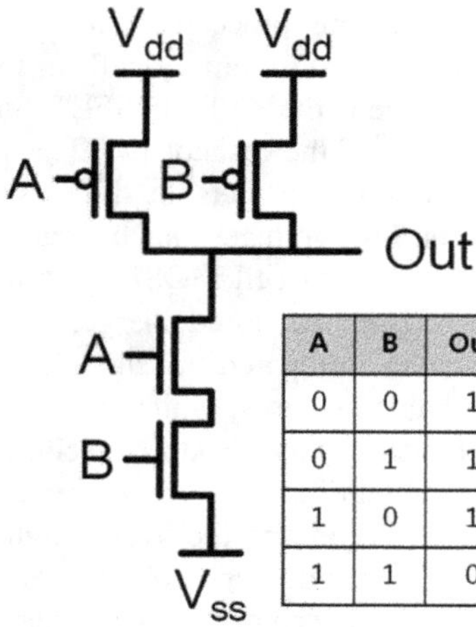

A	B	Out
0	0	1
0	1	1
1	0	1
1	1	0

Fig. 1.3 Simplified diagram of a NAND gate and the truth table of NAND logic

can be simply approximated as follows. In n-MOSFET, the high and low (or equal) gate voltage (V_g) relative to the source region (V_s) corresponds to the on- and off-states of the n-MOSFET, respectively. In p-MOSFET, the opposite is the case, where low (or equal) and high V_g relative to V_s correspond to the on- and off-states, respectively. The source voltage or ground potential connected to one of the n- and p-MOSFETs is transferred to the output depending on the combinations of two input voltages, which are applied to the gates of the four MOSFETs. Readers thus can easily understand that the output voltage is determined by the two inputs according to the truth table when the operation (on or off) of each MOSFET in the figure is traced according to the input voltages on each gate. One notable finding from such a circuit element is that there is no steady-state current flow; once the transient current flew to charge up the node capacitances, the source voltage and ground are never connected directly while the output node is connected to either the voltage source (logic state 1) or ground (logic state 0). The output is connected to the gates of following/neighboring logic circuits (inputs of them) so that there is no steady-state current flow from the output node to the following circuitry. They are separated by n- or p-MOSFET at the off-state and gate insulator, so that there is no steady-state power consumption. This is the most notable feature of the circuits composed of CMOSFETs compared to other circuits, and was one of the reasons why the CMOSFETs have dominated the logic chips. However, this will be the case only in the ideal case where there is no leakage through the gate dielectric and source-drain junctions, which is certainly not the case in reality. Particularly when the gate length is shortened to a very small value ($\ll$100 nm) for a high on-current, the short channel effect, i.e., increase in the off current even at the $V_g <$ threshold voltage, becomes very serious. It has been widely recognized that one of the most promising ways to suppress the short channel effect is to increase the gate capacitance, either by decreasing the SiO_2 gate dielectric thickness or adopting high-k dielectrics. Decreasing the gate dielectric thickness to $< \sim 1$ nm results in a significant gate leakage current whereby the power consumption cannot be tolerated. Even the logic state can be influenced erroneously with the overly high gate leakage (input can be directly transferred to output). Therefore, the high-k gate dielectric technology is indispensable and the metal gate technology is being pursued along with the high-k gate dielectric technology in an effort to eliminate the gate poly-Si depletion effect.

Another key asset of the CMOSFET in logic circuit is its ability to amplify signal levels due to the non-linear response of FET to the input signals. The input signal variation, which is generally induced by various noises and parasitic circuit components, can be successfully filtered once it passes through the MOSFET. Every future switching device, whether based on Si or other materials (ca. carbon), they must have this functionality in order to be used in the high performance logic circuits. Further details on this can be found in the recent book by Waser [12].

Reference

1. Hwang CS (2002) J Appl Phys 92:432
2. Kim SK, Kim KM, Kim WD, Hwang CS, Jeong J (2004) Appl Phys Lett 85:4112
3. Kim SK, Choi G-J, Lee SY, Seo M, Lee SW, Han JH, Ahn H-S, Han S, Hwang CS (2008) Adv Mater 20:1429
4. Kim SK, Lee SW, Han JH, Lee B, Han S, Hwang CS (2010) Adv Func Mater 20:2989
5. Ahonen M, Pessa M, Suntola T (1980) Thin Solid Films 65:301
6. Hausmann D, Becker J, Wang S, Gordon RG (2002) Sci 298:402
7. Seo M, Kim SK, Min Y-S, Hwang CS (2011) J Mater Chem Adv Art 21:18497
8. Waser R, Dittmann R, Staikov G, Szot K (2009) Adv Mater 21:2632
9. Shannon CE (1949) Bell Syst Tech J 28:59
10. Lee M-J, Park Y, Suh D-S, Lee E–H, Seo S, Kim D-C, Jung R, Kang B-S, Ahn S-E, Lee CB, Seo DH, Cha Y-K, Yoo I-K, Kim J-S, Park BH (2007) Adv Mater 19:3919
11. Oh JH, Park JH, Lim YS, Lee HS, Oh YT, Kim JS, Shin JM, Park JH, Song YJ, Ryoo KC, Lim DW, Park SS, Kim JI, Kim JH, Yu J, Yeung F, Jeong CW, Kong JH, Kang DH, Koh GH, Jeong GT, Jeong HS, Kim K (2006) IEEE Int Electron Devices Meet (IEDM) Tech Dig 1
12. Waser R (2003) Nanoelectronics and information technology. Introduction to part3, Wiley-VCH, 323

Part II
Fundamentals

Chapter 2
ALD Precursors and Reaction Mechanisms

Roy G. Gordon

2.1 Introduction

ALD has become accepted as a reliable tool for production of thin films in the microelectronics industry. Dielectrics and metal electrodes for DRAM capacitors are now routinely produced by ALD, which is the only technique capable of uniformly coating inside the narrow structures required by current technology. Transistors will soon be produced as 3D structures, with requirements for conformal coatings over their increasingly complex surfaces. Metal circuits between transistors are also pushing toward narrower and deeper structures, such as through-silicon vias. In order to extend the use of ALD into these new areas, a greater variety of materials will need to be deposited.

Successful use of ALD requires suitable chemical precursors used under reaction conditions that are appropriate for them. There are many requirements for ALD precursors: sufficient volatility, thermal stability, and self-limited reactivity with substrates and with the films being deposited. The precursor vapor should not etch or corrode the substrate or deposited film. In addition, it is easier to produce the required vapors if the precursor is liquid at room temperature, or if it is a solid with a melting point below its vaporization temperature, or if it is soluble in an inert solvent with vapor pressure similar to that of the precursor. Ideally, the precursors should be non-flammable, non-corrosive, non-toxic, non-hazardous, and simple and inexpensive to make. In applications of ALD to microelectronics, precursors with high purity may be required.

It is not easy to find ideal precursors that satisfy all of these requirements. When the properties of precursors are inadequate, satisfactory ALD results cannot be achieved.

R. G. Gordon (✉)
Department of Chemistry and Chemical Biology, Harvard University, Cambridge, MA 02138, USA
e-mail: gordon@chemistry.harvard.edu

C. S. Hwang (ed.), *Atomic Layer Deposition for Semiconductors*,
DOI: 10.1007/978-1-4614-8054-9_2,

If the precursor does not have sufficient vapor pressure at a temperature at which it is thermally stable, then the surface reactions cannot be saturated within a practical cycle time. Uniform step coverage will be limited to features with low aspect ratios. Impurities may also be incorporated into the films because some ligands are not removed during an exposure (defined as vapor pressure multiplied by the time of vapor contact) that is too low to complete the surface reactions.

If the thermal stability of a precursor is too low, then decomposition takes place during the surface reactions, and the reactions are usually not self-limited. Step coverage will be non-uniform, and impurities from the ligand decomposition products may remain in the films.

Self-limited surface reactions may not exist because of a number of non-ideal behaviors. The surface reactions may be reversible, so they do not remain saturated when the supply of reactant vapor is removed. If a precursor etches the film, a stable saturated condition will not be reached. If a precursor is not reactive enough with a substrate, nucleation of film is delayed many cycles, and growth happens only sporadically and non-reproducibly on defects on the substrate.

Because of the ingenuity and hard work of many scientists, useful, if not ideal, precursors have been found for most of the non-radioactive elements. Figure 2.1 shows this wide range of elements that have been incorporated into films deposited by ALD. Still missing from this repertoire are the heavier alkali metals sodium, potassium, rubidium, and cesium, for which few volatile compounds are known. The heavier halogens (chlorine, bromine, and iodine) and hydrogen are usually considered deleterious impurities to be avoided in ALD films. Non-reactive elements, such as the noble gases, are not included in ALD films. The highly toxic element, thallium, has also been neglected.

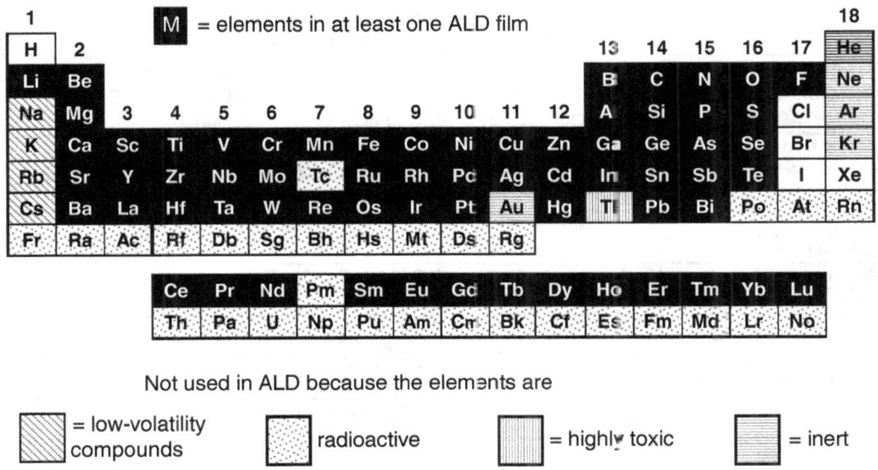

Fig. 2.1 Elements in ALD films

ALD of pure elements has been achieved for the 19 elements that are underlined in Fig. 2.2. Many different types of compounds have been made successfully by ALD. These compounds are shown in Fig. 2.2 as combinations of the element whose symbol is at the center of each box with one or more of the elements whose symbols lie on the periphery of the box. ALD materials are organized by the type or properties of the compound in Fig. 2.3. An impressive breadth of applicability for ALD has been achieved by researchers from around the world. As a result, the

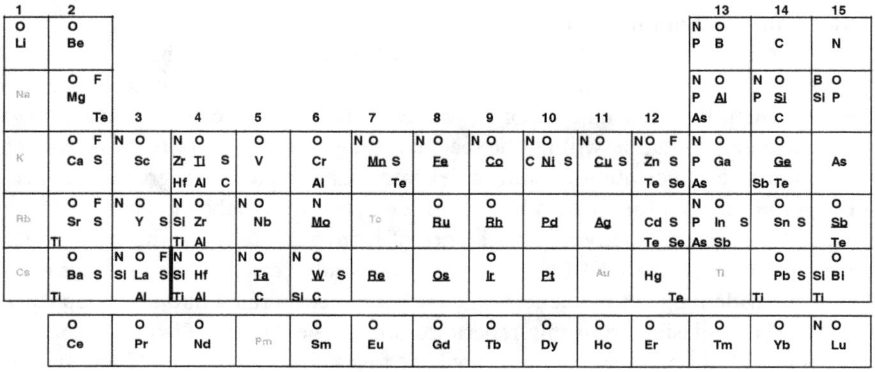

Fig. 2.2 Combinations of elements in ALD films. ALD films have been made with combinations of 2 or more elements within a box. Underlined elements have been deposited as pure, single elements. Updated from R. Puurunen, J. Appl. Phys. 97, 121301 (2005)

Oxide dielectrics	Al_2O_3, TiO_2, ZrO_2, HfO_2, Ta_2O_5, Nb_2O_5, Sc_2O_3, Y_2O_3, BeO, MgO, B_2O_3, SiO_2, GeO_2, La_2O_3, CeO_2, PrO_x, Nd_2O_3, Sm_2O_3, EuO_x, Gd_2O_3, Dy_2O_3, Ho_2O_3, Er_2O_3, Tm_2O_3, Yb_2O_3, Lu_2O_3, $SrTiO_3$, $BaTiO_3$, $PbTiO_3$, $PbZrO_3$, Bi_xTi_yO, Bi_xSi_yO, $SrTa_2O_6$, $SrBi_2Ta_2O_9$, $YScO_3$, $LaAlO_3$, $NdAlO_3$, $GdScO_3$, $LaScO_3$, $LaLuO_3$, $LaYbO_3$, $Er_3Ga_5O_{13}$
Oxide conductors or semiconductors	In_2O_3, In_2O_3:Sn, In_2O_3:F, In_2O_3:Zr, SnO_2, SnO_2:Sb, SnO_2:Al, SnO_2:N, Sb_2O_3, ZnO, ZnO:Al, ZnO:B, ZnO:Ga, RuO_2, RhO_2, IrO_2, Ga_2O_3, VO_2, V_2O_5, WO_3, W_2O_3, NiO, CuO_x, FeO_x, CrO_x, CoO_x, MnO_x
Other ternary oxides	$LaCoO_3$, $LaNiO_3$, $LaMnO_3$, $La_{1-x}Ca_xMnO_3$
Nitride dielectrics or semiconductors	BN, AlN, GaN, InN, Si_3N_4, Ta_3N_5, Cu_3N, Zr_3N_4, Hf_3N_4, LaN, LuN
Metallic nitrides	TiN, Ti-Si-N, Ti-Al-N, TaN, NbN, MoN, WN_x, WN_xC_y, Co_xN, Sn_xN
II-VI semiconductors	ZnS, ZnSe, ZnTe, CaS, SrS, BaS, CdS, CdTe, MnTe, HgTe
II-VI based phosphors	ZnS:M (M=Mn,Tb,Tm); CaS:M (M=Eu, Ce, Tb, Pb); SrS:M(M=Ce,Tb, Pb)
III-V semiconductors	GaAs, AlAs, AlP, InP, GaP, InAs
Fluorides	CaF_2, SrF_2, MgF_2, LaF_3, ZnF_2
Elements	Ru, Pt, Ir, Pd, Rh, Ag, Cu, Ni, Co, Fe, Mn, Ta, W, Mo, Ti, Al, Si, Ge, Sb
Other semiconductors	PbS, SnS, In_2S_3, Sb_2S_3, Cu_xS, $CuGaS_2$, WS_2, SiC, $Ge_2Sb_2Te_5$
Others	La_2S_3, Y_2O_2S, TiC_x, TiS_2, TaC_x, WC_x, $Ca_3(PO_4)_2$, $CaCO_3$, organics

Fig. 2.3 ALD materials by type. Updated from M. Ritala and J. Niinisto, in Chemical vapor deposition (Royal Society of Chemistry, 2009)

number of papers applying known ALD processes now greatly outnumbers those reporting new ones. Nevertheless, considerable challenges remain for finding precursors with more nearly ideal properties for use in ALD. Recently, many reviews of ALD and its applications have been published [1–18].

This review will first summarize the various types of chemicals that have been used successfully as precursors for ALD processes. Then the kinds of surface reactions that they undergo will be classified. Despite the wide variety of chemical types involved, it will become clear that only a few kinds of surface reactions account for the many ALD processes that have been discovered.

2.2 Types of ALD Precursors for Metals

In a few cases, the elements themselves, rather than compounds, can serve as ALD precursors. The metals zinc, cadmium, and mercury are the most common examples. Most other metals have vapor pressure that is too low, unless they are evaporated from high-temperature effusion sources in high vacuum, as in molecular beam epitaxy (MBE) systems. While MBE can be an excellent tool for laboratory investigations, it is not as easily scaled to large-area production of films as are the conventional ALD processes based on more volatile precursors.

Halide compounds, including fluorides, chlorides, and iodides, have been used as precursors for ALD for those elements shown with a dark background in Fig. 2.4. Halides generally display excellent thermal stability. Widespread industrial use is made of titanium tetrachloride, $TiCl_4$, to make titanium nitride, TiN, electrodes for DRAM, and hafnium tetrachloride, $HfCl_4$, to make HfO_2 gate dielectric for transistors. Vapor pressures of halides vary widely. Tungsten hexafluoride, WF_6, is a gas at room temperature. Copper(I) chloride, CuCl, requires

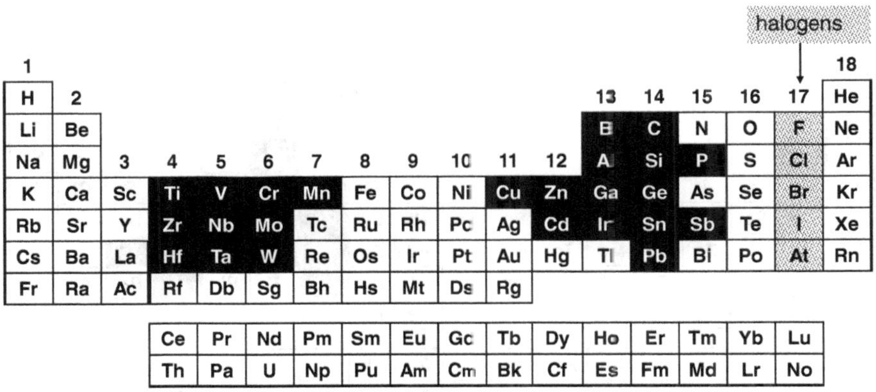

Fig. 2.4 Elements with halide ALD precursors. Halides are compounds MX_n, n = 1, 2, 3, 4, 5 or 6, of an element M and a halogen X = F, Cl, Br or I

high temperature (340 °C) to produce a useful amount of vapor [19]. The main disadvantage of metal halides is the corrosive nature of the precursors and their reaction byproducts. They can etch or corrode substrates, the deposited films, ALD equipment, and vacuum pumps. For example, a $NbCl_5$ precursor partly etches away Nb_2O_5 film as it is formed, so non-uniform thicknesses are observed [20]. Halide impurities can also be retained in the films, with deleterious consequences to their properties.

Metal alkyls, which have metal-carbon bonds, are another popular class of precursor for ALD. Trimethylaluminum, $(CH_3)_3Al$, is probably the precursor most often used for depositing aluminum oxide because it displays nearly ideal properties.[1] Diethylzinc, $(C_2H_5)_2Zn$ [21] and dimethylzinc, $(CH_3)_2Zn$ [22], are usually used for depositing zinc oxide. These metal alkyls need to be handled carefully because they ignite spontaneously if exposed to air (pyrophoric behavior), and can explode if contacted by water. Safer non-pyrophoric aluminum alkyl-alkoxides, such as dimethylaluminum isopropoxide [23], are also effective for ALD of aluminum oxide. Non-pyrophoric complexes of dimethylzinc with Lewis bases, such as dimethylsulfide, are also known [24–26]. Figure 2.5 shows those elements with metal alkyl precursors suitable for ALD. Other metal alkyls lack sufficient volatility or thermal stability for use in ALD.

Cyclopentadienyl ligands (Fig. 2.6) tend to form more stable metal compounds because they form 5 carbon-metal bonds for each ligand [27–30]. In comparison, each metal atom is bonded to only one carbon atom in each ligand of a metal alkyl. Small metal atoms, such as magnesium, are bound to cyclopentadiene rings with just small hydrogen atoms attached. For large atoms, such as lanthanum, bulky isopropyl groups are attached to the cyclopentadienyl ring so that the resulting compound remains monomeric and is thus slightly volatile. Examples of metal cyclopentadienyl compounds are listed in Fig. 2.7. A wide variety of metal cyclopentadienyls, shown in Fig. 2.8, have been used in ALD. Because of the strong metal-carbon bonding, cyclopentadienyl ligands can be difficult to remove completely, sometimes contaminating ALD films with carbon impurity.

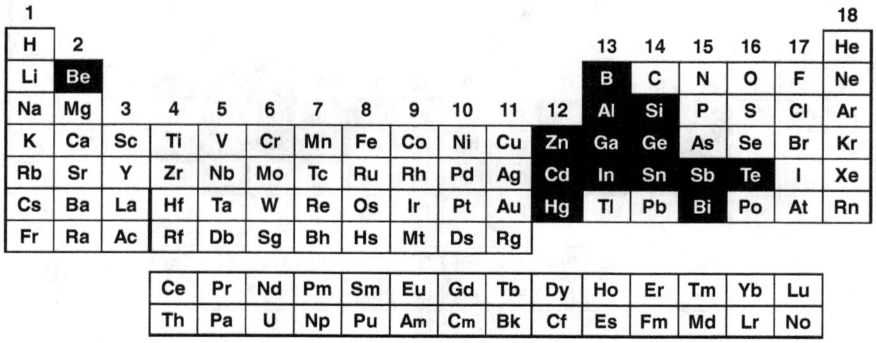

Fig. 2.5 Elements with alkyl ALD precursors

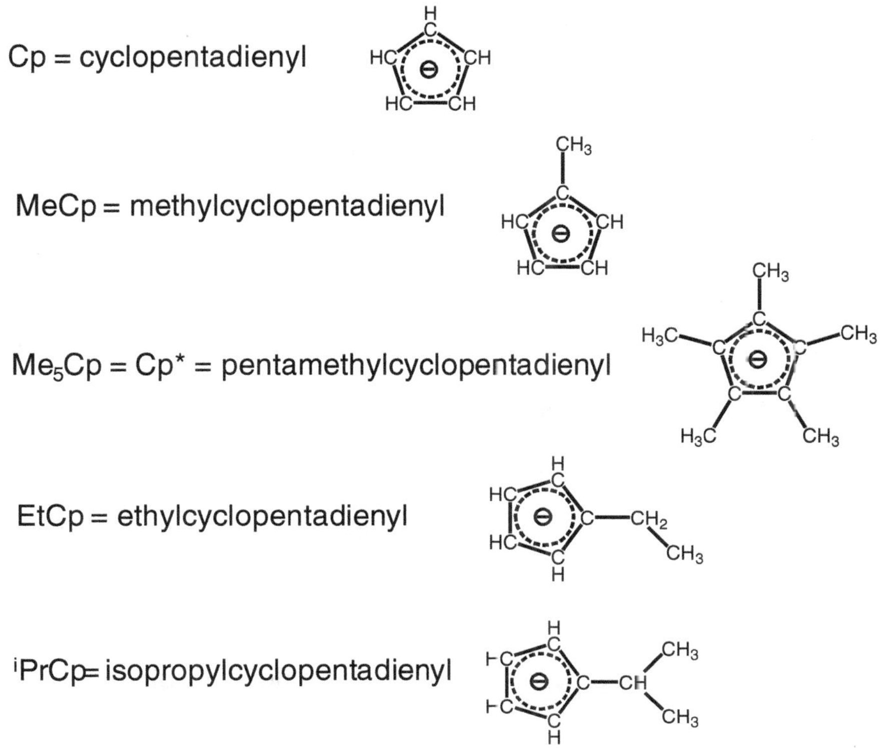

Fig. 2.6 Cyclopentadienyl ligands

Cyclopentadienyl ligands can also be mixed with alkyl groups attached to the same metal atom. A popular example of such heteroleptic precursors is (methylcyclopentadienyl)(trimethyl)platinum: [31, 32].

MeCpPtMe3

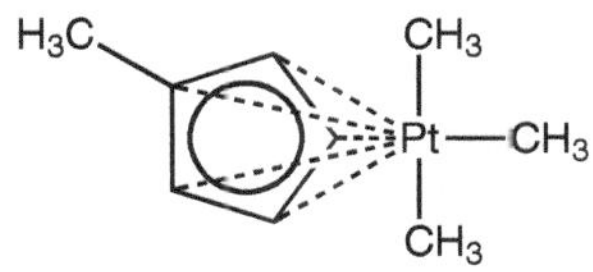

Alkoxide compounds have a metal bonded to oxygen, which is then attached to carbon, as illustrated in Fig. 2.9. Alkoxides are known for most of the metals in the periodic table, but only a few of them, identified in Fig. 2.10, have all the properties needed for ALD. Other alkoxides lack sufficient thermal stability and/or volatility [33]. Some alkoxide precursors are listed in Fig. 2.11. A popular alkoxide precursor is titanium isopropoxide [34, 35]:

Cp_2Ni = bis(cyclopentadienyl)nickel(II)

$(EtCp)_2Ru$ = bis(ethylcyclopentadienyl)ruthenium(II)

$(Me_5Cp)_2Sr$ = bis(pentamethylcyclopentadienyl)strontium

$(^iPrCp)_3La$ = tris(isopropylcyclopentadienyl)lanthanum

Cp_2Me_2Zr = (dicyclopentadienyl)(dimethyl)zirconium

$(MeCp)(Me)_3Pt$ = (methylcyclopentadienyl)(trimethyl)platinum(IV)

Fig. 2.7 Examples of cyclopentadienyl precursors

$Ti(OiPr)_4$

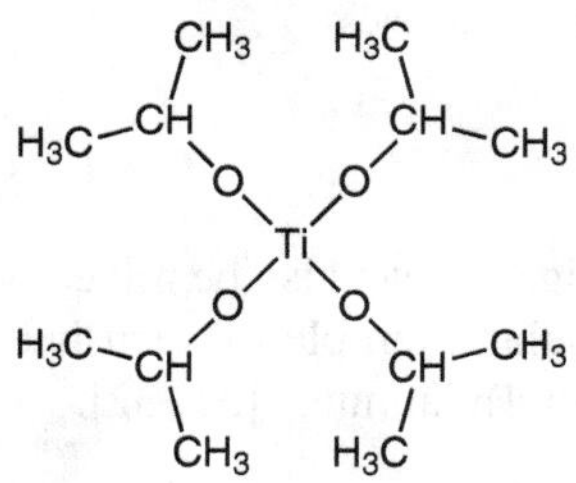

One way to make alkoxides more thermally stable is to attach a dialkylamine group at the end of an alkyl chain of the alkoxide, as in the 1-dimethylamino-2-methyl-2-propanolate (dmamp) ligand:

$Ni(dmamp)_2$

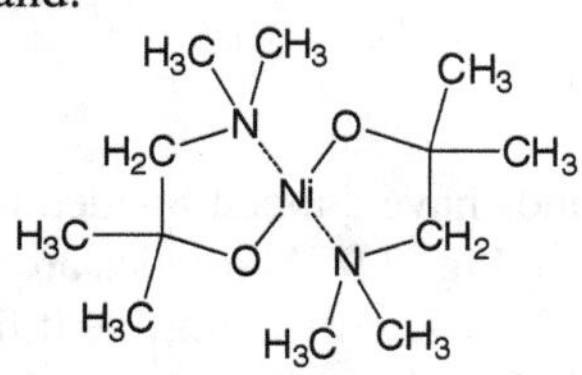

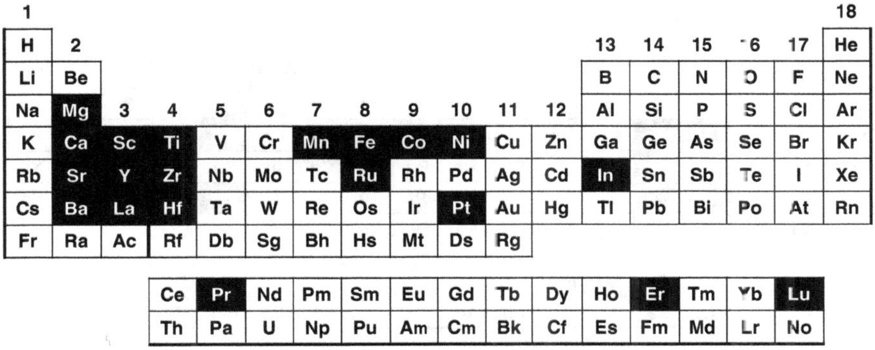

Fig. 2.8 Cyclopentadienyl ALD precursors

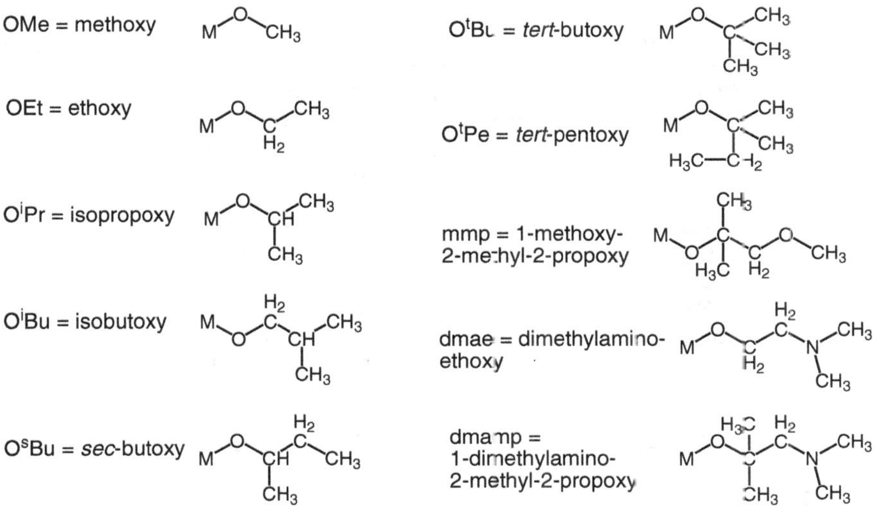

Fig. 2.9 Alkoxide Compounds

The attachment of the metal to each ligand by two bonds stabilizes the molecule by what is called the "chelate effect". It is, however, more difficult to remove these chelated ligands using reducing agents such as hydrogen or ammonia, so impurities of carbon, oxygen, and nitrogen from the ligands are found to contaminate the nickel metal films. Water vapor does cleanly cleave the ligands to deposit pure nickel oxide [36].

Metal β-diketonates have 2 metal-oxygen bonds for each ligand, as pictured in Fig. 2.12. The chelate effect makes metal β-diketonates more thermally stable than metal alkoxides. By varying the atoms or atom groups attached to the 3 carbon atoms, the properties of metal β-diketonates can be adjusted. Small groups, such as methyl, are suitable for small metals, while bulky groups are needed to maintain a

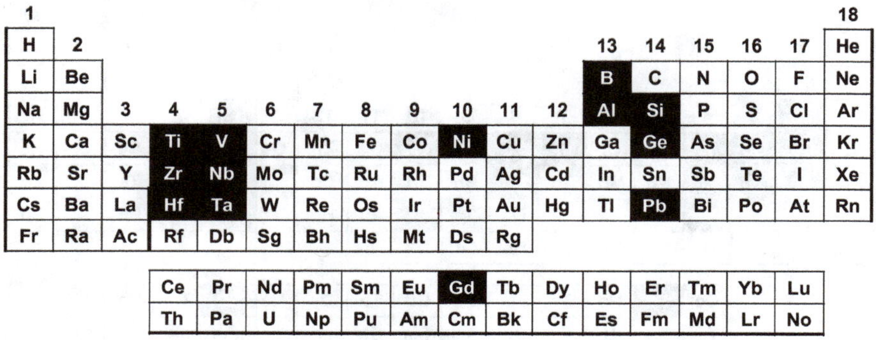

Fig. 2.10 Elements with alkoxide ALD precursors

$Al(OEt)_3$ = tris(ethoxy)aluminum = aluminum ethoxide
$AlMe_2(O^iPr)$ = isopropoxydimethylaluminum
$B(OMe)_3$ = tris(methoxy)boron = trimethylborate
$Hf(O^tBu)_4$ = tetra(*tert*-butoxy)hafnium = hafnium *tert*-butoxide
$Hf(mmp)_4$ = tetra(1-methoxy-2-methyl-2-propoxy)hafnium
$Nb(OEt)_5$ = penta(ethoxy)niobium = niobium ethoxide
$Ni(dmamp)_2$ = bis(1-dimethylamino-2-methyl-2-propoxy)nickel(II)
$Pb(O^tBu)_2$ = bis(*tert*-butoxy)lead(II) = lead(II) *tert*-butoxide
$Si(OEt)_4$ = tetra(ethoxy)silane = tetraethylorthosilicate = TEOS
$Si(O^tBu)_3OH$ = tris(*tert*-butoxy)silanol = TBOS
$Si(O^tPe)_3OH$ = tris(tert-pentoxy)silanol = TPSOL
$Ta(OEt)_5$ = penta(ethoxy)tantalum = tantalum ethoxide
$Ti(OMe)_4$ = tetra(methoxy)titanium = titanium methoxide
$Ti(OEt)_4$ = tetra(ethoxy)titanium = titanium ethoxide
$Ti(O^iPr)_4$ = tetra(isopropoxy)titanium = titanium isopropoxide
$VO(O^iPr)_3$ = tris(isopropoxy)oxovanadium = vanadyl isopropoxide

Fig. 2.11 Alkoxide compounds used in ALD

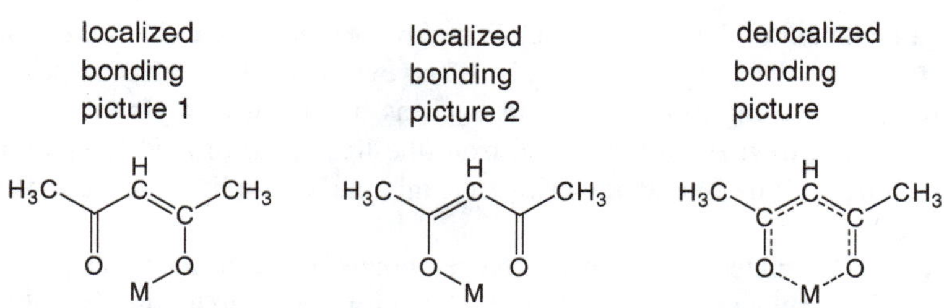

Fig. 2.12 Three ways to represent a metal acetylacetonate (acac)

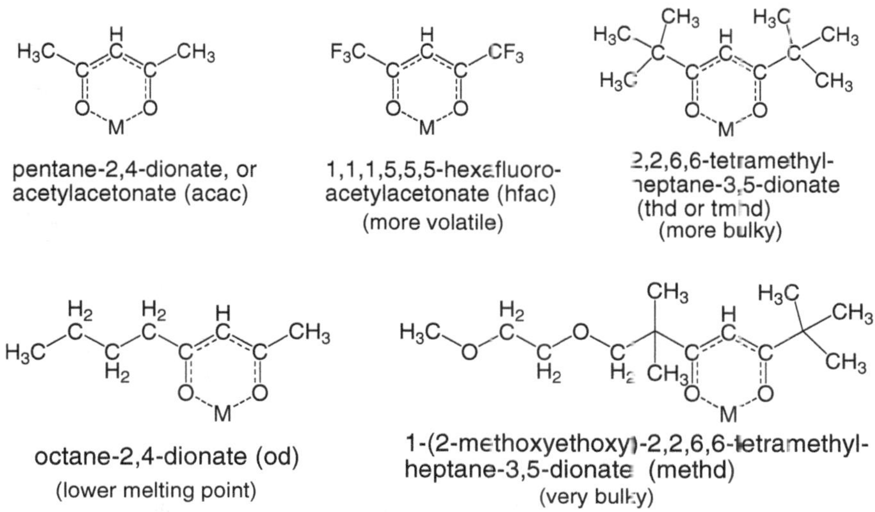

Fig. 2.13 Beta-diketonate compounds

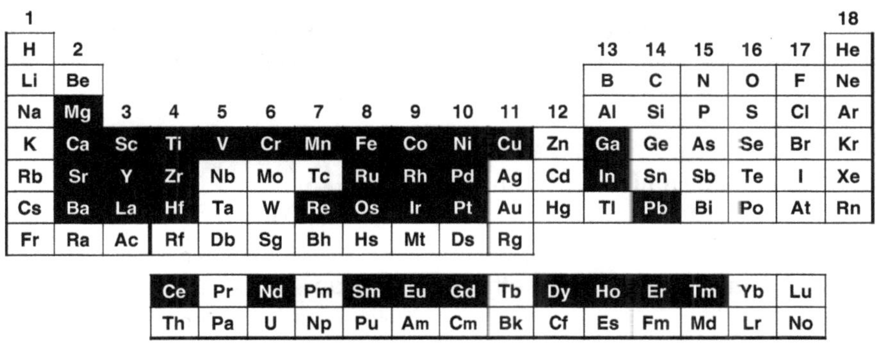

Fig. 2.14 Elements with beta-diketonate ALD precursors

monomeric molecular structure for large metals, as diagrammed in Fig. 2.13. Fluorine substitution increases the vapor pressure. The melting point can be decreased by substitution of longer hydrocarbon chains. By these adjustments, metal β-diketonate precursors have been demonstrated for many metals (Fig. 2.14) [37–40]. Typical metal β-diketonate precursors are listed in Fig. 2.15.

Because metal β-diketonates are so stable, very reactive sources must often be used as the co-reactants. For example, strong oxidants, such as ozone, are typically required to deposit metal oxides from metal β-diketonates, and these strong oxidants can attack substrates. Byproducts of these oxidation reactions include carbon dioxide, which can cause carbonate contamination of oxides of large metals, such

Fig. 2.15 Beta-diketonate ALD precursors

$Ba(thd)_2$	$Mg(thd)_2$
$Ce(thd)_4$	$Mn(thd)_3$
$Co(acac)_2$	$Nd(thd)_3$
$Co(acac)_3$	$Ni(acac)_2$
$Co(thd)_3$	$Ni(thd)_2$
$Cr(acac)_3$	$Pb(thd)_2$
$Cu(hfac)_2$	$Pd(hfac)_2$
$Cu(thd)_2$	$Pd(thd)_2$
$Dy(thd)_3$	$Pt(acac)_2$
$Er(thd)_3$	$Ru(thd)_3$
$Eu(thd)_3$	$Ru(od)_3$
$Fe(acac)_3$	$Sc(thd)_3$
$Fe(thd)_3$	$Sm(thd)_3$
$Gd(thd)_3$	$Sr(thd)_2$
$Ho(thd)_3$	$Sr(methd)_2$
$Ir(acac)_3$	$Tm(thd)_3$
$La(thd)_3$	$Y(thd)_3$

as strontium and lanthanum. Metal nitrides can be difficult to make from metal β-diketonates because the strong metal-oxygen bonds are hard to disrupt with nitrogen sources.

Nitrogen is bonded to metals in metal amides (a single chemical bond between nitrogen and a metal) and metal imides (a double bond between nitrogen and a metal). Some typical amides and imides are illustrated in Fig. 2.16. Precursors of these types have been developed for some metals highlighted in Fig. 2.17, and

NMe_2 = dimethylamino = dimethylamido

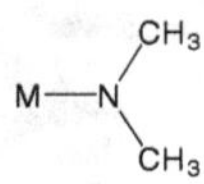

NEtMe = ethylmethylamino = ethylmethylamido

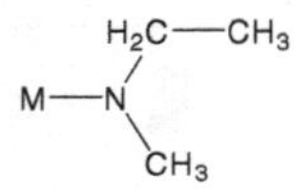

NEt_2 = diethylamino = diethylamido

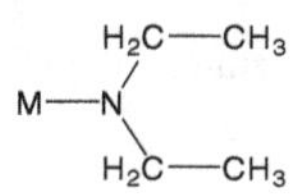

$N(SiMe_3)_2$ = bis(trimethylsilyl)amido = bis(trimethylsilyl)amino

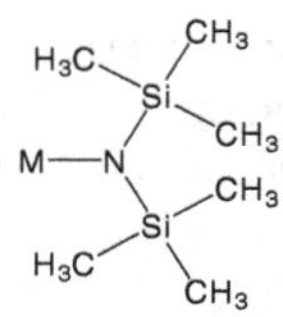

N^tBu = *tert*-butylimino = *tert*-butylimido

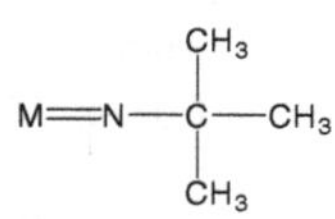

Fig. 2.16 Amide and imide ligands

1																	18
H	2											13	14	15	16	17	He
Li	Be											B	C	N	O	F	Ne
Na	Mg	3	4	5	6	7	8	9	10	11	12	**Al**	**Si**	P	S	Cl	Ar
K	Ca	Sc	**Ti**	V	Cr	Mn	Fe	Co	Ni	Cu	Zn	**Ga**	**Ge**	**As**	**Se**	Br	Kr
Rb	Sr	Y	**Zr**	**Nb**	**Mo**	Tc	Ru	Rh	Pd	Ag	Cd	In	**Sn**	**Sb**	**Te**	I	Xe
Cs	Ba	**La**	**Hf**	**Ta**	**W**	Re	Os	Ir	Pt	Au	Hg	Tl	Pb	**Bi**	Po	At	Rn
Fr	Ra	Ac	Rf	Db	Sg	Bh	Hs	Mt	Ds	Rg							

Ce	**Pr**	Nd	Pm	Sm	Eu	Gd	Tb	Dy	Ho	Er	Tm	Yb	Lu
Th	Pa	U	Np	Pu	Am	Cm	Bk	Cf	Es	Fm	Md	Lr	No

Fig. 2.17 Elements with amide and imide precursors for ALD

$Al(NMe_2)_3$ = tris(dimethylamido)aluminum
= $Al_2(NMe_2)_6$ = hexakis(dimethylamido)dialuminum
$Bi[N(SiMe_3)_2]_3$ = tris(bis(trimethylsilyl)amido)bismuth
$Hf(NMe_2)_4$ = tetrakis(dimethylamido)hafnium
$Hf(NEtMe)_4$ = tetra(ethylmethylamido)hafnium = TEMAH
$Hf(NEt_2)_4$ = tetrakis(diethylamido)hafnium = TDEAH
$La[N(SiMe_3)_2]_3$ = tris(bis(trimethylsilyl)amido)lanthanum
$Pr[N(SiMe_3)_2]_3$ = tris(bis(trimethylsilyl)amido)praseodymium
$Ta(NMe_2)_5$ = pentakis(dimethylamido)tantalum
$Ta(NEt_2)_5$ = pentakis(diethylamido)tantalum
$Ta(NtBu)(NEt_2)_3$ = (*tert*-butylimido)tris(diethylamido)tantalum
$Ti(NMe_2)_4$ = tetrakis(dimethylamido)titanium
$Ti(NEtMe)_4$ = tetra(ethylmethylamido)titanium = TEMAT
$W(NtBu)_2(NMe_2)_2$ = bis(*tert*-butylimido)bis(dimethylamido)tungsten
$Zn[N(SiMe_3)_2]_2$ = bis(bis(trimethylsilyl)amido)zinc
$Zr(NMe2)_4$ = tetrakis(dimethylamido)zirconium
$Zr(NEtMe)_4$ = tetra(ethylmethylamido)zirconium = TEMAZ
$Zr(NEt_2)_4$ = tetrakis(diethylamido)zirconium = TDEAZ

Fig. 2.18 Amide and imide precursors for ALD

particular examples are listed in Fig. 2.18. Dialkylamide precursors are popular for making zirconium and hafnium oxides and nitrides. The reactivity of metal dialkylamides is quite high, so deposition is possible at very low substrate temperatures, even down to room temperature. This low temperature capability can be used for coating plastics and biological specimens, or for patterning by lift-off of photoresist [41]. The thermal stability of dialkylamides is generally limited, so substrate temperatures must be kept below their decomposition temperatures. A widely used metal amide precursor is tetrakis(ethylmethylamido)hafnium (TEMAH) [42, 43]:

Hf(NEtMe)$_4$ 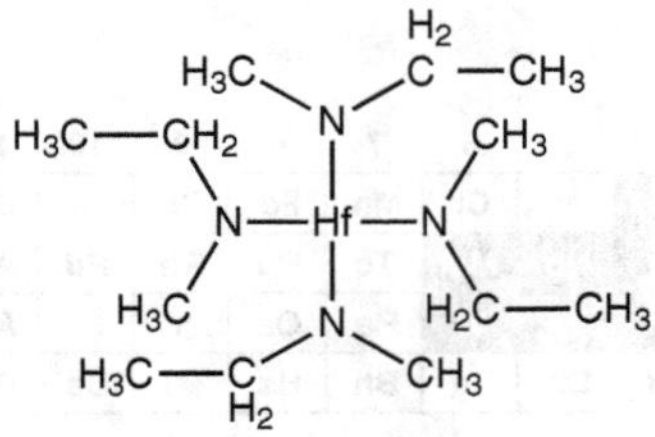

Metal amides may also be stabilized by the chelate effect. An example is a cyclic amide of tin(II) ("CAT"),

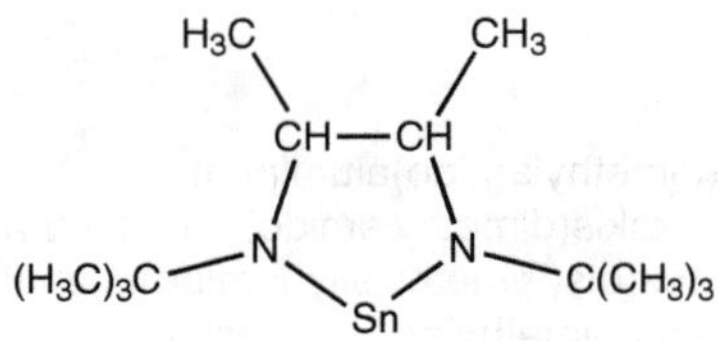

which is an effective ALD precursor for SnS and SnO_2 [44–46].

An example of a precursor with both imide bonds and amide bonds is provided by the useful precursor bis(*tert*-butylimido)bis(dimethylamido)tungsten(VI):

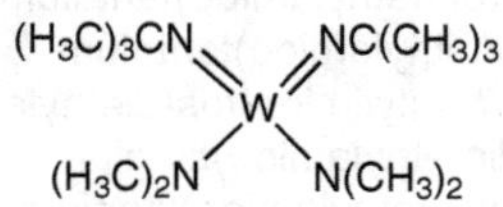

Metal amidinates have two metal-nitrogen bonds, as diagrammed in Fig. 2.19 [16]. Amidinates relate to amides analogously to how β-diketonates relate to alkoxides. The chelate effect results in metal amidinates being more thermally stable than metal amides, which have only one metal-nitrogen bond. By making

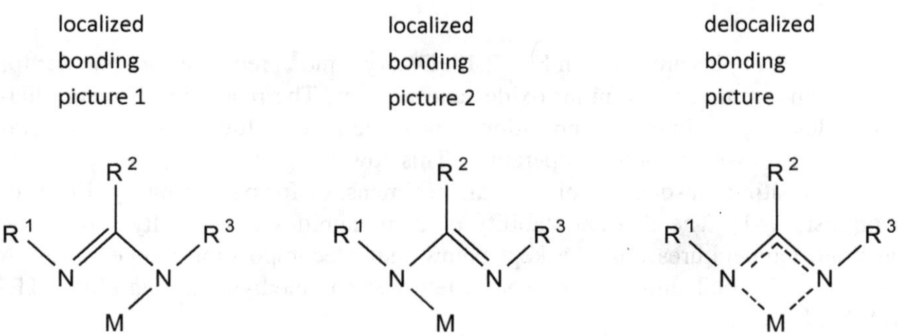

Fig. 2.19 Three ways to represent metal amidinate compounds. R^1, R^2 and R^3 are non-metals, usually alkyl groups C_xH_{2x+1}; other non-metals, such as silicon or nitrogen may be included

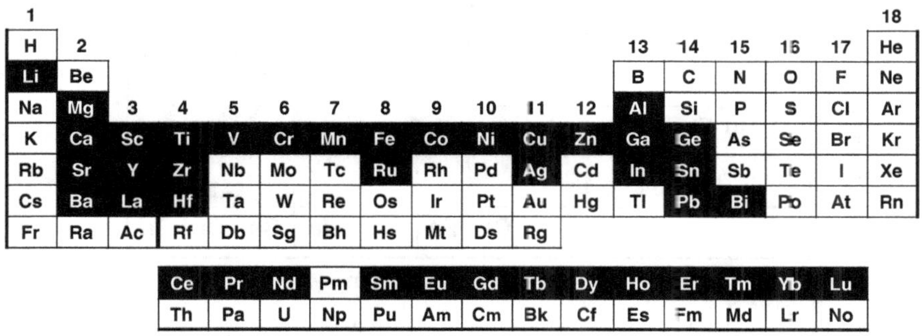

Fig. 2.20 Elements with amidinate ALD precursors

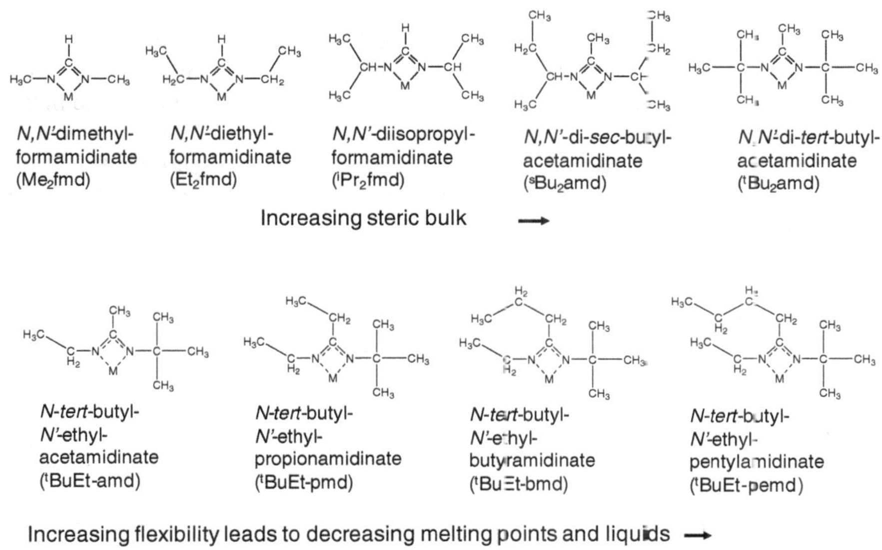

Fig. 2.21 Some amidinate ligands

appropriate choices of the amidinate ligands, volatile and reactive ALD precursors have been constructed for many metals, as shown in Fig. 2.20 [47–51].

The physical and chemical properties of metal amidinates can be adjusted by choice of the alkyl groups attached to the two nitrogens in an amidinate ligand (Fig. 2.21). The required size of the alkyl groups depends both on the size of the metal atom and the number of amidinate ligands attached to it (normally equal to the oxidation state of the metal, for a neutral compound).

To make the most volatile amidinates, the alkyl groups should be just large enough to prevent oligomerization of the compound. This trend is illustrated for metal acetamidinates with two ligands in Fig. 2.22, which display monomeric, dimeric, or polymeric structures, depending on the ligand size. Small metals and

Increasing ligand bulk →									
tert-butyl$_2$	m	m	m	m	m	m	d	d	d
isopropyl$_2$	m	m	m	d	d	d	d	p	p
^{t}Bu-Et	m	d	d	d				p	p
n-propyl$_2$		d							
	Ni	Co	Cr	Fe	Mg	Mn	Ca	Sr	Ba

Increasing size of metal atom →

m = volatile monomer | d = volatile dimer | p = non-volatile polymer

Fig. 2.22 Structures of metal(II) acetamidinates

bulky ligands form volatile monomers, as indicated in the upper left part of Fig. 2.22. The most volatile compound for each metal is the lowest monomeric one. The dimers below them are a little less volatile, and become monomeric in the vapor phase. Compounds with ligands that are too small for their large metals result in non-volatile polymers, as seen in the lower right part of Fig. 2.22. Bis(*N,N′*-di-*tert*-butylacetamidinato)nickel(II) is a widely used precursor for nickel nitride and nickel oxide films:

Ni(^{t}Bu$_2$amd)$_2$

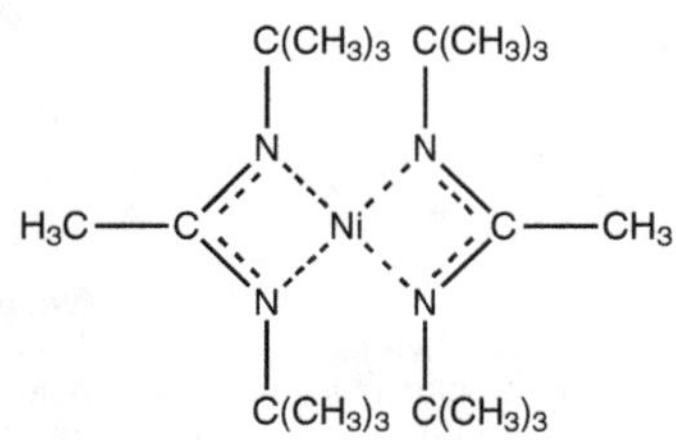

Because its surface reactions are slow, this same precursor is very useful in CVD, where it produces highly conformal coatings much more quickly than is possible by ALD [52, 53].

Copper(I) amidinates are always dimers, such as (*N,N′*-di-*sec*-butylamido)copper(I) dimer [47, 54, 55]:

Cu$_2$(*sec*-Bu$_2$amd)$_2$

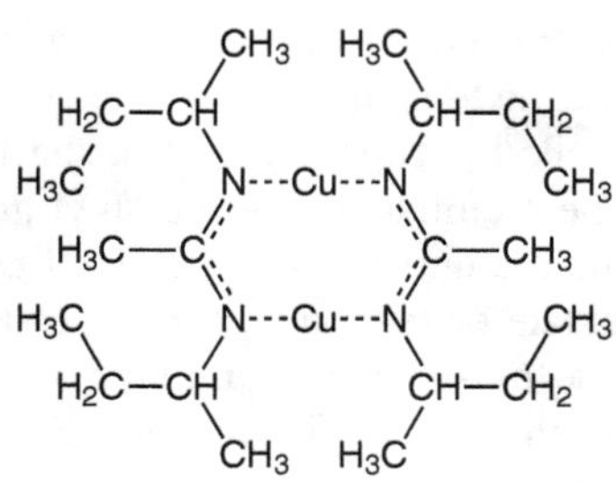

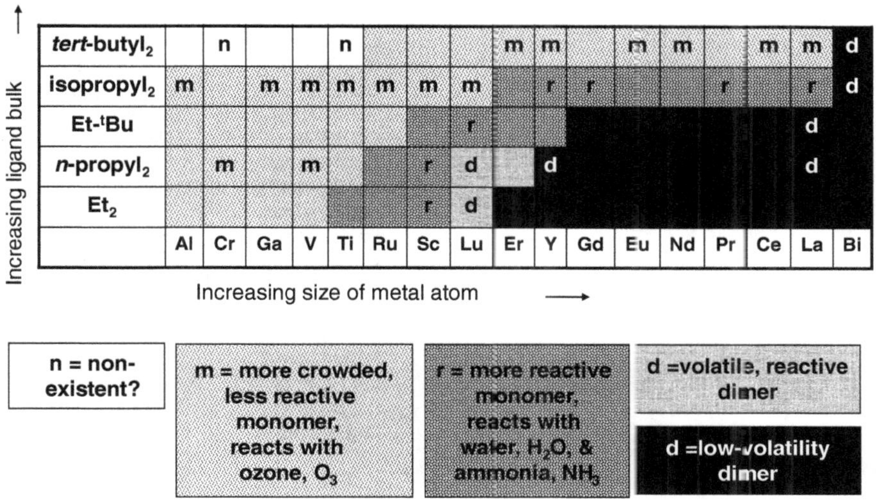

Fig. 2.23 Structures of metal(III) acetamidinates

For trivalent metals, if the alkyl groups are too large, reactivity can be decreased because of crowding of the ligands around the metal, or, in extreme cases (such as Ti(III) with bulky *tert*-butyl groups) the compound cannot be synthesized. These effects are illustrated in Fig. 2.23 for metals with three amidinate ligands. The crowded monomeric compounds (marked m) near the top and left do not react with water or ammonia. Thus to run ALD reactions with these reagents, less crowded monomers (marked r) are needed. Still smaller ligands give less volatile dimers. A very useful precursor for lanthanum is tris(*N,N′*-diisopropylformamidinato)lanthanum [56, 57], which is also the most volatile lanthanum compound known:

La(^{i}Pr$_2$fmd)$_3$

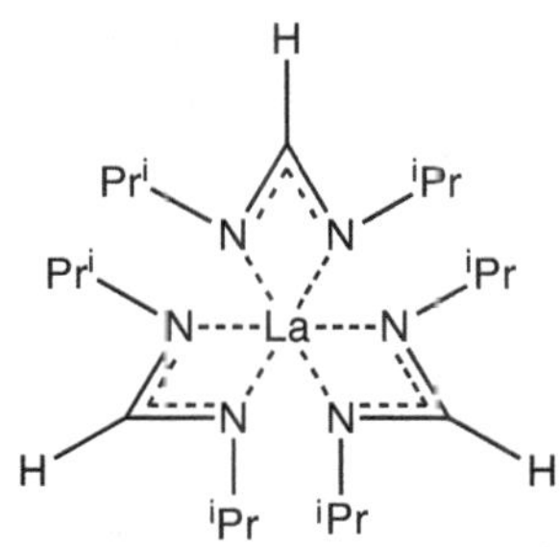

For tetravalent metal amidinates, only the smallest alkyl groups, such as methyl groups, can fit around even large metals, such as zirconium or hafnium [58]. An example is tetrakis(*N,N′*-dimethylformamidinato)zirconium, an extremely stable ALD precursor for zirconium oxide [59, 60]:

Fig. 2.24 Amidinate compounds used in ALD

$Ag_2(^tBu_2\text{-amd})_2$
$Ca(^tBu_2\text{-amd})_2$
$Co(^iPr_2\text{-amd})_2$
$Co(^tBuEt\text{-amd})_2$
$Cr(Et_2\text{-amd})_3$
$Cu_2(^iPr_2\text{-amd})_2$
$Cu_2(^sBu_2\text{-amd})_2$
$Er(^tBu_2\text{-amd})_3$
$Fe(^iPr_2\text{-amd})_2$
$Fe(^tBu_2\text{-amd})_2$
$Ga(Et_2\text{-amd})_3$
$Gd(^iPr_2\text{-amd})_3$
$Hf(Me_2\text{-fmd})_4$
$Hf(Me_2\text{-pmd})_4$
$Hf(Me_2\text{-bmd})_4$
$La(^iPr_2\text{-fmd})_3$
$La(^tBu_2fmd)_3$
$Lu(Et_2\text{-fmd})_3$
$Lu(Et_2\text{-amd})_3$
$Mg(^tBu_2amd)_2$
$Mg(^sBu_2\text{-amd})_2$
$Mn(^iPr_2\text{-pemd})_2$
$Mn(^tBu_2\text{-amd})_2$
$Ni(^tBu_2\text{-amd})_2$
$Pr(^iPr_2\text{-amd})_3$
$Ru(^tBu_2amd)_2(CO)_2$
$Sc(Et_2\text{-amd})_3$
$Sr(^tBu_2\text{-amd})_2$
$Ti(^iPr_2\text{-amd})_3$
$V(Et_2\text{-amd})_3$
$V(^iPr_2\text{-amd})_3$
$Y(^iPr_2\text{-amd})_3$
$Zn(^iPr_2\text{-amd})_2$
$Zr(Me_2\text{-fmd})_4$
$Zr(Me_2\text{-pmd})_4$
$Zr(Me_2\text{-bmd})_4$

$Zr(Me_2fmd)_4$

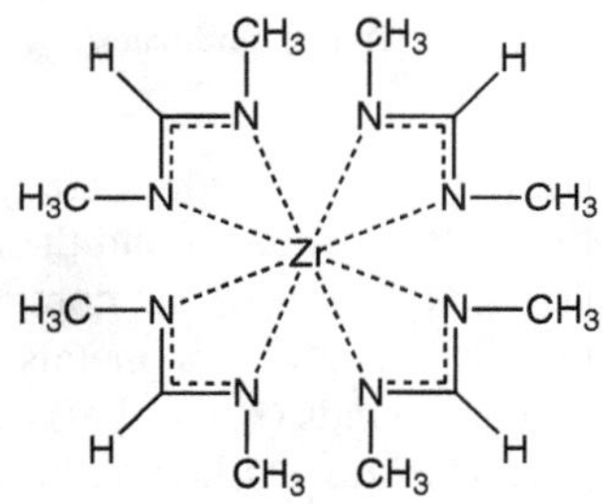

Particular examples of metal amidinate precursors are listed in Fig. 2.24. Ligand sizes on metal amidinates can generally be chosen so that they have high reactivity to many non-metal precursors, such as water vapor, ammonia, hydrogen sulfide, and even molecular hydrogen. Direct liquid injection can be used to vaporize liquid amidinates [61], as well as solid ones dissolved in inert hydrocarbon solvents.

2.3 Types of ALD Precursors for Non-metals

Metal oxides are the materials made most often by ALD. Water vapor is the most common source of oxygen, while alcohol vapors and elemental oxygen gas, O_2, have also been used. More reactive sources of oxygen include oxygen atoms, O, ozone, O_3, hydrogen peroxide, H_2O_2, and carboxylic acids, such as acetic acid, CH_3COOH. Oxygen atoms generated in a plasma are the most reactive oxygen sources. A disadvantage of O atoms is that they recombine easily on many surfaces, so it can be difficult to reach saturation coverage inside the deeper parts of narrow holes. An O plasma also contains O_2 and O_3, so their surface reactions should also be considered in interpreting data on plasma-enhanced ALD (PEALD). It should be noted that O_3 is always accompanied by an excess of O_2, and that H_2O_2 is always diluted by H_2O. Pure O_3 and H_2O_2 can be generated in small amounts, but these pure materials are far too explosive to be used in ALD. Thermal decomposition of O_3 and H_2O_2 can limit the uniformity of step coverage in holes with high aspect ratios, particularly at substrate temperatures over about 200 °C.

Ammonia, NH_3, is the most common source of nitrogen in ALD reactions. Dinitrogen, N_2, is normally unreactive under ALD conditions. Plasma-activated N_2 is much more reactive than N_2. Nitrogen atoms readily recombine into non-reactive N_2 on surfaces. Thus very long exposures may be needed to saturate ALD reaction with nitrogen plasma inside holes with high aspect ratios. Hydrazine, N_2H_4, is more reactive than NH_3 [62], but is highly toxic and explosive at high concentrations.

Phosphorus and arsenic are customarily delivered by their hydrides, phosphine, PH_3, and arsine, AsH_3. Despite their very poisonous properties, these gases are widely used in the deposition of III-V semiconductors, although mostly under CVD conditions, rather than strict ALD conditions. Metal phosphates have been made by ALD from a less dangerous phosphorus precursor diisopropylphosphate [63].

Fluorine precursors include hydrogen fluoride, which is corrosive and highly toxic. Fluorine may also be incorporated into ALD films from volatile metal fluorides, such as WF_6 and TiF_4 [64].

Carbon precursors for ALD include acetylene, C_2H_2, and formic acid, HCOOH. Carbon may also be incorporated from ligands of metal precursors because of thermal decomposition; however, such decomposition usually means that ideal ALD processes cannot be achieved and step coverage in narrow holes may be non-uniform.

Sulfides and selenides have normally been deposited from hydrogen sulfide, H_2S [30, 65–67], and hydrogen selenide, H_2Se, gases. Recently, bis(triethylsilyl)selenium, $(R_3Si)_2Se$, has been introduced as a less toxic source for ALD of selenides [68]. Its tellurium analog is effective as an ALD precursor for tellurides.

2.4 Types of ALD Reactions

ALD reactions usually transfer one atom from a surface-bound group to a vapor group, or from a vapor group to a surface-bound group (the reverse direction). The transferred atoms are hydrogen, oxygen, fluorine, or chlorine.

The vast majority of ALD reactions involve transfer of a hydrogen atom. The molecule or surface that brings the hydrogen into the reaction is called a Brønsted acid by chemists. The molecule or surface that receives the hydrogen atom is called a Brønsted base. Thus, these ALD reactions may be considered as examples of Brønsted acid–base reactions.

Normally, Brønsted acid–base reactions occur in solution (usually water), and many of the species involved are electrically charged ions. It is appropriate in this situation to describe these reactions as transfers of protons (positively charged hydrogen ions). The polar molecules of solvent (water) stabilize the ions.

In vapor-based ALD processes, which take place in the absence of any stabilizing solvent, ions are normally not present. (An exception would be a plasma applied directly to the surface.) Thus in discussing ALD, it would be quite artificial to discuss reactions involving non-existent ions. Therefore we will use language appropriate to reactions between neutral species, in which the transferred species are also neutral atoms. For example, we will speak of hydrogen transfers, rather than proton transfers. This approach has an advantage of simplicity, in that balancing of charges in reactions is automatic. In contrast, when using the language of proton transfer, one must include an extra step of keeping track of the electrons and balancing the charges in the reaction.

2.5 ALD Reactions that Transfer Hydrogen

The simplest source of hydrogen atoms is a remote hydrogen plasma. By "remote" we mean that the surface is far enough away from the excitation source of the plasma so that electrons and ions have recombined, leaving only neutral hydrogen atoms (along with many H_2 molecules). One reason for keeping the direct plasma away from the substrate surface is to avoid damage by impact of energetic ions. Another reason is that if the surface reaction actually required ions, it would not occur inside small holes, into which the plasma conditions cannot extend. Thus an ion-induced surface reaction would not have any step coverage, negating one of the important advantages of ALD.

Remote hydrogen plasma has been used to deposit very reactive metals, such as titanium and tantalum from their chlorides [69, 70]. Assuming that the plasma has already provided hydrogen atoms to the surface, the next reaction step would be the following:

Hydrogen atoms from the plasma then remove chlorine from the surface as hydrogen chloride gas and re-hydrogenate the surface.

The surface is now returned to the state in which it is ready for the first reaction to begin the ALD cycle again.

It is also noteworthy that the H_2 molecules in the remote plasma do not participate in this reaction; that is why plasma activation of the hydrogen is necessary. In fact, molecular H_2 is non-reactive to almost all potential metal precursors.

One important exception is copper amidinate, which is readily reduced to copper metal by H_2 under ALD conditions [47, 54, 55, 71, 72]. The first step is dissociative chemisorption of copper amidinate on a copper surface [73]:

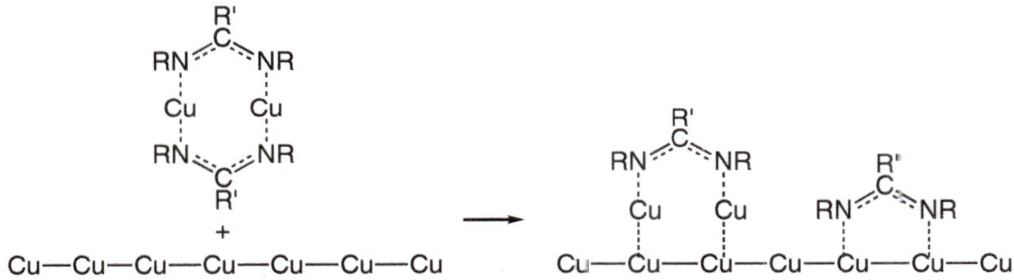

Molecular hydrogen then transfers to the amidinate ligands to release amidine vapor and return the surface to pure copper:

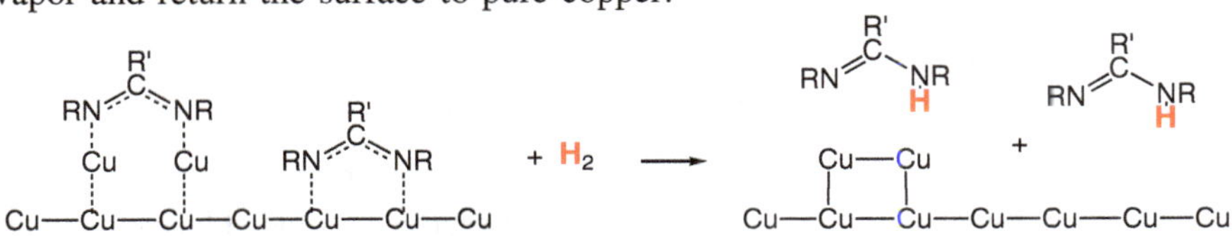

Similar reduction reactions allow ALD of other first-row transition metals (manganese, iron, cobalt, and nickel) from their amidinate precursors [51].

Many metal oxides can be deposited by reaction of water with suitably reactive metal precursors. One example is the reaction of tetrakis(dimethylamido)hafnium with water to make hafnium dioxide [42, 74]. The overall reaction is the following:

$$Hf(NMe_2)_4 + 2H_2O \;=>\; HfO_2 + 4\ HNMe_2$$

Chemisorption of the precursor occurs by hydrogen transfer to the dimethylamido ligands to release dimethylamine gas:

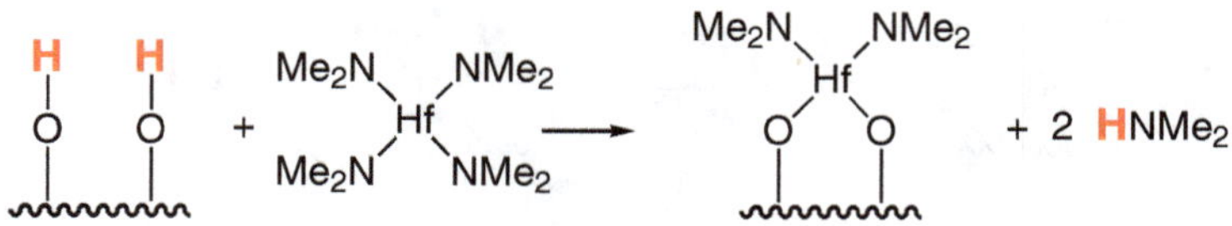

Hydrogen is next transferred from water to the remaining surface-bound dimethylamide ligands:

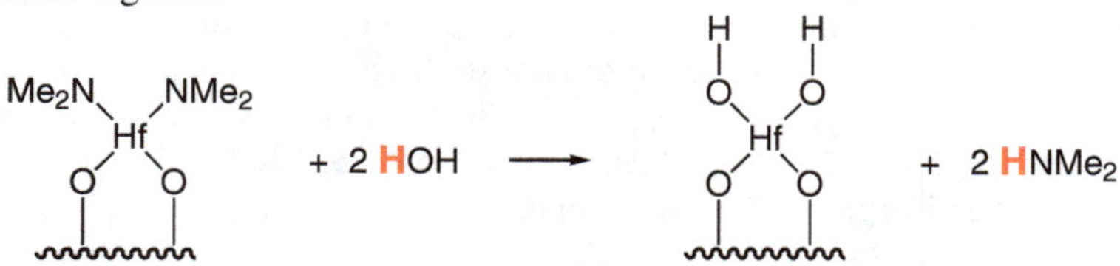

The surface is thus returned to the hydroxylated state in which it is ready to begin the next cycle.

Nitrides can be deposited by hydrogen transfer reactions from ammonia. A typical example is titanium(IV) tetrachloride reacting with ammonia to form titanium(III) nitride [75–77].

$$TiCl_4 + NH_3 => TiN + \text{gaseous products}$$

In the first step, hydrogen atoms transfer from surface NH groups to chlorine atoms on the titanium precursor:

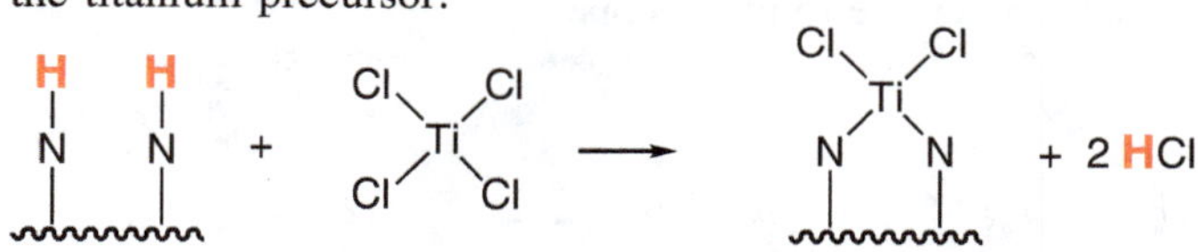

Hydrogen atoms then transfer from ammonia to surface-bound chloride ligands:

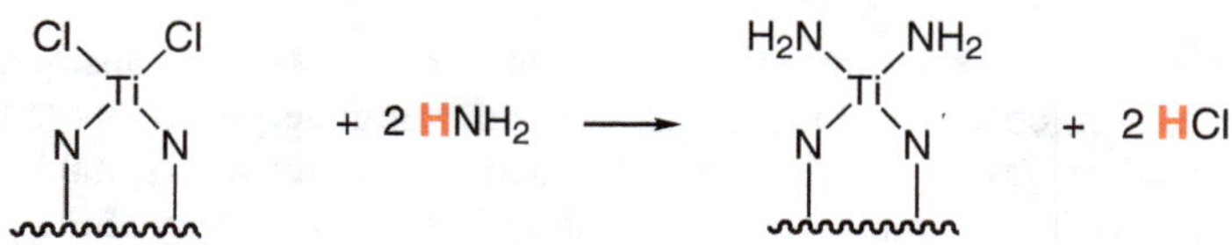

After this step, the titanium remains in the +4 oxidation state, corresponding to a product Ti_3N_4. In order for titanium to reach the +3 oxidation state in the known product, TiN, some as yet uncharacterized reduction and elimination reactions must occur:

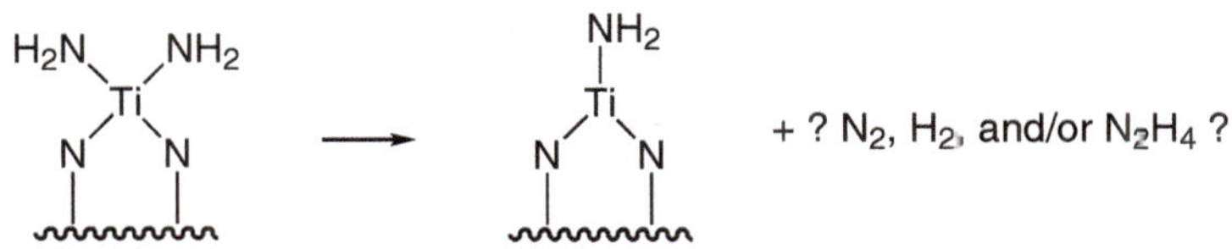

The driving force for this reaction is the fact that the hypothetical Ti(IV) nitride, Ti_3N_4, is unstable. However, the byproducts and mechanism of this reaction are unknown.

Tungsten nitride can be deposited by reaction of ammonia with a tungsten precursor bis(*tert*-butylimido)bis(dimethylamido)tungsten(VI) [78, 79]. Hydrogen first transfers from surface NH groups to dimethylamido groups on the precursor to free dimethylamine gas as a byproduct:

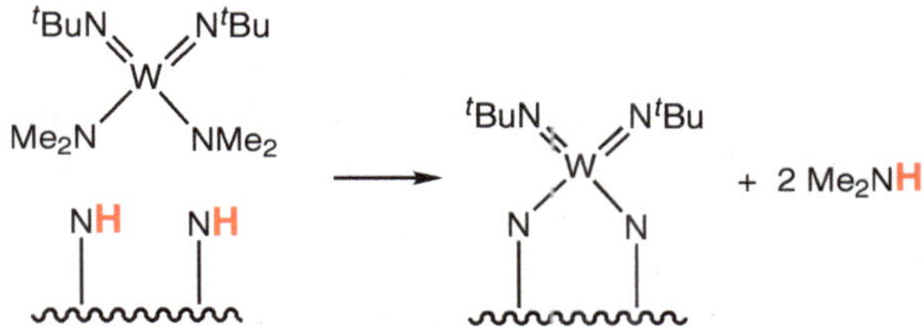

Hydrogen then transfers either from ammonia [80] or from a methyl group in a *tert*-butylimido group [79]:

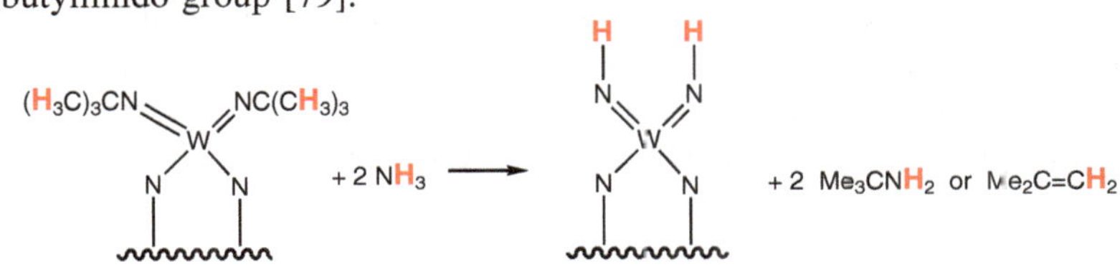

The product at this stage is the hypothetical, unstable WN_2, which must lose nitrogen to form the known stable product WN.

ALD of metal silicates can be achieved by hydrogen transfer reactions from tris(alkoxy)silanols, which serve as sources of both silicon and oxygen in the product film [81, 82]. As an example, we consider ALD of hafnium silicate. In the first step, hydrogen atoms transfer from surface hydroxyls to dimethylamide ligands of the hafnium precursor:

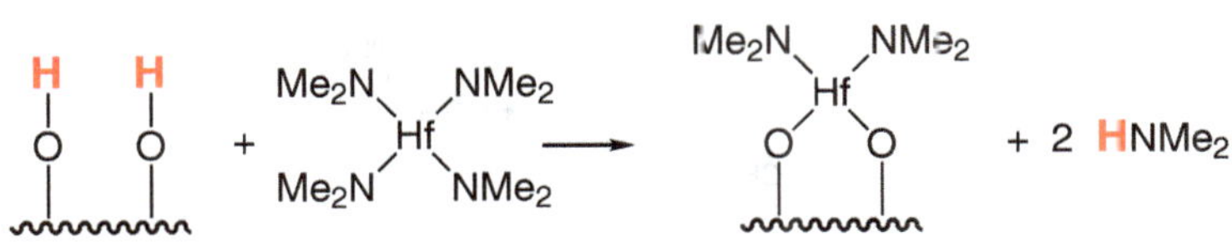

Hydrogen atoms then transfer from two silanols to the remaining surface-bound dimethylamides, releasing them as dimethylamine vapor:

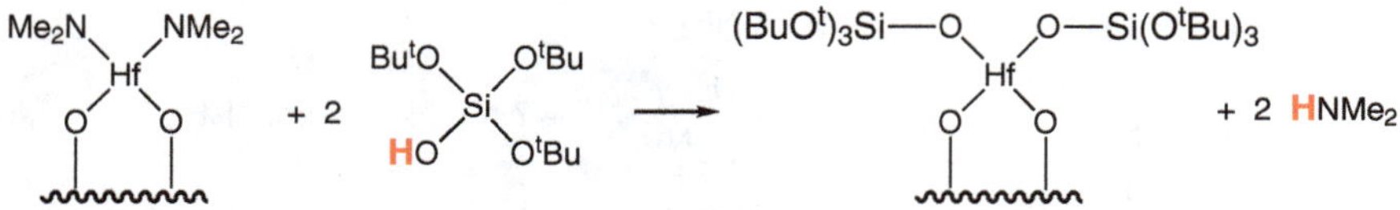

Regeneration of surface hydroxyls is achieved by β-hydrogen transfer from tertiary butyl groups:

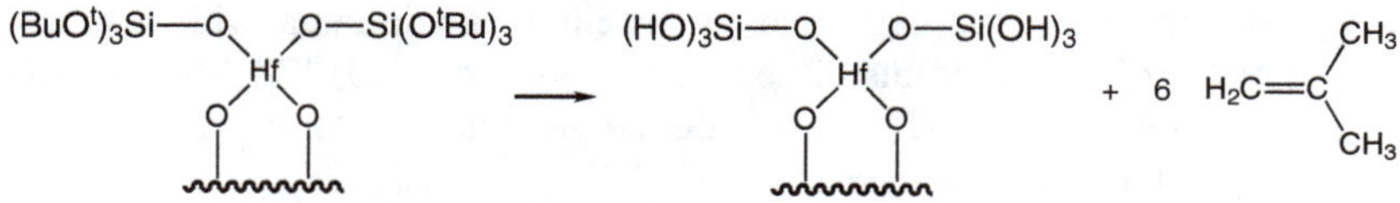

This last step recreates a hydroxylated surface, which is ready to repeat the first step, while releasing isobutene vapor as a byproduct. These reactions produce a hafnium silicate film with 2 to 3 times more silicon atoms than hafnium atoms [82]. The hafnium content can be increased by replacing some of the silanol cycles with water cycles, thereby depositing pure HfO_2 in those cycles [42].

In the case of ALD aluminum silicate using trimethylaluminum and tris(*tert*-butoxy)silanol, the product is almost pure SiO_2, with less than 1 % of Al_2O_3, and growth per cycle up to 12 nm, which is about two orders of magnitude higher than observed in all other known ALD reactions [83]. This unusually high growth rate appears to arise from the aluminum catalytically incorporating many silanol molecules. The first reaction is a usual hydrogen atom transfer from surface hydroxyl groups to methyl ligands of the trimethylaluminum precursor [84]:

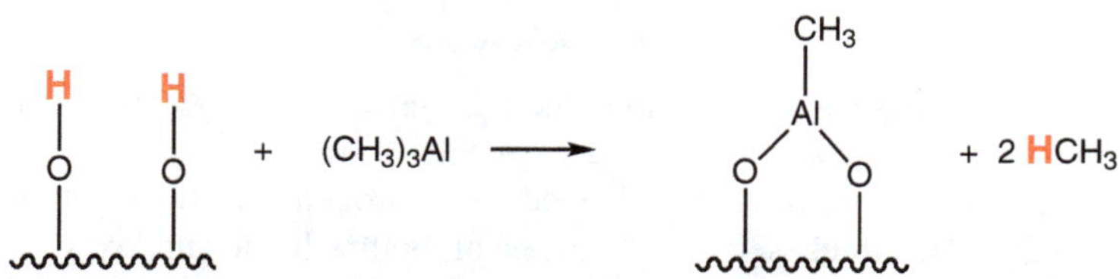

Next, hydrogen atoms transfer from silanols to methyl ligands that are still attached to aluminum on the surface:

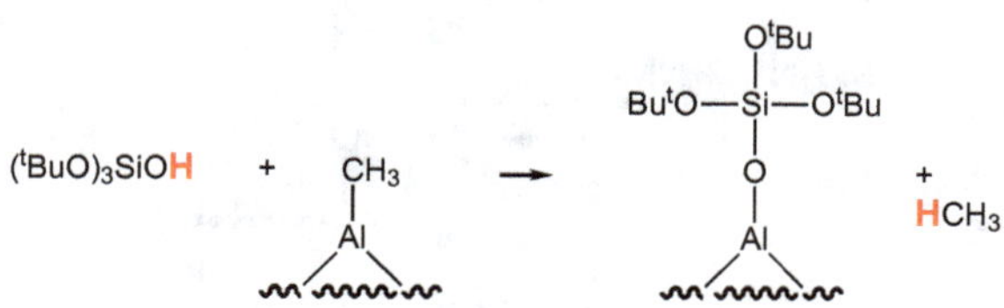

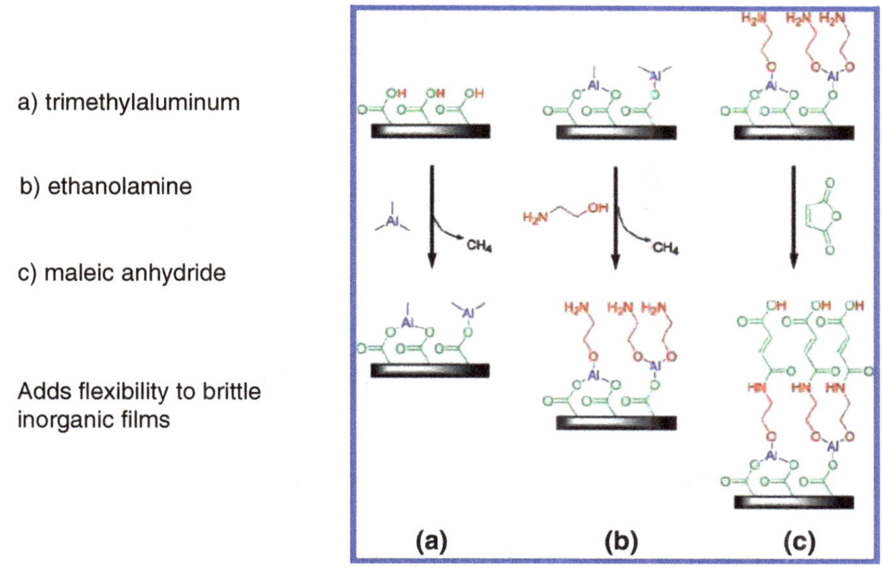

Fig. 2.25 Adding Organic Components to ALD Films (from S. M. George, Acc. Chem. Res. 42, 498 (2009), with permission)

Additional silanol precursors can then attach to the same aluminum atom by inserting themselves into the aluminum-oxygen bond, along with an internal hydrogen transfer from the silanol to a neighboring *tert*-butoxy group to liberate *tert*-butanol vapor:

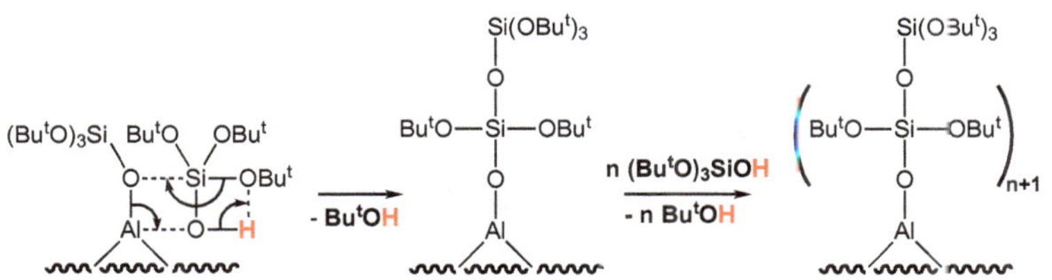

The result of these successive insertion reactions is a siloxane polymer tethered to the surface. The *tert*-butyl side-chains on this polymer are thermally unstable, so they eliminate isobutene by β-hydrogen transfer:

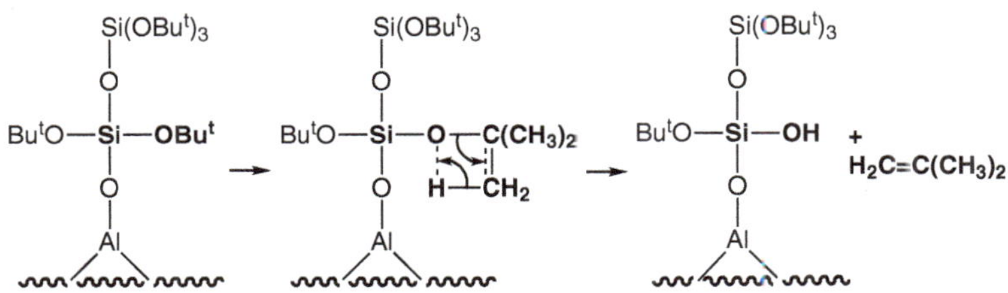

The siloxane polymer chains can cross-link by elimination of *tert*-butanol vapor:

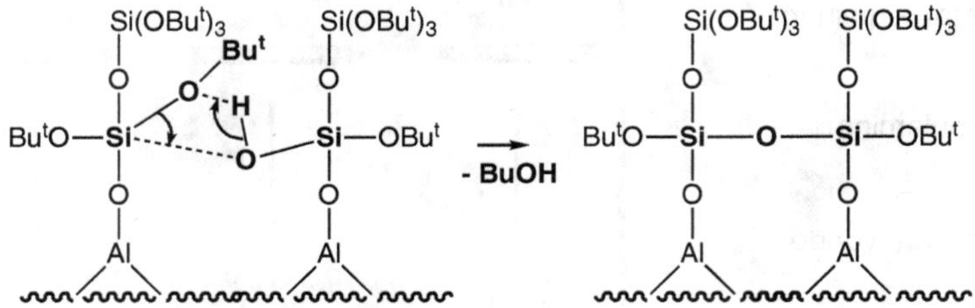

Neighboring hydroxyl side-groups can also cross-link by elimination of water vapor:

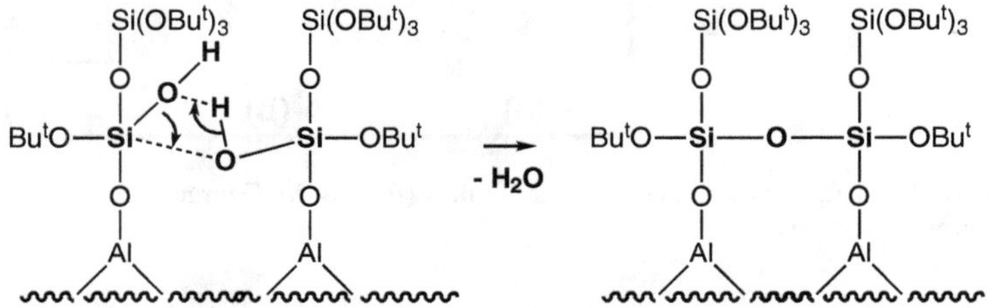

Complete crosslinking by these reactions produces a solid silica layer that is impervious to diffusion of more silanol up to the aluminum catalyst, so reaction stops.

Successive ALD reactions of trimethylaluminum, ethanolamine, and then maleic anhydride can incorporate organic chains along with aluminum oxide, as illustrated in Fig. 2.25 [85]. Each of these reaction steps involves a hydrogen transfer reaction.

2.6 ALD Reactions that Transfer Oxygen

Ozone, O_3, as an ALD reactant typically involves transfers of oxygen atoms. For example, trimethylaluminum reacts with ozone to make aluminum oxide by the overall reaction [86, 87].

$$2(CH_3)_3Al + 4\ O_3 => Al_2O_3 + 3\ CH_4 + 3\ H_2O + 3\ CO_2$$

In the first ALD half-reaction, hydrogen atoms transfer from surface hydroxyls to methyl ligands to form methane:

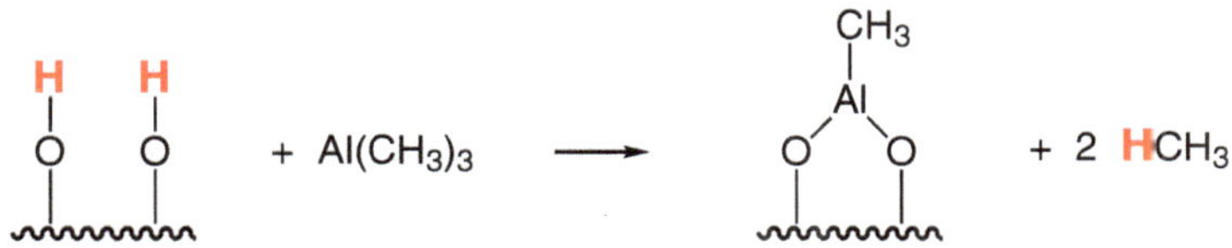

In the second half-reaction, ozone oxidizes the surface-bound methyl groups to carbon dioxide and water, and also transfers oxygen atoms to the aluminum on the surface, recreating a hydroxylated surface ready to begin the next cycle:

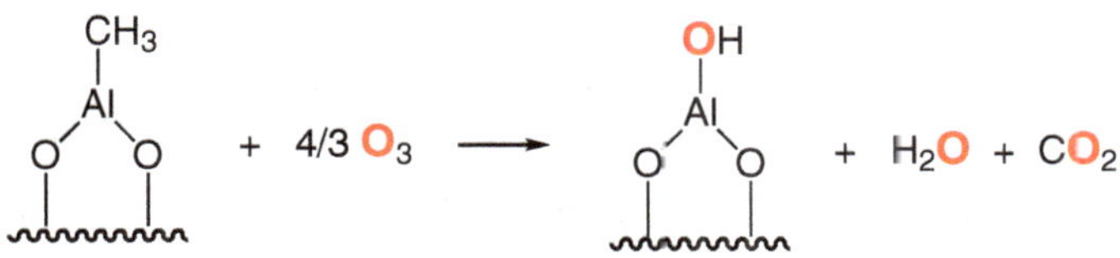

Under some conditions, water vapor may not be detected as a byproduct because it has reacted with other surface-bound methyl groups to produce methane.

Noble metals, such as platinum [31], rhodium [88], iridium [89], and ruthenium [90], may be deposited by oxygen or ozone reacting with precursors having carbon-containing ligands [91]. If a layer of noble metal already exists on the substrate, then oxygen can chemisorb onto the metal [31]:

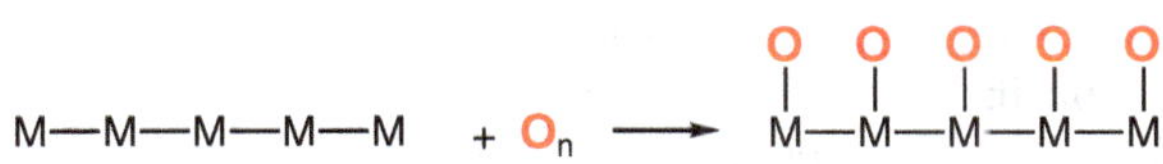

In the next half-cycle, adsorbed oxygen atoms combust the ligands to form carbon dioxide and water:

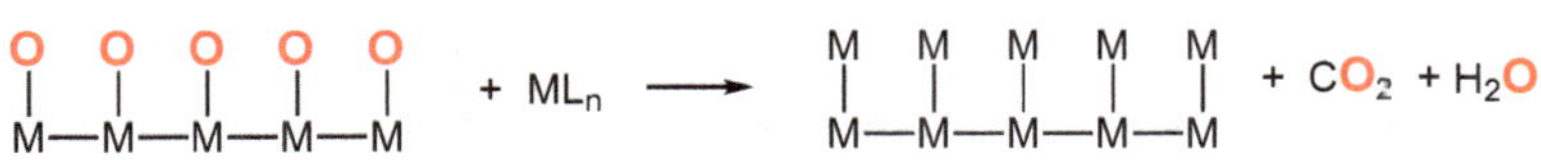

In order to supply enough oxygen atoms to complete this reaction, chemisorption of additional oxygen below the surface is required. Initiation of this ALD cycle is difficult on substrates that do not chemisorb oxygen atoms, such as metal oxides or non-noble metals. In such cases, little or no metal is deposited for many cycles, and even when growth begins, the metal atoms may remain in separated clusters until they become large enough to coalesce into a film.

2.7 ALD Reactions that Transfer Fluorine

Fluorine atom transfers are the key steps in deposition of tungsten and molybdenum metals. As an example, consider the reduction of tungsten hexafluoride by disilane [92]. The overall reaction is:

$$WF_6 + Si_2H_6 => W + 2\ SiF_3H + 2\ H_2$$

In the first half cycle, a fluorine atom moves from WF_6 gas to liberate SiF_2H groups chemisorbed on the surface by previous cycles:

$$\text{surface} - WSiF_2H + WFF_5 => \text{surface} - WWF_5 + SiFF_2H$$

After disilane is supplied in the next half-cycle, three fluorine atoms move from their attachment to tungsten on the surface to break apart the disilane:

$$\text{surface} - WWF_5 + Si_2H_6 => \text{surface} - WWSiF_2H + SiF_3H + 2\ H_2$$

Clearly, this is only a summary of a very complex series of reactions, which break 1 Si–Si bond, 5 W–F bonds and 4 Si–H bonds while forming a new W–Si bond, 5 new Si–F bonds, and 2 new H–H bonds. Starting this chemistry on a surface that is not already covered with tungsten is difficult, so nucleation delays are common on oxide substrates.

2.8 ALD Reactions that Transfer Chlorine

Chlorine atom transfer plays a key role in a process for ALD of germanium telluride [68]. Chlorine atoms on a previously chlorinated germanium surface transfer to trialkylsilicon groups released from the bis(trialkylsilyl)tellurium precursor vapor [93]:

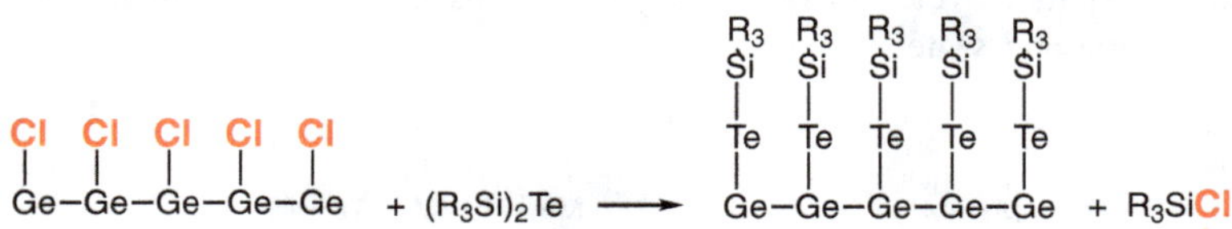

Chlorine atoms on the germanium precursor vapor remove surface trialkylsilyl groups during the next half-cycle:

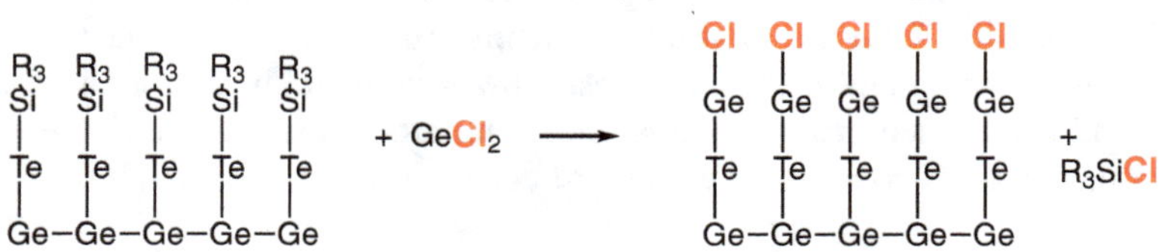

The main driving force for these reactions is the formation of the stronger Si–Cl bonds after breaking the weaker Ge–Cl bonds. A wide variety of selenides and tellurides, as well as antimony metal [94], have been made by similar reactions.

2.9 Conclusions

Successful ALD precursors and processes have been developed for many elements and compounds. Despite all that has been discovered so far, there still remain many areas for improvement in the range of compositions, purity, phase composition, structure, morphology, and properties of materials produced by ALD.

Acknowledgments I want to thank my students and post-doctoral fellows who have worked together with me in this interesting, challenging, and important field of research: Titta Aaltonen, Sean Barry, Jill S. Becker, Mike Coulter, Damon Farmer, Dennis M. Hausmann, Adam Hock, Jaeyeong Heo, Daewon Hong, Esther Kim, Kyoung-ha Kim, Jean Sebastien Lehn, Sang Bok Kim, Booyong S. Lim, Don Keun Lee, Sang Woon Lee, Huazhi Li, Zhefeng Li, Zhengwen Li, Xinye Liu, Yiqun Liu, Hisashi Nakagawa, Jun Ni, Wontae Noh, Dawen Pang, Helen H. Park, Jin-Seong Park, Antti Rahtu, P. Venkateswara Rao, Leonard N. J. Rodriguez, Philippe de Rouffignac, Prasert Sinsermsuksakul, Amethyst Smith, Seigi Suh, Hongtao Wang, Shenglong Wang, Xinwei Wang, Sheng Xu, and Andrew P. Yousef.

Support was provided by the US National Science Foundation, Dow Chemical, IBM, and Intel.

References

1. Puurunen RL (2005) Surface chemistry of atomic layer deposition: a case study for the trimethylaluminum/water process. J Appl Phys 97:121–301
2. Niinistö J, Kukli K, Heikkilä M, Ritala M, Leskela M (2009) Atomic layer deposition of high-k oxides of the group 4 metals for memory applications. Adv Eng Mater 11:223
3. Ritala M, Niinisto J (2009) Chemical vapour deposition: precursors, processes and applications. Jones AC, Hitchman ML (eds). Royal Society of Chemistry, Cambridge
4. George SM (2010) Atomic layer deposition: an overview. Chem Rev 110:111
5. Ritala M, Niinisto J (2009) Industrial applications of atomic layer deposition. ECS Trans 25:641
6. Sundaram GM et al (2009) Thin films for 3D: ALD for non-planar topographies. Solid State Technol 52:12
7. Tsai W, Goel N, Koveshnikov S, Majhi P, Wang W (2009) Challenges of integration of high-κ dielectric with III-V materials (Invited paper). Microelectron Eng 86:1540
8. Clancy P et al (2010) Analytical characterization of ALD thin film precursors. ECS Trans 28:349
9. Doll GL, Mensah BA, Mohseni H, Scharf TW (2010) Chemical vapor deposition and atomic layer deposition of coatings for mechanical applications. J Therm Spray Technol 19:510
10. Molle A, Baldovino S, Spiga S, Fanciulli M (2010) High permittivity materials for oxide gate stack in Ge-based metal oxide semiconductor capacitors. Thin Solid Films 518:S96
11. Puurunen RL, Kattelus H, Suntola T (2010) Atomic layer deposition in MEMS technology. Handbook of Silicon Based MEMS Materials and Technologies Elsevier. Boston, p. 433

12. Verghese M, Maes JW, Kobayashi N (2010) Atomic layer deposition goes mainstream in 22 nm logic technologies. Solid State Technol 53:18
13. Kim H, Lee H-B-R, Maeng WJ (2009) Applications of atomic layer deposition to nanofabrication and emerging nanodevices. Thin Solid Films 517:2563
14. Kittl JA et al (2009) High-*k* dielectrics for future generation memory devices (Invited Paper). Microelectron Eng 86:1789
15. Knoops HCM et al (2009) Atomic layer deposition for all-solid-state 3D-integrated batteries. ECS Trans 25:333
16. Edelmann FT (2009) Lanthanide amidinates and guanidinates: from laboratory curiosities to efficient homogeneous catalysts and precursors for rare-earth oxide thin films. Chem Soc Rev 38:2253
17. Godlewski M et al (2009) Barriers in miniaturization of electronic devices and the ways to overcome them—from a planar to 3D device architecture. Acta Phys Pol, A 116:S19
18. Hinkle CL, Milojevic M, Vogel EM, Wallace RM (2009) Surface passivation and implications on high mobility channel performance (Invited Paper). Microelectron Eng 86:1544
19. Torndahl T (2004) Growth of copper metal by atomic layer deposition using copper(I) chloride, water and hydrogen as precursors. Thin Solid Films 458:129
20. Knapas K, Rahtu A, Ritala M (2009) Etching of Nb_2O_5 thin films by $NbCl_5$. Chem Vap Deposition 15:269
21. Yousfi EB, Fouache J, Lincot D (2000) Study of atomic layer epitaxy of zinc oxide by in situ quartz crystal microgravimetry. Appl Surf Sci 153:223
22. Lukasiewicz M et al (2008) ZnCoO films obtained at low temperature by atomic layer deposition using organic zinc and cobalt precursors. Acta Physica Polonica 114:1235
23. Cho W et al (2003) Atomic layer deposition of Al_2O_3 thin films using dimethylaluminum isopropoxide and water. J Vac Sci Technol A: Vac, Surf Films 21:1366
24. Bayev AK, Podkovyrov AI (1993) Thermodynamics and associative interactions in dimethylzinc-dimethylsulfur system. Zhurnal Khimicheskoi Termodinamiki i Termokhimii 2:80
25. Lebedev BV, Kulagina TG (1993) Thermodynamics of the dimethyl(dimethyl sulfide)zinc complex at temperatures from 0 to 330 K. J Chem Thermodyn 25:591
26. Baev AK (1997) Thermodynamic study of structural changes in solutions of dimethyl-zinc with methyl compounds of Group VIB elements. Zh Neorg Khim 42:662
27. Niinistö J et al (2006) HfO[sub 2] Films grown by ALD using Cyclopentadienyl-type precursors and H_2O or O_3 as oxygen source. J Electrochem Soc 153:F39
28. Niinistö J et al (2007) Atomic layer deposition of HfO_2 thin films exploiting novel cyclopentadienyl precursors at high temperatures. Chem Mater 19:3319
29. O'Kane R et al (2007) Growth of HfO_2 by Liquid Injection MOCVD and ALD using new hafnium-cyclopentadienyl precursors. Chem Vap Deposition 13:609
30. Ihanus J et al (2002) Atomic layer deposition of SrS and BaS thin films using cyclopentadienyl precursors. Chem Mater 14:1937
31. Aaltonen T, Rahtu A, Ritala M, Leskela M (2003) Reaction mechanism studies on atomic layer deposition of ruthenium and platinum. Electrochem Solid-State Lett 6:C130
32. Aaltonen T, Ritala M, Sajavaara T, Keinonen J, Leskelae M (2003) Atomic layer deposition of platinum thin films. Chem Mater 15:1924
33. Jones AC et al (2006) MOCVD and ALD of high-j dielectric oxides using alkoxide precursors. Chem Vap Deposition 12:83
34. Xie Q, Jiang Y-l, Detavernier C, Deduytsche D, Meirhaeghe RLV (2007) Atomic layer deposition of TiO_2 from tetrakis-dimethyl-amido titanium or Ti isopropoxide precursors and H_2O. J Appl Phys 102:1
35. Xie Q et al (2007) Atomic layer deposition of TiO_2 from tetrakis-dimethyl-amido titanium or Ti isopropoxide precursors and H_2O. J Appl Phys 102:083521
36. Yang TS et al (2005) Atomic layer deposition of nickel oxide films using Ni(dmamp)[sub 2] and water. J Vac Sci Technol A: Vac, Surf Films 23:1238

37. Roeckerath M et al (2008) Characterization of lanthanum lutetium oxide thin films grown by atomic layer deposition as an alternative gate dielectric. Thin Solid Films 517:201
38. Kukli BK et al (2006) Atomic layer deposition and properties of lanthanum oxide and lanthanum-aluminum oxide films. Chem Vap Deposition 12:158
39. Nieminen M (2001) Formation and stability of lanthanum oxide thin films deposited from β-diketonate precursor. Appl Surf Sci 174:155
40. Kwon OS, Kim SK, Cho M, Hwang CS, Jeong J (2005) Chemically conformal ALD of $SrTiO_3$ thin films using conventional metallorganic precursors. J Electrochem Soc 152:C229
41. Biercuk MJ, Monsma DJ, Marcus CM, Becker JS, Gordon RG (2003) Low-temperature atomic-layer-deposition lift-off method for microelectronic and nanoelectronic applications. Appl Phys Lett 83:2405
42. Hausmann DM, Kim E, Becker J, Gordon RG (2002) Atomic layer deposition of hafnium and zirconium oxides using metal amide precursors. Chem Mater 14:4350
43. Becker JS, Kim E, Gordon RG (2004) Atomic layer deposition of insulating hafnium and zirconium nitrides. Chem Mater 16:3497
44. Heo J, Hock AS, Gordon RG (2010) Low temperature atomic layer deposition of tin dioxide, SnO_2. In: Proceedings of the AVS Atomic layer deposition conference
45. Heo J, Hock AS, Gordon RG (2010) Low temperature atomic layer deposition of tin oxide. Chem Mater 22:4964
46. Sinsermsuksakul P, Hartman K, Kim S, Heo J, Sun L, Park HH, Chakraborty R, Buonassisi T, Gordon RG, (2013) Appl Phys Lett 102:0539011; Sinsermsuksakul P, Chakraborty R, Kim SB, Heald SM, Buonassisi T, Gordon RG (2012) Chem Mater, 24:4556; Heo J, Kim SB, Gordon RG (2012) J Mater Chem 22:4599; Heo J, Kim SB, Gordon RG (2012) Appl Phys Lett 101:113507/1; Heo J, Kim SB, Gordon RG (2012) Appl Phys Lett 101:113507; Heo J, Kim SB, Gordon RG (2012) J Mater Chem 22:4599
47. Li Z, Barry ST, Gordon RG (2005) Synthesis and characterization of copper(I) amidinates as precursors for atomic layer deposition (ALD) of copper metal. Inorg Chem 44:1728
48. De Rouffignac P, Park J-S, Gordon RG (2005) Atomic layer deposition of Y2O3 thin films from Yttrium Tris(N, N'-diisopropylacetamidinate) and water. Chem Mater 17:4808
49. Gordon RG (2004) New precursors for atomic layer deposition (ALD) of metals and diffusion barriers. In: Proceedings of the AVS atomic layer deposition conference
50. Lim BS, Rahtu A, Park J-S, Gordon RG (2003) Synthesis and characterization of volatile, thermally stable reactive transition metal amidinates. Inorg Chem 42:7951
51. Lim BS, Rahtu A, Gordon RG (2003) Atomic layer deposition of transition metals. Nat Mater 2:749
52. Li Z, Gordon RG, Pallem V, Li H, Shenai DV (2010) Direct-liquid-injection chemical vapor deposition of nickel nitride films and their reduction to nickel films. Chem Mater 22:3060
53. Li Z, Gordon RG, Li H, Shenai DV, Lavoie C (2010) Formation of nickel silicide from direct-liquid-injection chemical-vapor-deposited nickel nitride films. J Electrochem Soc 157:H679
54. Li Z, Rahtu A, Gordon RG (2006) Atomic layer deposition of ultrathin copper metal films from a liquid copper(I) amidinate precursor. J Electrochem Soc 153:C787
55. Li Z, Gordon RG (2005) Atomic layer deposition of copper from Cu(I) Amidinates. In: Proceedings AVS atomic layer deposition conference
56. Liu Y, Xu M, Heo J, Ye PD, Gordon RG (2010) Heteroepitaxy of single-crystal $LaLuO_3$ on GaAs(111)A by atomic layer deposition. Appl Phys Lett 97:162910/1
57. Wang H et al (2009) Atomic layer deposition of lanthanum-based ternary oxides. Electrochem Solid-State Lett 12:G13
58. Gordon RG, Lehn J-S, Liu Y, Kim KH, Li Z, Coulter M, Li H, Pugh R, Shenai D (2007) New precursors for ALD of High-*k* dielectrics. In: Proceedings of the AVS atomic layer deposition conference
59. Lee B et al (2009) A novel thermally-stable zirconium amidinate ALD precursor for ZrO2 thin films. Microelectron Eng 86:272

60. Li H, Shenai DV, Lehn J-S (2010) Novel zirconium formamidinate precursor for the ALD of ZrO2. In: Proceedings of the materials research society symposium 1194E:A1105–1110. doi:10.1557/proc-1194-a05-10
61. Li Z, Lee DK, Coulter M, Rodriguez LNJ, Gordon RG (2008) Synthesis and characterization of volatile liquid cobalt amidinates. Dalton Trans 19:2592–2597
62. Burton BB, Lavoie AR, George SM (2008) Tantalum nitride atomic layer deposition using (tert-Butylimido)tris(diethylamido)tantalum and Hydrazine. J Electrochem Soc 155:D508
63. Gordon RG, Becker JS, Hausmann DM, Suh S (2005) US Patent 6,969,539
64. Pilvi T, Arstila K, Leskelä M, Ritala M (2007) Novel ALD process for depositing CaF_2 thin films. Chem Mater 19:3387
65. Hänninen T et al (1997) [$Ca(Thd)_2$(Tetraen)]: a monomeric precursor for deposition of CaS thin films. Chem Mater 9:1234
66. Kim JY, George SM (2010) Tin monosulfide thin films grown by atomic layer deposition using tin 2, 4-pentanedionate and hydrogen sulfide. J Phys Chem 114:17597
67. Sinsermsuksakul P, Hock AS, Gordon RG (2009) ALD of tin monosulfide, SnS. In: Proceedings of the AVS atomic layer deposition conference
68. Pore V, Hatanpää T, Ritala M, Leskelä M (2009) Atomic layer deposition of metal tellurides and selenides using alkylsilyl compounds of tellurium and selenium. J Am Chem Soc 131:3478
69. Lee J-H et al. (2002) IEDM technical digest. In: IEEE international electron devices meeting
70. Rossnagel SM, Sherman A, Turner F (2000) Plasma-enhanced atomic layer deposition of Ta and Ti for interconnect diffusion barriers. J Vac Sci Technol B: Microelect Nanomet Struct 18:2016
71. Li Z, Gordon RG, Farmer DB, Lin Y, Vlassak J (2005) Nucleation and adhesion of ALD copper on cobalt adhesion layers and tungsten nitride diffusion barriers. Electrochem Solid-State Lett 8:G182
72. Li Z, Gordon RG (2006) Thin, continuous, and conformal copper films by reduction of atomic layer deposited copper nitride. Chem Vapor Deposition 12:435
73. Ma Q, Guo H, Gordon RG, Zaera F (2011) Surface chemistry of copper(I) acetamidinates in connection with atomic layer deposition (ALD) processes. Chem Mater 23:3325
74. Hausmann DM, Gordon RG (2003) Surface morphology and crystallinity control in the atomic layer deposition (ALD) of hafnium and zirconium oxide thin films. J Cryst Growth 249:251
75. Mårtensson P, Juppo M, Ritala M, Leskelä M, Carlsson J-O (1999) Use of atomic layer epitaxy for fabrication of Si/TiN/Cu structures. J Vac Sci Technol B: Microelect Nanomet Struct 17:2122
76. Kim H (2003) Plasma-enhanced atomic layer deposition of tantalum thin films: the growth and film properties. Thin Solid Films 441:311
77. Kim H, Cabral C, Lavoie C, Rossnagel SM (2002) Diffusion barrier properties of transition metal thin films grown by plasma-enhanced atomic-layer deposition. J Vac Sci Technol B: Microelectron Nanomet Struct 20:1321
78. Becker JS, Gordon RG (2003) Diffusion barrier properties of tungsten nitride films grown by atomic layer deposition from bis(tert-butylimido)bis(dimethylamido)tungsten and ammonia. Appl Phys Lett 82:2239
79. Becker JS, Suh S, Wang S, Gordon RG (2003) Highly conformal thin films of tungsten nitride prepared by atomic layer deposition from a novel precursor. Chem Mater 15:2969
80. Mukhopadhyay AB, Musgrave CB (2007) The role of ammonia in atomic layer deposition of tungsten nitride. Appl Phys Lett 90:173120
81. Gordon RG, Becker J, Hausmann D, Suh S (2001) Alternating layer chemical vapor deposition (ALD) of metal silicates and oxides for gate insulators. Mater Res Soc Symp Proc 670:K2 4/1
82. Gordon RG, Becker J, Hausmann D, Suh S (2001) Vapor deposition of metal oxides and silicates. Possible gate insulators for future microelectronics. Chem Mater 13:2463

83. Hausmann D, Becker J, Wang S, Gordon RG (2002) Rapid vapor deposition of highly conformal silica nanolaminates. Science 298:402
84. Dillon AC, Ott AW, Way JD, George SM (1995) Surface chemistry of Al_2O_3 deposition using $Al(CH_3)_3$ and H_2O in a binary reaction sequence. Surf Sci 322:230
85. George SM, Yoon B, Dameron AA (2009) Surface chemistry for molecular layer deposition of organic and hybrid organic-inorganic polymers. Acc Chem Res 42:498
86. Lim J et al (2006) Dielectric properties of $(Zr_xTi_{1-x})O_2$ film on Ru/SiO_2/Si substrates deposited by the atomic layer deposition using [$Zr(OtBu)_4$ + $Ti(OtBu)_4$] cocktail source. ECS Trans 1:153
87. Li M-Y et al. (2007) Effect of process pressure on atomic layer deposition of Al_2O_3. J Electrochem Soc 154:H967
88. Aaltonen T, Ritala M, Leskelä M (2005) ALD of rhodium thin films from Rh(acac)[sub 3] and oxygen. Electrochem Solid-State Lett 8:C99
89. Aaltonen T, Ritala VM, Sammelselg I, Leskelä M (2004) Atomic layer deposition of iridium thin films. J Electrochem Soc 151:G489
90. Aaltonen T, Ritala M, Arstila K, Keinonen J, Leskelä M (2004) Atomic layer deposition of ruthenium thin films from $Ru(thd)_3$ and oxygen. Chem Vap Deposition 10:215
91. Hämäläinen J, Munnik F, Ritala M, Leskelä M (2008) Atomic layer deposition of platinum oxide and metallic platinum thin films from $Pt(acac)_2$ and ozone. Chem Mater 20:6840
92. Grubbs RK, Steinmetz NJ, George SM (2004) Gas phase reaction products during tungsten atomic layer deposition using WF[sub 6] and Si[sub 2]H[sub 6]. J Vac Sci Technol B: Microelectron Nanomet Struct 22:1811
93. Knapas K, Hatanpää T, Ritala M, Leskelä M (2010) In Situ Reaction Mechanism Studies on Atomic Layer Deposition of Sb_2Te_3 and GeTe from $(Et_3Si)_2Te$ and chlorides. Chem Mater 22:1386
94. Pore V et al (2011) Atomic layer deposition of antimony and its compounds using dechlorosilylation reactions of Tris(triethylsilyl)antimony. Chem Mater 23:247

Chapter 3
ALD Simulations

Simon D. Elliott

3.1 Introduction and Methods

Simulations provide a bridge between theory and experiment, exploiting the extraordinary computational power available today so as to carry out virtual experiments and test theories in complex scenarios. This chapter reviews how simulations have been brought to bear on the pulsed version of chemical vapor deposition (CVD) that is commonly called atomic layer deposition (ALD), and historically also atomic layer epitaxy. The review only covers simulations of the ALD process itself, and not properties of the materials deposited by ALD. Computational studies of non-pulsed CVD and of crystal growth [1] are out of scope, as are surface science simulations that do not make conclusions specific to ALD. For instance, although ALD has many parallels with heterogeneous catalysis, the huge body of modeling work in that field [2] cannot be covered in this short review.

The role of a simulation within a scientific investigation can simply be to *validate* a model by showing agreement with experiment (and without validation a model is merely a sophisticated hypothesis). It is better if a simulation can *explain* an experimental result by providing evidence for one model over another. Best of all is if a simulation can *predict* results, narrow down experimental options and lead directly to new insights without experimental input—but of course such prediction requires a quantitative accuracy that can rarely be achieved. Nevertheless, whatever the balance between computed and measured data, computational studies can be an efficient way to investigate novel processes and shorten process development times in the laboratory.

In terms of subject matter and popularity, ALD simulations follow the same trend as ALD experiments, with most recent work motivated by the needs of the electronics industry, particularly the ALD of high-permittivity ("high-*k*") dielectrics onto semiconductors as part of the gate stack in CMOS transistors. The

S. D. Elliott (✉)
Tyndall National Institute, University College Cork, Dyke Parade, Cork, Ireland
e-mail: simon.elliott@tyndall.ie

C. S. Hwang (ed.), *Atomic Layer Deposition for Semiconductors*,
DOI: 10.1007/978-1-4614-8054-9_3,

two largest sections of this review therefore cover the ALD of high-*k* oxides, both the reaction mechanisms themselves and how deposition onto semiconductor substrates occurs. The remaining sections outline what simulations can contribute to describing the particular properties that are of interest in experiment—precursor stability, film growth, conformality, and uniformity. Many of these properties are only accessible via multi-scale simulation, which is introduced next.

3.1.1 Multi-Scale Simulation

The ALD process is manifest across many length scales [3]. The pulsed flow of gases into meter-scale reactors, around millimeter-scaled geometries, leads to chemical reactions between atoms, which grow into nanometer thick films and coat micron-scaled pores or particles. It is clearly impossible to describe explicitly all of these length scales in one model, and most simulations are 'multi-scale' insofar as they involve coupling between selected length scales according to the property of interest. A successful multi-scale model minimizes the propagation of errors from one scale to the next [4]. The problem of timescale in ALD is perhaps even more acute than that of length scale, since a combination of fast (ps-ns) and slow reactions (μs-ms) contribute to film growth, and gases are pulsed and purged over second-long timescales [5].

3.1.2 Atomic-Scale Simulation

Atomistic calculations, in which discrete atoms have explicit locations in space, are now established as an important adjunct to experiment [6]. The electronic structure is determined by approximately solving the quantum mechanical Schrödinger equation, which yields the wavefunction, energy, gradients etc., and hence gives access to atomic structure, reaction energetics at zero temperature, and dynamics at finite temperatures. The more accurate the approximate solution, the more demanding the computational load and the smaller the system that can be simulated. The various methods of atomistic simulation thus form a hierarchy of accuracy *versus* cost and system size. Of these, first principles (or ab initio) methods are distinguished by requiring no empirical fitting parameters, so that genuinely novel situations can be investigated.

Hartree–Fock (HF) theory is the original ab initio technique, but today density functional theory (DFT) provides a higher level of accuracy at roughly the same computational cost [7, 8]. There is in fact a range of DFT approaches to approximately solving the electronic problem, but in this review we use the term 'DFT' as shorthand for Kohn–Sham calculations of the ground state using the local density approximation or generalized gradient approximation, often mixed with HF exchange (hybrid DFT). Physically, the trick in these DFT calculations is to find the density of mutually interacting electrons by computing non interacting pseudo-electrons subject to a potential that partially accounts for electron correlation and exchange.

DFT generally gives a good description of classical two-center covalent bonds, nonclassical multi-center bonds, metallic bonding in conductors, and polar/ionic bonding in semiconductors and insulators. However, 'self-interaction' in the DFT density means that there are systematic but unpredictable errors in band gaps, defect levels, excited state properties, van der Waals (dispersion) interactions, and curve crossing at transition states. For ALD, this means that DFT is highly accurate in determining changes in structure and bonding when organometallic molecules interact with various material surfaces, i.e., reaction mechanism. Computed properties such as vibrational spectra can help in the assignment of experimental data and thus provide direct evidence for mechanism [9–11]. On the other hand, volatility is difficult to simulate with DFT, because of the poor description of weak intermolecular forces. There is also an upper limit on the system size that can be simulated with DFT, although algorithmic and hardware advances mean that this limit is being pushed ever upwards: Today, DFT is the method of choice for systems of 100–1,000 atoms, occupying roughly one cubic nanometer.

The main source of uncertainty in first principles calculations is therefore the choice of model and its interpretation. This review will include a brief assessment of various models for describing ALD. The proposed model should be viewed as a hypothesis and the first principles simulation ensures that accurate structures and energies can be obtained for this hypothesis, without fitting to experiment.

3.1.3 Activation Energies

It is commonly stated that ALD is driven by kinetics, rather than thermodynamics, appealing presumably to the irreversible loss of by-products into the stream of exhaust gases. One could argue that certain reaction steps are reversible and thermodynamically controlled, such as the sampling of surface sites by frequently adsorbing/desorbing precursor molecules, and the densification of ad-layers to match the contour of the substrate. Nevertheless, the competition between slower side reactions and faster growth reactions is crucial for successful ALD, and hence much of the focus of atomic-scale simulations is on the activation energies for such reaction steps.

Computing Gibbs free energies of activation to so-called 'chemical accuracy' (<5 kJ mol^{-1}) remains an embarrassing difficulty in atomic-scale simulation [8]. Such bond-making/-breaking can only sensibly be attempted with ab initio methods (e.g., HF or approximate DFT), thereby limiting the system size to $<$1,000 atoms on today's computer hardware. Even so, error cancelation is poor when the wavefunction changes strongly at the transition state, so that chemical accuracy can only reliably be achieved by post-HF methods such as coupled cluster [12, 13], or configuration interaction [14], tractable for fewer than 100 atoms.

There is therefore an unavoidable trade-off between the accuracy of the quantum mechanical solution and the veracity of the atomic model. The latter depends on the number of atoms, the number of possible geometries, and how many alternative pathways can be tested. For example, if the model contains just one H and one ligand, then the number of possible H-transfer pathways is clearly unrealistically low compared to actual ligands on an OH-covered surface. In this sense, ergodicity (i.e., whether all relevant pathways can be sampled) is as big a problem as accuracy when computing activation energies. In principle, all that is needed to sample pathways with higher accuracy is computational brute force, which is now available in massively parallel computers. However, practical experience shows that ALD chemistries are so complicated that automated searches are rarely successful, and that the bias of the user in defining models is still a major constraint.

3.2 Simulating Reaction Mechanism During Homodeposition

Unlike standard CVD, the steady state during a single ALD pulse is when the surface is saturated and no further net reaction occurs. Instead, 'steady-state' is used here to refer to the constant growth rate that is achieved over cycles of product-on-product growth, which we may also term 'homodeposition' in order to distinguish it from the initial product-on-substrate cycles. ALD process development is clearly dependent on understanding first the mechanism of homodeposition, and after that, the mechanism of deposition onto various substrates (Sect. 3.3). The aim is to identify the elementary chemical reactions that transform precursor molecules into product films (both ALD and non-ALD reactions), as well as those that lead to undesired by-products or impurities. This knowledge should enable us to answer the key questions about a particular ALD process, namely, how does the surface become saturated in one pulse and how are these groups consumed during the next pulse? Quantitative answers to these questions yield predictions of growth rate, temperature dependence and pulse/purge durations, and help in choosing ligands and designing processes for new materials. As shown in the following sections, many simulations have achieved these goals.

In ideal ALD, precursors react only on the growing surface and not in the gas phase. The film growth reactions of a precursor for element M may therefore be written:

$$\text{gas-ML}_n \rightleftarrows \text{surf-(ML}_n) \rightleftarrows \text{surf-(ML}_x) \rightleftarrows \text{bulk-M}, \qquad (2.1)$$

where ML_n is a metal-containing precursor with n ligands L, ligands can be eliminated from the surface so that $0 \leq x \leq n$, and 'bulk-M' refers to an atom in a local environment like that of the as-deposited solid film. The position of the equilibria between the steps in Eq. 3.1 depends on the reaction energetics and on the availability of reagents and ligands during the pulse-purge cycle of ALD. The first step is molecular adsorption. The second step shows the nature of adsorption changing as ligands are eliminated. The final 'bulk' status may not be reached until after many ALD cycles. In a general sense, the whole sequence in Eq. 3.1 is the reactive adsorption of a precursor onto the surface, along with desorption of ligand remnants. Describing the ALD reaction mechanism means describing each of the steps in Eq. 3.1, and in particular, describing the various interactions between ML_x adsorbates and the substrate or growing surface.

Adsorption of the precursor molecule onto a substrate is the first stage in the chemical reactions of deposition. There are numerous first principles computational studies of adsorption, both of organometallic complexes and of molecules like H_2O, NH_3, and O_2 that are used as co-reagents in ALD, and it is not possible to review all these studies here. This survey is instead limited to papers where computation of reaction steps leads to a conclusion about the mechanism of ALD. As will be seen, most of these studies do consider molecular adsorption of organometallic precursor or co-reagent, and also the subsequent elimination of ligands.

3.2.1 Mechanism of Oxide ALD

For the mechanism of oxide ALD, possibly the earliest quantum chemical study is that by Siodmiak, Frenking, and Korkin on $TaCl_5 + H_2O$ [12]. The model consists of complexes totaling ten atoms or less, and despite the model being so small, reasonable reaction energies are obtained for adsorption, HCl elimination, and etching. Energy barriers for intra-ligand reaction pathways are also computed, even though these pathways are highly constrained relative to an actual surface.

The next mechanism to be computed with DFT was the prototypical ALD system $AlMe_3 + H_2O$ (Me $= CH_3$) by Widjaja and Musgrave [15]. An Al_2 cluster model is used and tested against an Al_4 cluster (up to 30 atoms), showing little nonlocal electronic effect on the reactions. Adsorption of $AlMe_3$ onto a surface hydroxyl group is found to form a Lewis acid–base adduct and a barrier was found for intra-cluster proton transfer producing CH_4. Calculations yield a similar energy profile for the corresponding H_2O half-reactions. A similar model is used to determine activation energies for kinetics, which are fed into continuum simulation of alumina ALD [5].

Widjaja and Musgrave follow a similar approach based on Hf_4 or Hf_8 clusters in their early study of HfO_2 from metalchloride precursors and H_2O [16]. In both half-reactions, the zero-temperature energetics show strongly favored adsorption of the precursor molecule, but a barrier and overall energy cost for subsequent elimination of HCl, confirmed later in a periodic model [17]. Inclusion of entropy effects at realistic ALD temperatures turns out to be decisive—reducing the adsorption free energy and making desorption of HCl more favorable, albeit still via a substantial barrier. The results for $ZrCl_4$ are nearly identical [18]. Similar results are obtained for $AlCl_3$, along with consideration of a non-growth side reaction [14]. Hydroxychloride side-products are also computed [19]. Since growth is a multi-step reaction in competition with side-reactions, it is not straightforward to correlate the computed energy minima and barriers with the experimental growth rate (e.g., the trend with increasing temperature). Nevertheless, this case illustrates that endothermic reaction steps are not necessarily an obstacle to film growth in ALD, presumably because certain reaction steps are far from chemical equilibrium. Calculating the activation and reaction energies for a single pathway has therefore no predictive power on its own.

The coverage dependence of water adsorption and surface acidity is added to the picture in cluster calculations by Deminsky [20] (confirmed later in a periodic model [21]). The authors note that "the mechanism and kinetics of the ALD process cannot be interpreted even qualitatively without taking into account stereochemical effects, in particular, the effects of surface coverage on the reactivity of chemisorbed surface species".

The $AlMe_3 + H_2O$ mechanism is investigated by Elliott and Greer using a periodic model [22]. The model imposes crystallinity and so the most stable surface of the most stable crystalline polymorph of alumina is chosen as the substrate. However, amorphous alumina is grown in ALD, which exhibits lower coordination numbers and it has not yet been proven whether the use of a crystalline model negatively affects the simulation results. The energy obtained for Lewis adduct formation agrees with that from the cluster model [15] but, from the various pathways investigated, proton transfer is found to occur most readily from a neighboring hydroxyl group. Substantial release of energy is seen to accompany increases in Al–O coordination at the surface as the CH_4 by-product desorbs, in contrast to the reluctance of HCl to desorb in the cases above.

An alternative alumina process is $AlMe_3 + O_3$, which is perhaps a model reaction for other ALD reactions with O_3 or O_2-plasma, but the reaction mechanism is much more complex. Ab initio molecular dynamics show that O radicals insert into metal–ligand bonds, ultimately transforming Al–CH_3 into Al–OH [23]. However, formate intermediates are detected with in situ IR, as confirmed by DFT assignments of the bands [24]. DFT-assigned spectra are also used to identify decomposition products for $La(amd)_3 + O_3$ [9]. Extensive computations of surface intermediates during homodeposition from $La(C_5H_5)_3$ to $Er(C_5H_5)_3$ are relevant for the corresponding oxide ALD processes with O_3 [25]. The La precursor is found to undergo surface catalyzed decomposition, whereas C_5H_6 elimination is favored for Er, partly due to the level of distortion in the oxide substrates, and this explains why Er_2O_3 growth is successful but La_2O_3 is not.

Sulfides have many similarities with oxides and so similar adsorption and elimination reactions are computed for CdS homodeposition from $CdMe_2 + H_2S$ [26]. The calculated reaction barriers for the ALD half reactions suggest that elevated temperatures are required for the film growth.

Because ALD depends on self-saturating surface reactions in the metal precursor pulse, it can be shown that the growth rate depends linearly on the degree to which ligands are eliminated during this pulse [27]. This motivates using the reaction energy for elimination as a simple metric for ALD growth rate (and perhaps even for the ALD growth temperature), allowing computational screening of a wide variety of ligands [28]. For this purpose, a cluster model is quick and efficient, as the influence of the surface is entirely neglected. In a similar vein, cluster models rationalize the elimination of protonated ligands from $Ti(N(CH_3)_2)_4$ *versus* $Ti(O^iPr)_4$ [29], from $Zr(MeCp)_2(Me)(OMe)$ [30] and from $Zr(Cp)_2(Me)_2$ [31]. Such a simple approach is successful because ligand elimination (often termed 'ligand exchange') is known to be an important element of the mechanism of oxide ALD.

Another important element is revealed by Olivier et al., who examine the tendency for Hf and Sn cations in oxide films to aggregate and increase their coordination number to oxygen, which they term 'densification' [32]. Based on this, the authors highlight the role of water as a co-reagent for oxide ALD, not only in providing reactive OH, but also in allowing densification via OH or O. Densification is important in ALD for the following reason. To be volatile, the metal center in an ALD precursor molecule must be kept from aggregating with metal centers in neighboring molecules, which is achieved via the coordination of bulky ligands. By contrast, the product film should be a dense solid, with highly coordinated metal atoms (bonded either to other metal atoms or to oxygen, sulfur, etc.). As indicated in Eq. 3.1 above, precursor adsorption and film growth therefore involves the change from low to high coordination of the constituent atoms. To date, the importance of densification as an element of ALD growth has not been widely recognized, although in hindsight, many of the published studies mentioned above show exothermic reactions correlating with increases in coordination number.

3.2.2 Mechanism of Nitride ALD

Transition metal nitrides such as TiN and TaN are metallic and so of interest in thin film form as diffusion barriers. The first DFT study of nitride growth is by Tanaka et al., [33] computing $TiCl_4 + NH_3$ onto a cluster representing amorphous SiO_2; following adsorption, surface reactions between adsorbates via both the Langmuir–Hinshelwood and Eley–Rideal mechanisms were considered. In a study of HfN growth from $Hf(NMe_2)_4 + NH_3$, it is found that the ALD mechanism is similar to that of HfO_2 from H_2O, although NH_3 is computed to be less reactive than H_2O and so O impurities could be a problem [34]. Thermal stability is a central question for transition metal amides/imides. The adsorption and decomposition of a whole sequence of such precursors show that breaking metal–ligand bonds is the favored pathway at low temperature, but also that C–H scission can lead to C impurities [35]. A comprehensive study of an amido/imido-Ta precursor with NH_3 shows that ammonia is a more facile source of N than the ligand, but also identifies decomposition pathways that may operate at elevated temperatures

(MOCVD) [36]. Thermodynamic modeling of TaN is carried out using the literature values [37].

In the growth of silicon nitride from $SiH_4 + NH_3$, the computed activation energies reveal why stoichiometric Si_3N_4 is favored in ALD but a Si-rich material can be obtained in higher temperature CVD [38]. Boronitride growth has also been computed with DFT, although the radical intermediates that are postulated may be more relevant for CVD rather than ALD [39].

3.2.3 Mechanism of Metal or Elemental ALD

There is an extra layer of mechanistic complexity in the ALD of metals, since electrons must be transferred to metal cations during deposition in order to reduce them to the neutral product. Of the elementary steps that are known for oxide ALD, only reductive elimination is a candidate for showing how this might occur, and so there is a need for other redox reaction steps to be considered.

One of the earliest simulations of ALD is the investigation by Hirva and Pakkanen in 1989 of elemental silicon from silanes to chlorosilanes, in both radical and molecular mechanisms [40]. While chemisorption to bare Si surfaces is computed to be favorable, releasing HCl from the surface is more difficult. Although not a redox process, a related study by Mochizuki et al. considers GaAs growth from $GaCl_3 + H_2$ using high level ab initio methods [41]. The deposition of elemental carbon as diamond from fluorinated compounds is simulated by Hukka et al., here focusing on radical pathways that may be more relevant for high-temperature CVD rather than ALD [42].

The ab initio simulation of metal ALD is pioneered in three papers by Mårtensson et al. that consider adsorption of Cu_2Cl_2 on a Cu (1 1 1) model surface [43], reaction with hydrogen [44] and barriers [45]. Given the range of oxidation states available for Cu, disproportionation is one of the likely redox steps, but is computed here to be unfavorable for Cu(I)Cl. The Cl ligands are found to eliminate via combination with surface-H. The reduction potential of metals is emphasized in the method of Orimoto et al. for screening metal precursors, illustrated with DFT calculations on copper(II) β-diketonates that compare well with measured values [46]. DFT is also used for explicit calculation of the reaction energy and barrier for reducing a Ta(V) precursor with H radicals, as a model for TaCN growth with H_2/Ar plasma [47]. Care is needed in interpreting these results: Because of the changing numbers of electrons in redox reactions, there is the possibility of poor error cancelation, leading to uncontrolled errors in the DFT energetics.

Substrate effects—nucleation, adhesion, and island growth—are crucial in metal ALD. First principles dynamics give insights into Cu agglomeration and its temperature-dependence [48, 49]. Poor adhesion of Cu films is attributed to spontaneous precursor decomposition during adsorption onto metals, but not onto nitrides, based on the example of Cu(hfac)(vtms) on Ti, Ta, and W [50, 51] . The adsorption of the same precursor onto Si is also computed and compared to experiments at a range of temperatures [52]. On a hydroxylated SiO_2 substrate, chemisorption of $Cu_2(amd)_2$ is shown to proceed by elimination of amdH, followed by release of amidine during the H_2 pulse, while transfer of ligands to the substrate can lead to contamination with C. These conclusions are reached via the powerful combination of DFT and in situ IR spectroscopy [53]. Although not specific to ALD, a DFT study of the growth mode of Pt onto $SrTiO_3$ is also relevant [54]. There seem to be no DFT studies to date specifically devoted to noble metal ALD, despite the fact that open questions remain about the mechanism—perhaps this is because of the complexity of the redox mechanism relative to the Bronsted acid–base reactions of oxide ALD [55].

3.3 Simulating ALD onto Silicon, Silica, and Other Substrates

Some of the most exciting applications of ALD are the controlled deposition of sub-nanometre thin films onto a substrate, which we term 'heterodeposition' to distinguish it from the steady state 'homodeposition' of Sect. 3.2 above. For the electronics industry, the substrate is often a semiconductor, although the actual surface may be oxidized or cleaned, and there is also interest in ALD onto porous dielectrics and metal electrodes. Most non-electronics applications of ALD also depend on successful heterodeposition. If heterodeposition can be fully understood, it opens up the potential for interface control at the atomic level, which is particularly important if the target films are just a few nanometres thick. Of course, this level of understanding is still far off and relatively little is known at present about how ALD precursors interact with various substrates.

In terms of growth rate, heterodeposition has been classified as linear growth, inhibited growth, or substrate-enhanced growth [56], followed by the transition to homodeposition or steady-state ALD. In the simplest case, substrate effects are limited to the earliest few ALD cycles, until a single monolayer of interfacial layer is formed, which is then followed by homodeposition. Detection of the interface reactions is only possible if in situ monitoring of adequate sensitivity is employed. The situation may be complicated if substrate and product mix to form a more extended inter-layer with its own distinct growth chemistry, as often the case when using an aggressive oxidizing agent such as ozone. Alternatively, the product may homodeposit onto islands that ripen into a closed layer only after many ALD cycles (Sect. 3.5.2).

One may envisage two chemical scenarios during heterodeposition. One possibility is that the same ALD mechanism operates as in homodeposition, modified only by the different coverage of reactive species (e.g., hydroxyl groups) on the substrate. Alternatively, non-ALD side-reactions may take place between precursor and substrate, either contributing extra product or etching the substrate away. First principles simulations clearly have a role to play in investigating what chemical interactions are possible between precursor and substrate. The situation is complicated by the strong dependence of heterodeposition on the preparation history of the substrate.

The (1 0 0) surface of diamond-structured silicon is the substrate used in the electronics industry, and a popular atomistic model is consequently the Si_9 cluster. This consists of a Si_2 surface dimer, on top of three (1 0 0)-oriented 'layers' of four + two + one Si atoms, respectively. The subsurface layers are terminated with 12 H atoms, to give roughly tetrahedral Si coordination and no dangling bonds. The atoms of the surface dimer have one dangling bond each (if bare) or can be bound to H (simulating the situation after HF cleaning) or OH (after washing with H_2O). Tests indicate that activation energies from Si_9 are converged with respect to cluster size [57, 58], apparently due to the open structure of Si(1 0 0), since the same is not true for Si_{10} as a model for more densely packed Si(1 1 1) [57].

3.3.1 Heterodeposition of Zirconia and Hafnia

Perhaps the earliest atomistic simulation of heterodeposition from two precursors is of $B(OMe)_3$ and $POCl_3$ onto silica [59]. However, the most common subject is the ALD of high-*k* ZrO_2 or HfO_2 onto silicon-based substrates, because of its technological importance in nano-CMOS. Brodskii et al. use periodic and cluster models to compute the reactions of $ZrCl_4$ with hydroxylated silicon: Elimination of HCl in the Zr-pulse is found to be endothermic, but subsequent hydrolysis of $ZrCl_2$ fragments in the H_2O pulse is slightly exothermic with a low activation barrier [60]. The adsorption of $ZrCl_4$ onto various substrates is compared: Onto bare, H-terminated hydroxylated [18] and hydrated [61] Si surfaces and onto hydroxylated Ge [62]. A common theme is the tendency of the HCl by-product to remain complexed and not desorb [63, 64]. A similar pathway for the formation of a HCl intermediate from adsorbed $HfCl_4$ on Si(1 0 0) is computed by Estève [65] and on SiO_2 by Dkhissi et al. [66]. Slower kinetics are predicted when $Si(OMe)_4$ is introduced as co-precursor with $HfCl_4$ for hafnium silicate heterodeposition [67, 68].

Jeloaica et al. compare three common precursors ($AlMe_3$, $ZrCl_4$, and $HfCl_4$) and find that they adsorb favorably on OH-terminated SiO_2, show similar barriers to H-transfer, but differ in ligand elimination energetics [69]. Calculations are presented for the same three precursors on GaAs, showing the inhibiting effect of passivation with sulfur [70], and for $HfCl_4$ on hydroxylated GaAs [71]. Fenno et al. also compare Cl and Me ligands for ZrO_2 and HfO_2 heterodeposition, this time onto H-terminated Si, and also find CH_4 elimination to be much more favorable than HCl elimination [72]. Contamination with H_2O is predicted to lead to interfacial Si–O-Zr/Hf. Switching to amide ligands, decomposition reactions are computed to result in interfacial Si–C bonds [35], or Si–N and Si–CN bonds [13], which are important insights that can guide precursor choice for specific interface properties.

3.3.2 Heterodeposition of Alumina

Halls and Raghavachari use DFT and the Si_9 model to reveal why $AlMe_3$ nucleates poorly on H-terminated Si: Both adsorption and elimination of CH_4 are barely energetically favored [73]. A side reaction leading to O incorporation is identified and the reaction pathway is computed [74]. By contrast, if hydroxyl groups are present after treatment with H_2O, then both ALD half reactions are computed to be thermodynamically favorable and kinetically uninhibited [75]. Indeed, many Lewis acidic metal precursors are found to adsorb favorably on OH-terminated SiO_2 and the Si–OH groups are sufficiently acidic for subsequent elimination [69]. Treatment with O_3 is shown to lead to Si–O–Al or Al–O–CH_3 but not SiO_2 [24]. Alternatively, a basic oxygen atom in the precursor [e.g., $AlMe_2(O^iPr)$] facilitates adsorption onto H-terminated Si [76]. More complex surface models are now becoming accessible to DFT calculations, as illustrated by a recent study of multiple $AlMe_3$ on Si–OH that shows an increase in activation energies as the surface becomes saturated [77].

There is interest currently in replacing Si with higher mobility materials for the transistor channel and, consequently, the ALD of high-*k* dielectrics onto Ge, GaAs, etc. is being studied. For $AlMe_3$ onto GaAs, Lu et al. compute adsorption and elimination reactions with hybrid DFT and find similar energetic trends as on other OH-terminated surfaces [78]. Ren et al. find that these reactions become less favored when GaAs is passivated with sulfur [70]. Surface structures of $AlMe_3$ fragments on InAs(0 0 1) and $In_{0.53}Ga_{0.47}As$(0 0 1) are computed with DFT and compared with scanning tunneling microscopy [79]. $AlMe_3$ can remove native oxides from GaAs surfaces and DFT calculations point out the role of Ga-OH *vs* As-O in this process, as well as proposing a redox reaction [80]. This illustrates that $AlMe_3$ is a multi-purpose reagent for surface preparation, functioning as a Lewis acid (due to under-saturated Al), as a Brønsted base (CH_3^- to CH_4), and also as a reducing agent (CH_3^+ plus two electrons). Adsorption of $AlMe_3$ onto H-terminated Ge is found to be more favorable, but still hindered, when compared with OH-terminated Si [81], [82]. Using Ge-SH as a model for sulfur-passivated Ge, DFT shows favorable adsorption of $AlMe_3$ and no removal of S [58].

Moving to C-based substrates, the ALD of Al_2O_3 onto the open edges of a graphene nanoribbon is studied with DFT for a range of temperatures and pressures, revealing selectivity for the zigzag edge during the H_2O pulse [83]. For organic molecules, Al_2O_3 ALD onto self-assembled monolayers (SAMs) is computed to be favorable in terms of adsorption and elimination on amine- and hydroxyl-terminations, but not on methyl-terminated molecules [84].

3.3.3 Heterodeposition of Titanium-Based Materials

Titanium oxides and nitrides are important materials and their heterodeposition onto silicon-based substrates can easily be studied by extension of the models for hafnia and zirconia. Hu et al. investigate $TiCl_4 + H_2O$ ALD onto various models of the SiO_2 surface and find a strong dependence of the growth rate and morphology on coverage and arrangement of reactive groups [85]. This is contrasted with the $TiI_4 + H_2O$ process, where binding energies are computed to be higher and impurity levels to be lower [86]. On H-terminated Si, direct chlorination is found to be the kinetically favored reaction of $TiCl_4$ [87]. High-temperature reactions of a Ti(III) precursor with SiO_2 are also computed with DFT and found to agree with solid-state NMR [88].

An early study of the adsorption of NH_3 onto bare Si [89] is followed with simulation of reaction steps for heterodeposition of TiN on SiO_2 from $TiCl_4 + NH_3$ [33]. DFT simulations and IR spectra are integrated in a substantial study of TiN onto silicon from $Ti(NMe_2)_4$ by Rodriguez-Reyes et al., investigating also amides of other transition metals [35]. The precursor is proposed to adsorb onto bare Si via N and then undergo N–Ti scission, based on a lower barrier relative to N–C scission [90]. This leads in turn to weakening of C–H, which is consistent with data from temperature to programmed desorption [91] Several decomposition pathways are computed to lead to Si–C bonds, explaining the high carbon content at the interface. On an ammonia-saturated surface, the dominant reaction is found to be transamination, which again matches IR results [10].

Like heterodeposition onto Si, ALD onto SAMs also depends on their chemical termination. Amidotitanium precursors are computed to form adducts with amine-terminated SAMs, but H-bonds with thiol- or hydroxyl-terminated SAMs, while the amidozirconium analogs always form dative-bonded adducts [92].

3.3.4 Heterodeposition of Other Oxides and Sulfides

The pyridine-catalyzed deposition of SiO_2 from an alternating sequence of $SiCl_4 + H_2O$ on a Si substrate is computed using MP2, which is a post-HF ab initio method for improved description of van der Waals forces [93]. Kang et al. consider the same system without catalysis using DFT and find that the standard mechanistic steps occur in both half-cycles: Adsorption, elimination of HCl, and movement to bridging sites [11]. The overall barrier matches the experimental value remarkably well. Reaction and activation energies for ALD steps onto hydroxylated Si are also computed for ZnO from $Zn(C_2H_5)_2 + H_2O$ [94], Y_2O_3 from $Y(C_5H_5)_3 + H_2O$ [95], and MgO from $Mg(C_5H_5)_2 + H_2O$ [96]. PbS growth is shown with DFT to be unfavorable on a methyl-terminated SAM, so that ALD takes place selectively on the areas not covered with SAM [97].

Simulations are now increasingly being reported of the mechanism of oxide ALD onto more complex substrates. Popovici et al. compute that $HfCl_4$ reactivity is enhanced on a Ti–OH substrate while depositing hafnium titanate [98]. For hafnium aluminate, Nyns et al. use DFT calculations to show that changes relative to the respective binary oxides are due to different dehydroxylation behavior of the ternary substrate [99]. The reactive adsorption of $AlMe_3$ onto anatase-TiO_2

surfaces is shown with DFT to produce atomically rough films [100] and to proceed only at defects and edges where water is adsorbed, explaining the experimentally observed island growth [101].

3.4 Thermal Stability of ALD Precursors

The study of ALD must also include the study of low-temperature or surface-catalyzed CVD. CVD reactions are generally responsible for an upper limit to the ALD temperature window and associated impurity levels in homodeposition [36], and can explain heterodeposition by certain precursors and resulting interfaces with substrates [10, 13, 35, 50, 51, 90, 91]. New ALD precursors should be designed so as to be thermally stable, minimizing unwanted CVD-like decomposition during storage and delivery of the precursor. As shown by the examples below, this is an area where ab initio simulation can contribute, although finding unknown decomposition reactions is much harder than confirming a known ALD mechanism.

A DFT assessment of decomposition reactions of various amido/imido Mo precursors reveal why the di-isopropylamido complex is the least stable, as determined by calorimetry [102]. Calculations by Zydor and Elliott show that intramolecular α-H transfer in Zr and Hf precursors $M(MeCp)_2(Me)_2$ accounts for experimentally observed decomposition and can be avoided by altering the ligands or increasing the electrophilicity of the metal [103]. The effect on electrophilicity is also calculated for related precursors with bridged cyclopentadienyl ligands [104]. Using DFT corrected for dispersion, the fragmentation modes of a Cu(I) guanidinate precursor are computed by Coyle et al. to be carbodiimide deinsertion and β–H elimination, with the latter pathway agreeing with results from mass spectrometry and IR spectroscopy [105]. Migration of β–H is also the focus of a first principles study of Co and Ni amidinate precursors, giving relative stabilities in agreement with experiment [106–108].

3.5 Simulating Film Growth

3.5.1 Kinetic Rate Equations

In principle, if the ALD mechanism is completely known, then the growth of films can be simulated without resorting to explicit atomistic dynamics. Reaction energetics for each step of the mechanism can be converted into a kinetic rate equation via the Arrhenius expression. All of the growth steps can then be 'lumped' into a master equation or selected in a random sequence (Sect. 3.5.3). The former approach can be straightforwardly applied to ALD from elemental reactants, as shown by Suntola [108]. Even for $HfCl_4 + H_2O$, however, Alam and Green show that just two differential equations are sufficient to describe ALD growth rates as a function of hydroxyl coverage and provide a good fit to experiments [109]. A kinetic scheme of 15 rate equations for the same system is obtained from ab initio simulation by Deminsky et al. [20]. Hemeryck et al. implement ab initio derived rate equations in a mesoscopic model for the deposition of bimetallic multilayers [110].

As an aside, it is important to distinguish between the kinetic rate of the growth reactions and the film growth rate (or growth per cycle). The kinetic rates determine the competition between reactions and hence the time needed for the surface to become saturated and its chemical make-up. The film growth rate is the concentration of product deposited in one cycle via the saturated surface. Finally, the throughput is the growth per unit time, taking the duration of pulses and purges into account.

3.5.2 Analytical Models

Since ideal ALD is a cyclic process, with a well-defined surface at the end of one precursor pulse acting as initial surface for the next pulse, simple models based on these characteristics can lead to quantitative predictions. For instance, Ylilammi derives analytical expressions for ALD growth rate as a function of density of adsorption sites and ligand coverage [111]. Park et al. develop equations that show how growth rate is only determined by the extent of the surface atomic layer coverage exchange at each pulse [112]. A graphical representation of growth provides a simple way to relate maximum growth rate of binary oxides to saturating coverages, which in turn depend on ligand size and on process conditions [22, 27, 113]. The same idea underlies the formula for growth rate given by Puurunen and illustrated for a series of test cases [114–116]. This is successfully applied to estimating hydroxyl coverage [117] and growth rate [118] from ligand size in binary oxide ALD. By making some assumptions about mechanism, the same strategy is applied to noble metal ALD, yielding analytical expressions for product ratios and growth rate in terms of saturating coverages [55].

Experimental growth rates are generally quoted as thickness increments per ALD cycle, whereas these phenomenological models are based on chemical reactions, and so predict molar or mass increments per cycle. Direct comparison between a model and thickness measurements is therefore only possible if the molar or mass density of the film is known, which can vary considerably with precursor chemistry and reactor conditions. Unfortunately, film densities are rarely measured along with thicknesses.

The ALD of ternary oxides provides a more complicated situation in which to apply these formulae. The question is how the stoichiometry of the ternary oxide is related to growth during each precursor pulse, and hence to the precursor pulsing ratio, since experiments often show a non-linear relationship. Kim et al. propose a model that accounts for changes in surface coverage during adsorption/elimination and apply it to strontium titantate [119]. Lie et al. develop a 'surface-utilization' model and fit this to the measured stoichiometries of the quaternary system of mixed lanthanum, strontium and iron oxides deposited by ALD from thd-precursors and O_3 [120].

The evolution of overall thickness and roughness is affected by the morphology of the growing film, for instance, whether layer-by-layer or island growth. Nilsen et al. employ a geometrical description of various morphologies to derive analytical formulae with just a few system-dependent parameters; the model successfully accounts for substrate-inhibited growth [121], amorphous films [122] or crystallites of various shapes [123]. A similar geometric model is used to describe the ALD of W onto Co nanoparticles and, by fitting to experiment, this yields the amount of surface oxide [124]. A different approach is adopted by Yim et al. to describe nucleation of Ru onto various substrates, yielding expressions for the coverage of nuclei as a function of number of cycles and homodeposition rate [125]. On a SAM substrate, the growth of tungsten nitride carbide is found to follow fractal scaling laws, while that of HfO_2 does not [126].

3.5.3 Stochastic Models

As an alternative to imposing ALD conditions and deriving formulae, one may investigate to what extent film growth is a result of the random sequence of growth events. Such a model has been proposed by Puurunen [127], successfully showing how roughness and film closure can evolve in time. Ahn et al. included surface relaxation after random deposition and applied the model to strontium titanate growth [128].

A more sophisticated approach is to model film growth by the sequential execution of elementary reaction steps, stochastically selected according to their relative kinetics. In this kinetic Monte Carlo (KMC) approach, time is coarse-grained, proceeding in irregular jumps from one reaction event to the next [129]. The timescales of seconds needed for ALD pulses and purges can therefore be simulated if the events are sufficiently rare. There is loose coupling with the atomic scale, as this is the source of the reaction events and their probabilities (i.e., activation energies). The most common implementation is Lattice KMC, where events take place on a fixed lattice of cells in space, and a group of atoms are associated with each cell. The bulk crystallographic lattice of the product film is often chosen as the simulation lattice and including a few thousand such cells means that areas of the order of 10 nm $\times$ 10 nm can be simulated.

In most multi-scale strategies that use activation energies as input, the relative magnitudes are more important than the absolute values (unless one is interested in absolute timescales). Even here however, the error in DFT activation energies is not constant across various classes of reactions, so that trends and relative values may not be accurate. For this reason, many KMC strategies adjust and fit activation energies, rather than use values directly from DFT. In this case, the main value of the DFT simulations is in discovering plausible pathways.

Potential outputs from KMC simulations as a function of time are the amount of film deposited and gaseous by-products (in terms of moles, mass, and thickness), the stoichiometry or purity of the film, and the evolution of nanostructure (morphology, interface to substrate, etc.)—all predictions that can be compared with experiment. By artificially switching particular reaction steps on and off in the simulation, their effect on film growth can be investigated and the proposed mechanism can be validated.

The first KMC model of ALD is developed by Deminsky et al. as part of a multi-scale simulation of $HfCl_4$ or $ZrCl_4 + H_2O$ and good agreement is obtained with experiment [20, 130]. The temperature-dependence of residual chlorine in the film can be traced back to steric hindrance between adsorbates. Heterodeposition of high-*k* oxides onto Si is simulated by lattice KMC by Mazaleyrat et al. (Al_2O_3) [131] and Dkhissi et al. (HfO_2) [66, 132], again based on DFT data, revealing the effect on film growth of substrate hydroxylation, temperature, and pulse durations. Neizvestny et al. use KMC to show the relationship between growth rate and island nucleation, with recommendations for improving film quality at the interface [133].

3.5.4 Accelerated Molecular Dynamics

KMC is able to access long timescales by separating rare growth events from frequent non-growth events (such structural relaxation). The same concept underlies the explicit all-atom molecular dynamics study of Al_2O_3 ALD by Hua et al. that use analytic potentials to relax atomic geometries for 500 ps, combined with deposition rules on the timescale of ALD pulses [134]. The simulations predict that growth rate and roughness depend strongly on initial surface composition and process temperature. As accelerated molecular dynamics approaches continue to be developed [135], it is likely that simulations of this sort will be able to extend our understanding of the atomistic processes underlying ALD.

3.6 Conformality and Uniformity

The majority of ALD simulations focus on the atomic-scale (e.g., DFT) and the nanometer length scale (e.g., KMC). However, the unique attractiveness of ideal ALD comes from the constancy with which substrate features are coated, regardless of their geometry and proximity to the gas inlet, on any length scale up to meters. This property of ALD will become more important for the semiconductor industry in the future, as performance improvements will depend on exploiting the third dimension in everything from fin-shaped nanotransistors to micron scaled through silicon vias. It is also a critical property in the application of ALD to large surface area substrates. The key questions are whether film growth does indeed conform to the geometry of a given feature and whether growth is uniform at all points within the reactor. This requires consideration of how variations in gas flow affect the film growth reactions, which again is a multi-scale modeling problem. In these approaches, the detailed surface chemistry is often 'lumped' into a single parameter, the sticking coefficient, which can be estimated, fitted to computed energetics or derived from growth experiments in channel-type reactors [136] or in high aspect ratio holes [137].

In the first studies that couple gas transport and surface reactions for ALD, Gobbert et al. postulate generic adsorption and desorption reactions for two precursors during the pulse and purge cycles [138, 139]. This reveals that growth is not usually limited by gas transport, but rather, by the reaction kinetics at the surface, confirming the importance of ALD chemistry. Estimates are made of pulse/purge durations that would give optimum wafer throughput, trading off rapid pulsing against undersaturation of the surface. The dependence on Knudsen number is also considered [140].

Metal and dielectric coatings of high aspect ratio trenches allow higher capacitance density, e.g., in DRAM, and so there is acute interest in how to achieve conformality in such demanding features. Gordon et al. assume perfect sticking to derive scaling laws for film conformality as a function of precursor dose, flow rate and the aspect ratio of the feature [141]. Dendooven et al. propose a more general model, valid for any value of sticking coefficient, modified by the Langmuir isotherm to account for the evolving surface coverage, and this yields realistic thickness profiles for undersaturation [142]. A similar model uses a sticking coefficient that falls with precursor injection time, an indirect measure of precursor coverage [143]. The performance of ALD in coating other shapes, such as sharply scalloped corners, has also been modeled [144]. Representative half-reactions for $AlMe_3 + H_2O$ are used in a lattice Monte Carlo simulation of film growth and are coupled with Knudsen diffusion-based transport with a pore, revealing how film composition, roughness, and thickness depend on depth within the pore [145].

Tortuous geometries provide an extra twist in the case of plasma-assisted ALD, since radicals can recombine at the surfaces over which they pass; indeed the prevailing assumption is that this technique is unsuited to high aspect ratio features. Parametrized Monte Carlo simulations by Knoops et al. reveal that conformal growth can in fact be achieved if the radical recombination probability is sufficiently low (e.g., on metal oxide substrates), but not if it is high (e.g., on metals) [146]. In a contrasting approach, the physical characteristics of a remote O_2-plasma (including surface recombination) are built into an ALD gas transport model by Tinck and Bogaerts, allowing the effect of chamber pressure, coil power, and substrate orientation to be assessed [147]. One result is that radicals are found to be the main species actually reaching the substrate.

Uniformity of deposition across a wafer (or across a batch of wafers) is crucial in semiconductor processing and so the next level of ALD modeling couples the feature scale with the reactor scale. Gas flow in the reactor provides the boundary conditions for gas injection into features distributed across the surface of the substrate. The coupling should be bidirectional, since product gases emerge from features and affect reactor pressure and temperature. Such models can aid in reactor design.

Fluid dynamics in the reactor can be simulated in a similar way as in standard MOCVD, namely continuum modeling in a realistic three-dimensional geometry, and DFT energetics are included in such models, e.g., for GaN [148, 149], or Si [150]. However, rather than steady state MOCVD, the pulse/purge cycle of ALD means that gas flow is 'transient', which leads to potential difficulties in solving the convection–diffusion equations for each gas-phase species (because chemical interconversion causes them to be stiffly coupled) [151]. Nevertheless, early work by Ylilammi shows how surface kinetics can be built into a model of mass transport in a flow-type ALD reactor, yielding quantitative estimates of growth rate and wafer uniformity for the example of Ta_2O_5 [152].

Simulations coupling fluid dynamics at the reactor-scale and feature-scale (simple sticking and Knudsen diffusion) are described by Lankhorst et al. for a multi-wafer vertical batch reactor and fitted to HfO_2 growth from an amide precursor and water [153]. The authors quantify the timescales that operate in the process and how these depend on the aspect ratio of the features. ALD of HfO_2 is also the subject for Stout et al., this time from the chloride precursor and water, using a Monte Carlo model for transport in trenches and vias [154]. Readsorption of the HCl product is found to contribute to non-uniform impurity levels of Cl.

3.7 Conclusion and Outlook

ALD modeling is a specialized niche represented at present by around 150 papers, most of them published in the last decade. The main questions have concerned growth reactions, precursor decomposition, substrate effects, growth rate, conformality in features, and uniformity across a reactor. These topics have been successfully addressed by judiciously combining models at different length and time scales. Nevertheless, many ambitious targets remain; for instance, there are no models to date that explicitly show how precursor chemistry and reactor conditions dictate nanomorphology. Predicting precursor volatility is another open question for simulation.

Simulations have mostly addressed the ALD of oxides and nitrides, with relatively little on metals and chalcogenides. By far the most common subject to date has been computing reaction steps in the deposition of high-*k* dielectrics (HfO_2, ZrO_2, Al_2O_3) onto silicon-based substrates, clearly driven by the interest from the semiconductor industry that has propelled ALD into prominence over this period. A few simulations have considered inorganic ALD onto organic molecules, but there has been no modeling yet of the issues unique to molecular layer deposition.

For mechanistic questions, DFT is the method of choice and this is mostly used to compute reaction and activation energies for adsorption and ligand elimination from a single precursor molecule interacting with an idealized reactive surface. The simulations reveal a palette of competing reactions, subject to fascinating stereochemical effects. Recently, more attention is being paid to interactions between adsorbate fragments on the surface, leading to steps such as densification that can have a major influence on growth. Detailed models of non-ALD decomposition reactions during heterodeposition onto various substrates are also emerging, opening up the potential for using ALD precursors to achieve atom-by-atom control of interfaces. Nevertheless, an open challenge is the reliable simulation of redox reactions and plasma-based ALD processes.

One of the strengths of DFT and other first principles approaches is that quantitative data (such as reaction energetics) can be obtained without fitting to experiment. However, whether a single reaction step is endothermic or exothermic is not in itself evidence of whether it plays a role in deposition. Only a series of simulations provide useful insights—e.g., assessing which decomposition reaction is most likely—and so the choice of model and its interpretation remains the greatest source of uncertainty. First principles results have turned out to be most powerful when they can add atomic-scale explanation to experimental data, particularly in situ techniques such as IR spectroscopy. More generally, simulations must aim to move beyond simply validating data already available from experiment, and instead explain, predict, and contribute to innovations in ALD processing.

Acknowledgments Financial support from Science Foundation Ireland under "ALDesign", 09.IN1.I2628, is gratefully acknowledged.

References

1. Jonsson H (2000) Annu Rev Phys Chem 51:623
2. Sautet P, Delbecq F (2010) Chem Rev 110:1788
3. Elliott SD (2007) In Fanciulli MSG (ed) Rare earth oxide thin films: growth, characterization, and applications, Springer-Verlag, Berlin. Platz 3, D-14197 Berlin
4. http://www.tyndall.ie/hike/method/index.html
5. Shankar S, Simka H, Haverty M (2008) J Phys: Condens Matter 20:064232
6. Hafner J (2000) Acta Mater 48:71
7. Kohn W, Becke AD, Parr RG (1996) J Phys Chem 100:12974
8. Zhao Y, Truhlar DG (2008) Acc Chem Res 41:157
9. Kwon J, Dai M, Halls MD, Langereis E, Chabal YJ, Gordon RG (2009) J Phys Chem C 113:654
10. Rodriguez-Reyes JCF, Teplyakov AV (2007) J Phys Chem C 111:16498
11. Kang JK, Musgrave CB (2002) J Appl Phys 91:3408
12. Siodmiak M, Frenking G, Korkin A (2000) J Phys Chem A 104:1186
13. Li K, Li S, Li N, Dixon DA, Klein TM (2010) J Phys Chem C 114:14061
14. Heyman A, Musgrave CB (2004) J Phys Chem B 108:5718
15. Widjaja Y, Musgrave CB (2002) Appl Phys Lett 80:3304

16. Widjaja Y, Musgrave CB (1931) J Chem Phys 2002:117
17. Mukhopadhyay AB, Musgrave CB, Sanz JF (2008) J Am Chem Soc 130:11996
18. Widjaja Y, Han JH, Musgrave CB (2003) J Phys Chem B 107:9319
19. Mukhopadhyay A, Musgrave C (2006) Chem Phys Lett 421:215
20. Deminsky M (2004) Surf Sci 549:67
21. Mukhopadhyay AB, Sanz JF, Musgrave CB (2006) Chem Mater 18:3397
22. Elliott SD, Greer JC (2004) J Mater Chem 14:3246
23. Elliott SD, Scarel G, Wiemer C, Fanciulli M, Pavia G (2006) Chem Mater 18:3764
24. Kwon J, Dai M, Halls MD, Chabal YJ (2008) Chem Mater 20:3248
25. Nolan M, Elliott SD (2010) Chem Mater 22:117
26. Tanskanen JT, Bakke JR, Bent SF, Pakkanen TA (2010) J Phys Chem C 114:16618
27. Elliott SD (2005) Comput Mater Sci 33:20
28. Elliott SD (2007) Surf Coat Technol 201:9076
29. Xie Q, Jiang Y-L, Detavernier C, Deduytsche D, Van Meirhaeghe RL, Ru G-P, Li B-Z, Qu X-P (2007) J Appl Phys 102:083521
30. Elam JW, Pellin MJ, Elliott SD, Zydor A, Faia MC, Hupp JT (2007) Appl Phys Lett 91:253123
31. Ren J, Cui C, Zhou G, Liu Y, Hu Y, Wang B (2011) Thin Solid Films 519:3716
32. Olivier S, Ducéré J-M, Mastail C, Landa G, Estève A, Rouhani MD (2008) Chem Mater 20:1555
33. Tanaka T (2002) Thin Solid Films 409:51
34. Xu Y, Musgrave C (2005) Chem Phys Lett 407:272
35. Rodríguez-Reyes JCF, Teplyakov AV (2008) J Appl Phys 104:084907
36. Somani S, Mukhopadhyay A, Musgrave C (2011) J Phys Chem C 115:11507
37. Blanquet E, Nuta I, Brizé V, Boichot Rl, Mantoux A, Violet P, Daniele S (2010) ECS Trans 33(2):321
38. Mui C (2004) Surf Sci 557:159
39. Arvidsson I, Larsson K (2007) Diam Relat Mater 16:131
40. Hirva P, Pakkanen T (1989) Surf Sci 220:137
41. Mochizuki Y, Takada T, Usui A (1994) Appl Surf Sci 82–83:200
42. Hukka T(1996) Surface Science 359, 213
43. Mårtensson P (1998) Appl Surf Sci 136:137
44. Mårtensson P (1999) Appl Surf Sci 148:9
45. Mårtensson P (2000) Appl Surf Sci 157:92
46. Orimoto Y, Toyota A, Furuya T, Nakamura H, Uehara M (2009) Ind Eng Chem Res 48:3389
47. Xie Q, Deduytsche D, Musschoot J, Meirhaeghe RLV, Detavernier C, Ding S-F, Qu X-P (2011) Microelectron Eng 88:646
48. Han B, Wu J, Zhou C, Li J, Lei X, Norman JAT, Gaffney TR, Gordon R, Roberts DA, Cheng H (2008) J Phys Chem C 112:9798
49. Wu J, Han B, Zhou C, Lei X, Gaffney TR, Norman JAT, Li Z, Gordon R, Cheng H (2007) J Phys Chem C 111:9403
50. Machado E, Kaczmarski M, Braida B, Ordejón P, Garg D, Norman J, Cheng H (2007) J Mol Model 13:861
51. Machado E, Kaczmarski M, Ordejón P, Garg D, Norman J, Cheng H (2005) Langmuir 21:7608
52. Pirolli L, Teplyakov AV (2006) Surf Sci 600:3313
53. Dai M, Kwon J, Halls MD, Gordon RG, Chabal YJ (2010) Langmuir 26:3911
54. Asthagiri A, Sholl DS (2002) J Chem Phys 116:9914
55. Elliott SD (2010) Langmuir 26:9179
56. Puurunen RL (2005) J Appl Phys 97:121301
57. Halls MD, Raghavachari K (2004) J Phys Chem A 108:2982
58. Delabie A, Sioncke S, Rip J, Van Elshocht S, Caymax M, Pourtois G, Pierloot K (2011) J Phys Chem C 110815122459076

59. Gun'ko VM (1993) Kinet Katal 34:463
60. Brodskii V, Rykova E, Bagatur'yants A, Korkin A (2002) Comput Mater Sci 24:278
61. Petersen M (2004) Comput Mater Sci 30:77
62. Chen W, Zhang D, Ren J, Lu H, Zhang J, Xu M, Wang J, Wang L (2005) Thin Solid Films 479:73
63. Han JH, Gao G, Widjaja Y, Garfunkel E, Musgrave CB (2004) Surf Sci 550:199
64. Jeloaica L, Esteve A, Dkhissi A, Esteve D, Djafari Rouhani M (2005) Comput Mater Sci 33:59
65. Estève A (2003) Comput Mater Sci 27:75
66. Dkhissi A, Mazaleyrat G, Estève A, Rouhani MD (2009) Phys Chem Chem Phys 11:3701
67. Ren J, Zhang Y-T, Zhang DW (2007) J Mol Struct (Thoechem) 803:23
68. Ren J, Hu Y, Wang J, Jiang H, Zhang D (2008) Thin Solid Films 516:2966
69. Jeloaica L, Estève A, Djafari Rouhani M, Estève D (2003) Appl Phys Lett 83:542
70. Ren J, Zhou G, Hu Y, Jiang H, Zhang D (2008) Appl Surf Sci 254:7115
71. Lu H-L, Xu M, Ding S-J, Chen W, Zhang DW, Wang L-K (2006) Appl Phys Lett 89:162905
72. Fenno RD, Halls MD, Raghavachari K (2005) J Phys Chem B 109:4969
73. Halls MD, Raghavachari K (2003) J Chem Phys 118:10221
74. Halls M, Raghavachari K, Frank M, Chabal Y (2003) Phys Rev B 68:161302
75. Halls MD, Raghavachari K (2004) J Phys Chem B 108:4058
76. Ghosh MK, Choi CH (2006) J Phys Chem B 110:11277
77. Kim D-H, Baek S-B, Seo H-I, Kim Y-C (2011) Appl Surf Sci 257:6326
78. Lu H-L, Chen W, Ding S-J, Xu M, Zhang DW, Wang L-K (2005) J Phys: Condens Matter 17:7517
79. Clemens JB, Chagarov EA, Holland M, Droopad R, Shen J, Kummel AC (2010) J Chem Phys 133:154704
80. Hegde G, Klimeck G, Strachan A (2011) Appl Phys Lett 99:093508
81. Yu S, Qing-Qing S, Lin D, Han L, Shi-Jin D, Wei Z (2009) Chin Phys Lett 26:053101
82. Shi Y, Sun Q-Q, Dong L, Liu H, Ding S-J, Zhang W (2009) Chin Phys Lett 26:053101
83. Xu K, Ye PD (2010) J Phys Chem C 114:10505
84. Xu Y, Musgrave CB (2004) Chem Mater 16:646
85. Hu Z, Turner CH (2006) J Phys Chem B 110:8337
86. Hu Z, Turner CH (2007) J Am Chem Soc 129:3863
87. Ghosh MK, Choi CH (2008) Chem Phys Lett 461:249
88. Wasslen YA, Tois E, Haukka S, Kreisel KA, Yap GPA, Halls MD, Barry ST (1976) Inorg Chem 2010:49
89. Widjaja Y, Mysinger MM, Musgrave CB (2000) J Phys Chem B 104:2527
90. Rodriguez-Reyes JCF, Teplyakov AV (2007) J Phys Chem C 111:4800
91. Rodríguez-Reyes JCF, Teplyakov AV (2008) J Phys Chem C 112:9695
92. Haran M, Engstrom JR, Clancy P (2006) J Am Chem Soc 128:836
93. Okamoto Y (1999) J Phys Chem B 103:11074
94. Ren J (2009) Appl Surf Sci 255:5742
95. Ren J, Zhou G, Hu Y, Zhang DW (2009) Appl Surf Sci 255:7136
96. Lu H-L, Ding S-J, Zhang DW (2009) J Phys Chem A 113:8791
97. Lee W, Dasgupta NP, Trejo O, Lee J-R, Hwang J, Usui T, Prinz FB (2010) Langmuir 26:6845
98. Popovici M, Delabie A, Van Elshocht S, Clima S, Pourtois G, Nyns L, Tomida K, Menou N, Opsomer K, Swerts J, Detavernier C, Wouters D, Kittl JA (2009) J Electrochem Soc 156:G145
99. Nyns L, Delabie A, Pourtois G, Van Elshocht S, Vinckier C, De Gendt S (2010) J Electrochem Soc 157:G7
100. Mäkinen V, Honkala K, Häkkinen H (2011) J Phys Chem C 115:9250
101. Terranova U, Bowler DR (2011) J Mater Chem 21:4197
102. Miikkulainen V, Suvanto M, Pakkanen TA (2008) Chem Vap Deposition 14:71

103. Zydor A, Elliott SD (1879) J Phys Chem A 2010:114
104. Black K, Aspinall HC, Jones AC, Przybylak K, Bacsa J, Chalker PR, Taylor S, Zhao CZ, Elliott SD, Zydor A, Heys PN (2008) J Mater Chem 18:4561
105. Coyle JP, Johnson PA, DiLabio GA, Barry ST, Müller J (2010) Inorg Chem 49:2844
106. Li J, Wu J, Zhou C, Han B, Lei X, Gordon R, Cheng H (2009) Int J Quantum Chem 109:756
107. Li J-YW, Wu J-P, Zhou C-G; Yao, S-J; Han B (2008) Acta Chimica Sinica 66:165
108. Suntola T (1989) Mater Sci Rep 4:261
109. Alam MA, Green ML (2003) J Appl Phys 94:3403
110. Hémeryck A, Petrantoni M, Estève A, Rossi C, Djafari Rouhani M, Landa G, Estève D (2010) J Phys Chem Solids 71:125
111. Ylilammi M (1996) Thin Solid Films 279:124
112. Park H (2000) Appl Surf Sci 158:81
113. Elliott SD, Pinto HP (2004) J Electroceram 13:117
114. Puurunen RL (2003) Chem Vap Deposition 9:249
115. Puurunen RL (2003) Chem Vap Deposition 9:327
116. Puurunen R (2005) Appl Surf Sci 245:6
117. Nyns L, Delabie A, Caymax M, Heyns MM, Van Elshocht S, Vinckier C, De Gendt S (2008) J Electrochem Soc 155:G269
118. Alekhin AP, Lapushkin GI, Markeev AM, Sigarev AA, Toknova VF (1091) Russ J Gen Chem 2010:80
119. Kim J-H, Kim J-Y, Kang S-W (2005) J Appl Phys 97:093505
120. Lie M, Nilsen O, Fjellvåg H, Kjekshus A (2009) Dalton Trans 3:481
121. Nilsen O, Mohn CE, Kjekshus A, Fjellvåg H (2007) J Appl Phys 102:024906
122. Nilsen O, Karlsen O, Kjekshus A, Fjellvåg H (2007) Thin Solid Films 515:4527
123. Nilsen O, Karlsen O, Kjekshus A, Fjellvåg H (2007) Thin Solid Films 515:4538
124. Wilson CA, Goldstein DN, McCormick JA, Weimer AW, George SM (2008) J Vac Sci Technol A: Vac Surf Films 26:430
125. Yim S–S, Lee D-J, Kim K-S, Kim S-H, Yoon T-S, Kim K-B (2008) J Appl Phys 103:113509
126. Martin Hoyas A, Schuhmacher J, Whelan CM, Fernandez Landaluce T, Vanhaeren D, Maex K, Celis JP (2006) J Appl Phys 100:114903
127. Puurunen RL (2004) Chem Vap Deposition 10:159
128. Ahn J-HK, Kwon S-H, Kim J-H, Kim J-Y, Kang S-W (2010) J Mater Sci Technol 26:371
129. Battaile CC, Srolovitz DJ (2002) Annu Rev Mater Res 32:297
130. Deminsky M, Knizhnik A, Belova I, Umanskii S, Rykova E, Bagatur'yants A, Potapkin B, Stoker M, Korkin A (2004) Surf Sci 549:67
131. Mazaleyrat G, Estève A, Jeloaica L, Djafari Rouhani M (2005) Comput Mater Sci 33:74
132. Dkhissi A, Estève A, Mastail C, Olivier S, Mazaleyrat G, Jeloaica L, Djafari Rouhani M (1915) J Chem Theory Comput 2008:4
133. Neizvestny I, Shwartz N, Yanovitskaja Z, Zverev A (2006) Comput Mater Sci 36:36
134. Hua Z, Shi J, Turner CH (2009) Mol Simul 35:270
135. Voter AF, Montalenti F, Germann TC (2002) Annu Rev Mater Res 32:321
136. Aarik J, Siimon H (1994) Appl Surf Sci 81:281
137. Rose M, Bartha JW (2009) Appl Surf Sci 255:6620
138. Gobbert M (2002) Thin Solid Films 410:129
139. Gobbert MK, Prasad V, Cale TS (1031) J Vac Sci Technol B: Microelectron Nanometer Struct 2002:20
140. Gobbert MK, Cale TS (2007) Surf Coat Technol 201:8830
141. Gordon RG, Hausmann D, Kim E, Shepard J (2003) Chem Vap Deposition 9:73
142. Dendooven J, Deduytsche D, Musschoot J, Vanmeirhaeghe RL, Detavernier C (2009) J Electrochem Soc 156:P63
143. Kim J-Y, Ahn J-H, Kang S-W, Kim J-H (2007) J Appl Phys 101:073502
144. Cleveland ER, Banerjee P, Perez I, Lee SB, Rubloff GW (2010) ACS Nano 4:4637
145. Adomaitis RA (2010) J Cryst Growth 312:1449

146. Knoops HCM, Langereis E, van de Sanden MCM, Kessels WMM (2010) J Electrochem Soc 157:G241
147. Tinck S, Bogaerts A (2011) Plasma Sources Sci Technol 20:015008
148. Sengupta D, Mazumder S, Kuykendall W, Lowry S (2005) J Cryst Growth 279:369
149. Mukinovic M, Brenner G, Khanderi J, Spoellmann S, Fischer RA, Tafipolsky M, Cadenbach T, Schmid R (2005) Chem Vap Deposition 11:306
150. Cavallotti C (2010) Chem Vap Deposition 16:329
151. Van Veldhuizen S, Vuik C, Kleijn CR (2007) Surf Coat Technol 201:8859
152. Ylilammi M (1995) J Electrochem Soc 142:2474
153. Lankhorst AM, Paarhuis BD, Terhorst HJCM, Simons PJPM, Kleijn CR (2007) Surf Coat Technol 201:8842
154. Stout PJ, Adams V, Ventzek PLG (2006) J Vac Sci Technol B: Microelectron Nanometer Struct 24:2372

Part III
ALD for Memory Devices

Chapter 4
Mass-Production Memories (DRAM and Flash)

Cheol Seong Hwang, Seong Keun Kim and Sang Woon Lee

4.1 Introduction

Computers process information and store data. Their workloads can be duly classified into computing-centric and data-centric. In the computing-centric workload, the supreme performance of a microprocessor, which is composed of logic gates, is of primary importance. Issues related with logic gate formation will be discussed in chapters 7 and 8. On the other hand, memory plays a more important role in data-centric workload [1]. As the world is more closely connected by internet and the amount of end-user generated data is increasing exponentially (such as video images uploaded onto YouTube), the usual computer workload shifts more to the data-centric variety. The amount of indexed online data was estimated to increase from ~5 exabytes in 2002 to ~280 exabytes in 2009, a ~56-fold increase in only 7 years [2]. However, the computing power increase estimated by the increase in the number of logic gates and its operation speed according to Moore's law is only ~16-fold for the same period, suggesting the current trend of exploding data. Most computer users nowadays "search and see" rather than "process" information. This trend means that the importance of memory in any form of computers is ever-increasing. As of the end of 2011, the shipping of flash memory, which is purely a memory-related device, exceeds that of dynamic random access memory (DRAM), which is more closely related with

C. S. Hwang (✉)
Department of Materials Science and Engineering and Inter-university Semiconductor Research Center, Seoul National University, Seoul 151-744, Korea
e-mail: cheolsh@snu.ac.kr

S. K. Kim
Electronic Materials Research Center, Korea Institute of Science and Technology, Seoul 136-791, Republic of Korea

S. W. Lee
Department of Chemistry and Chemical Biology, Harvard University, Cambridge, MA 02138, USA

C. S. Hwang (ed.), *Atomic Layer Deposition for Semiconductors*,
DOI: 10.1007/978-1-4614-8054-9_4,

microprocessor operation. This trend clearly shows a very large change in the worldwide computing environment; memory becomes more and more important.

This chapter describes the memory devices and ALD-related fabrication processes. As described in the Introduction chapter, memory devices can be classified into 2 groups: the mass-production memories, including DRAM, Flash, and static RAM (SRAM); and emerging memories, such as phase-change RAM (PcRAM), magnetoresistive RAM (MRAM), and resistive RAM (ReRAM), or niche markets such as ferroelectric RAM (FeRAM). At the moment, MRAM and ReRAM are of little relevance for ALD or are too premature to be considered from the ALD point of view, so they are not discussed in this chapter. From the fabrication point of view, SRAM is identical to the logic gates, so it is also not discussed here.

4.1.1 Role of DRAM and Flash Memories in a Computer

Dynamic random access memory (DRAM) works as the main memory in every modern computer, from high-end server computers to simple hand-held devices. Computing in any computer requires two key information sets; programs, and the data to program with. Both sets are stored as a form of 'bits' in the core memory part of a computer. Any computed output that comes from these actions is also stored as a data set within different parts of the memories that comprise the computer. Up to now, the conventional hard disk has been the primary memory element for all data storage (program and user data). However, the mechanical motions of the magnetic hard disk limit the data access time for both read and write operations, making them incompatible with the modern high speed microprocessors. Microprocessors actually require a considerably large amount of memory for device operation. The primary temporal memory element to cope with this issue is an embedded SRAM, which is fabricated concurrently with the logic circuits. However, due to its large cell size, the memory density of SRAM is limited to a rather low level, and a massive amount of data must inevitably be stored away in the DRAM portion during computer operation. In addition, the program that controls microprocessor operation must also be retrieved from the hard disk and loaded into the DRAM. Therefore, DRAM must have a memory density large enough to satisfy these requisites, which is typically at least a few gigabytes for modern computers to fluently operate. In everyday life, we sometimes see the computer freezing its functions momentarily as the hard disk operates when a computer with an insufficient main memory is used. This is usually due to the dumping of an abnormally large data set that cannot be handled by the limited DRAM when information is transferred in and out of the hard disk. It is not difficult for us to infer from these everyday experiences that the main memory itself plays a role as important as high speed processors for modern computers in terms of performance. It is, therefore, evident that main memories should have high density, high speed, robust data storage, and an almost unlimited number of writing/reading cycles, which can all be achieved only from DRAM at present. If DRAM technology persist its evolving trend for faster operation, it might be able to replace the embedded SRAM with embedded DRAM in future high performance microprocessors. This has actually been attempted in the past but the success was limited mainly due to the process incompatibility between logic circuits and memory cells. Logic circuits generally require a low temperature process (<500 °C) after the gate formation, which is composed of Ni-silicides, but DRAM cell fabrication requires a much higher temperature (at least 600 °C) during capacitor fabrication processes. In addition, the largely different topology between the logic area, where there are no capacitors, and the DRAM area, where there are capacitors, has been a big hurdle for the overall fabrication process. This is a research field that will be pursued in future.

Hard disks are now being replaced by the NAND flash memories, which are another topic of this chapter. However, the much higher cost of flash memory compared to magnetic hard disks is slowing down the replacement speed. Since the NAND flash is a solid-state device (sometimes it is called a solid-state disk, although it actually has nothing to do with disks), the data retrieve speed is far superior to that of conventional hard disks. However, as explained below, the writing speed is much slower in comparison, even comparable with that of the hard disk, due to its characteristic writing scheme. Therefore, flash memory is not expected to replace DRAM.

If the main memory density increases to a very large degree, say to a tera-bit size, and it is nonvolatile, i.e., the data are still stored under the absence of electric power, and has other features that DRAM has, no extra-memory element may be needed for computers. This corresponds to the dream memory, which is literally possible 'in dreams' only and nonexistent at present. DRAM has the potential to be the closest candidate to fulfill this dream goal in various aspects, except for its nonvolatility. The reason why DRAM cannot have nonvolatility is explained in detail in this section. Perhaps phase change RAM (PcRAM) is closer to this dream goal in principle, but PcRAM may need to be nurtured for quite a long time to realize this goal. PcRAM is currently positioned as a subsidiary memory between the DRAM and disc-type memory devices, which subsidize its overall memory functionality (so-called storage class memory). In other applications, it may replace NOR-type flash memories, which are mainly used as the code memory elements. MRAM, especially with the spin torque transfer function (STTRAM), appears to be the closest to DRAM in terms of performance (high speed, write/read endurance). However, as discussed in Chap. 2, its requirement for a high operating current, larger cell area, and extreme difficulty in fabrication at the mass-production level prohibit its replacing DRAM at the present time.

Flash memory features a unique double gate-structured MOSFET that works as the memory cell as well as the selection device. This is a very useful feature as a memory in general, because the memory and switching elements do not take up separate space or area, meaning that the memory density can be higher compared to ca. DRAM. However, this same feature also gives rise to a demerit in terms of reading and writing operations. In general, random access, especially during writing, in flash is very difficult, indicating that it cannot work as a main memory in principle. Therefore, flash memory is now being explored as a nonvolatile solid-state memory that may eventually replace magnetic hard disks. Among the two types of flash, NOR and NAND, the basic function of NOR flash was explained in Chap. 2. However, NOR type flash memories actually have a lower significance in industry due to its larger cell size ($\sim$10 F^2, where F is minimum feature size) compared to the NAND type device (cell size $\sim$5 F^2) and even DRAM (6–8 F^2). Therefore, this chapter focuses only on NAND flash memory.

This chapter, therefore, first introduces the operation principles of DRAM and NAND flash memories, and then deals with the thin film processing issues related with ALD. The writing principle of DRAM was already explained in Chap. 2 when the active matrix type memory operation was explained. Therefore, here the discussions begin with the read scheme of DRAM. As the reading of data in a DRAM cell corresponds to a destructive read out, the data must be written back to the cell immediately after read out (refresh). This can also be understood easily from the operation of the sense amplifier. Once we understand this reading/refresh scheme, the reason why there are two types of cell architectures, the so-called folded (8 F^2 cell) and open bit-line (6 F^2 cell) structures, and what the merits and demerits of these two schemes are can also be understood.

Understanding the writing and reading schemes of NAND flash requires a little more knowledge on MOSFET operation and electron tunneling. However, as long as basic understandings on circuit operation are concerned, the discussions are strictly limited to a qualitative level, so that many readers who did not major electrical engineering can still understand the key assets without any large difficulties.

The ALD parts review the progress of ALD processes for various functional ALD layers, which either contribute to device operation in the final products (called active layers), or are removed during the subsequent processes (called sacrificial layers) in memory devices.

Making the ever-shrinking pattern size, where "pitch" is a more appropriate term in memory, whereas feature size generally represents the smallest dimension in logic devices, has always been the biggest hurdle to overcome at each technology node. Even with the very recent immersion lithographic technology combined with the 193 nm ArF excimer laser exposure tool, making patterns with a feature size <30 nm relies on supplementary processes. These include deposition/etching back for making a smaller space or contact hole, and a double patterning technique for denser and smaller patterns. For these resolution enhancement technologies, several sacrificial layers that are mostly compatible with the photoresist (PR) are necessary. Finely patterned PR maintains its fidelity only up to $\sim$200 °C. Therefore, the generally low processing temperature of ALD is well suited to processes where thin films are deposited on top or sides of PR patterns. These techniques will keep contribute to the next generation lithography with the extreme ultra-violet lithography tool.

Several ALD materials are used as the sacrificial hard masks for dry etching processes, which means that they must have reasonable density and chemical stability with very tight thickness control. These requirements are another challenge for ALD films, which require further developments in the precursors and process conditions.

4.1.2 How is the Data in DRAM Read Out?

Figure 4.1 shows the schematic diagram of a sense amplifier and two connected memory cells, one to a bit line (B/L) and the other to a bit line bar (B/L bar), in the folded bit line architecture. The sense amplifier is composed of two cross-coupled latches which are made of two NMOSFETs and two PMOSFETs.

It can be immediately noted that the cell arrangement in this architecture is not optimal in terms of the highest density, as can be understood from the fact that there is space for one more memory cell to be placed at the crossing node between word line 1 (W/L1) and the B/L bar, and that it is empty. The reason for this seemingly inefficient arrangement can be understood from the following. Here, let us assume that the high potential stored in the memory cell, indicated as D1 in Fig. 4.1, is about to be read out. In order to do that, first, the potentials of the B/L and B/L bar are set to $V_{cc}/2$ (V_{cc} is the operation voltage), which is usually accomplished by shorting, separating, and subsequently floating the B/L and B/L bar, which were initially connected to the voltage source (V_{cc}) and ground, respectively. The potential of the two crossing lines, PLAG and NLAG, which are connected to the source regions of the two PMOSFETs and NMOSFETs, respectively, are also set to $V_{cc}/2$. Therefore, at this stage, the V_{GSP1} (= $V_{source} - V_{gate}$ of PMOS1) and V_{GSP2} (= $V_{source} - V_{gate}$ of PMOS2) are zero. Then, W/L1 was pulled up to turn on the cell transistor with D1. It must be noted that the cell capacitor with D1 had a high potential, which is higher than $V_{cc}/2$ but lower than V_{cc}, so the stored charge (or potential) was transferred into the B/L and increased its potential from $V_{cc}/2$ to $V_{cc}/2 + \Delta V/2$, where $\Delta V \sim V_{cc}C_c/2(C_c + C_b)$, which is called charge sharing, while the potential of B/L bar remained unaffected. C_c and C_b are the capacitances of the cell capacitor and B/L respectively, where $C_b \sim 10\ C_c$. At the same time, the potential of PLAG increases to V_{cc} and that of NLAG decreases to 0, in an attempt to turn on the MOSFETs in the sense amplifier. From this operation, V_{GSP1} becomes $V_{cc}/2$ while V_{GSP2} becomes $V_{cc}/2 - \Delta V/2$, because the gate voltage of PMOS2 was increased by $\Delta V/2$ due to the

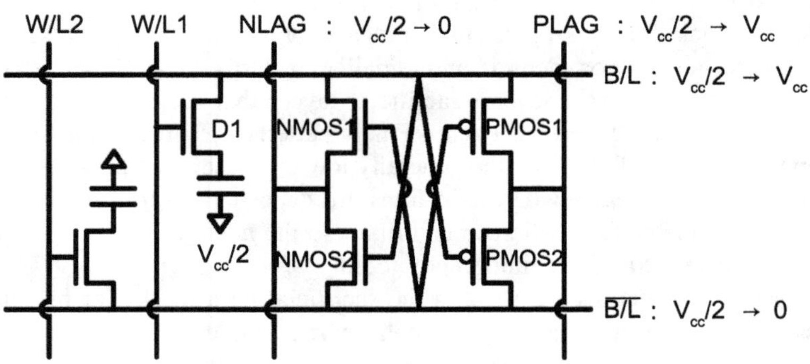

Fig. 4.1 Schematic diagram of sense amplifier and connected two memory cells in the folded bit line architecture

charge sharing between the cell capacitor and B/L. V_{GSN1} and V_{GSN2} were regarded as 0 and $-\Delta V/2$, respectively, at the moment. As was explained in Chap. 2, a PMOSFET turns on when the gate potential is lower than that of source, while the opposite is the case for an NMOSFET. Therefore, at this stage of circuit operation, PMOS1 is more turned on than PMOS2 because $V_{GSP1} > V_{GSP2}$. Here, the threshold voltages of all MOSFETs are controlled to be $<V_{cc}/2$. In other words, both PMOS1 and PMOS2 would be turned equally on when the potential of PLAG increases from $V_{cc}/2$ to V_{cc}, and then both the B/L and B/L bar potentials would be equally increased from $V_{cc}/2$ to V_{cc}. However, due to the involvement of the cell capacitor, PMOS1 turns on more than PMOS2, and most of the PLAG potential is transferred to the B/L. In other words, the transferred charge (or voltage) from the cell capacitor interferes with the operation of PMOS2 and the PLAG voltage can hardly be transferred to B/L bar. At this point of operation, the B/L potential is V_{cc}, which was the voltage originally used to write data 1 into the cell D1. It must be noted here that because the W/L1 potential is still high, the increased voltage of B/L (V_{cc} now) is transferred back to the cell capacitor with D1, which corresponds to the refreshing of the data destroyed just a moment ago. Of course, a full V_{cc} cannot be written into the cell capacitor due to the presence of a V_{th} in MOSFETs and many other parasitic effects. However, neglecting these quantitative issues and focusing on the more qualitative aspects, it can be understood that reading in DRAM is also writing back. Even though the variation in the B/L potential by charge sharing is quite small ($\sim$100 mV, while V_{cc} is $\sim$2 V), the operation of the circuit (reading) returns the B/L potential to its original high value (V_{cc}) when writing is performed, and the circuit knows that data 1 was written into that memory cell.

Next, let us consider what is going on in the NMOSFETs and the lines connected to it (B/L bar and NLAG). The initial potential of the B/L bar ($V_{cc}/2$) played an important role in operating the two PMOSs in a correct way by providing a reference voltage ($V_{cc}/2$) to the gate of PMOS1 when charge sharing occurred between the cell capacitor and B/L. It must be stressed that maintaining the reference voltage as accurate as possible is very important because the ΔV is only $\sim$100 mV. This issue is discussed in detail below. When charge sharing occurs, the B/L voltage increases from $V_{cc}/2$ to $V_{cc}/2 + \Delta V/2$, and the gate voltage of NMOS2 also increases from $V_{cc}/2$ to $V_{cc}/2 + \Delta V/2$, while that of NMOS1 still remains at $V_{cc}/2$. As the NLAG voltage decreases to 0 V, both NMOSs are about to be turned on. However, due to the different gate voltages, NMOS2 turns on more and the ground potential of NLAG is transferred mainly to B/L bar. Then, this lowering of the B/L bar voltage also decreases the gate voltages of PMOS1 and NMOS1, which actually further strengthens the intended operation of the circuit; PMOS1 turns on completely (now $V_{GSP1} = V_{cc}$) while NMOS1 turns off completely (now $V_{GSN1} = 0$). At the same time, PMOS2 turns off completely (because $V_{GSP2} = 0$: consider that the B/L potential is now V_{cc}) while NMOS2 turns on completely (now $V_{GSN2} = -V_{cc}$). Therefore, after the circuit completes its read operation, the status of all elements returns to its previous state when the data were written. This is a very clever design!

It can also be noted that the memory cell connected to W/L2 and B/L bar is not influenced by this reading action because W/L2 remained at a low voltage, and the data in that cell are retained, uninfluenced in principle. It is now possible to understand why an additional memory cell cannot be placed at the junction of W/L1 and B/L bar. If a memory cell was placed here, two problems occur; first, when W/L1 is boosted up, this additional cell is also connected to the sensing circuit that is attempting to read the data in D1 cell. Therefore, the data in this additional cell are disturbed by reading of the other cell that shares the same W/L1. Second, depending on the written voltage level of this additional cell, the reference voltage of B/L bar for reading D1 cell is disturbed, and the circuit does not work appropriately. Therefore, this additional cell cannot be placed at that position. This is why this type of array architecture is called a 'folded bit line architecture'; memory cells are arranged in a folded way along the B/L direction.

It can be easily understood that a low potential (data 0) stored in the same cell can be read out in a similar way. The reading recovers the zero potential of B/L after the read operation, so a data 0 is written back to the memory cell. It can also be understood that the B/L works as the reference voltage line when the memory cells connected to B/L bar are read out (W/L2 pulled up but W/L1 remained low). The detailed descriptions for reading data 0 are left for the readers' practice.

The four MOSFETs in a sense amplifier must have very well-controlled characteristics to operate accurately (identical V_{th}), since the sensing margin (ΔV) is quite small compared to the reference voltage. Another key factor is an identical C_b for B/L and B/L bar. As mentioned above, the reference voltage of both lines is produced by shorting and separating the two lines. If the C_b values are different for the two lines, the potential will deviate from the ideal value inversely proportional to the C_b values. The major factor that influences C_b is the lithographical error that occurs during the fabrication of the two narrow and long B/Ls. In the folded B/L architecture, they are formed on the same side of the sense amplifier, and the proximity between them minimizes error. However, in the open B/L architecture, discussed below, there is a higher chance of mismatching C_b.

Figure 4.2 shows the schematic diagrams for cell arrangement in folded and open B/L architectures. It can be seen that the B/L and B/L bar locates on opposite sides of the sense amplifier in the open B/L architecture, which increases the risk of lithographical error during fabrication. Therefore, the open B/L architecture has a weaker resistance to noises. However, if we consider the cell arrangements in the open B/L architecture, it is clear that the cells can be arranged in a regular matrix way, meaning that the memory cell density could be higher than the folded B/L architecture. However, the memory cell density in the open B/L architecture is not simply double of its counterpart because larger number of sense amplifiers is necessary. One sense amplifier serves four and two B/Ls, respectively, in folded and open B/L architectures. In overall, the overhead by the increased sense amplifier area is overcompensated by the increased memory cell numbers and the cell area efficiency is higher for the open B/L architecture. Due to this merit, almost all gigabit density DRAMs now adopt the open B/L architecture, even though fabrication is a bit trickier than the other case.

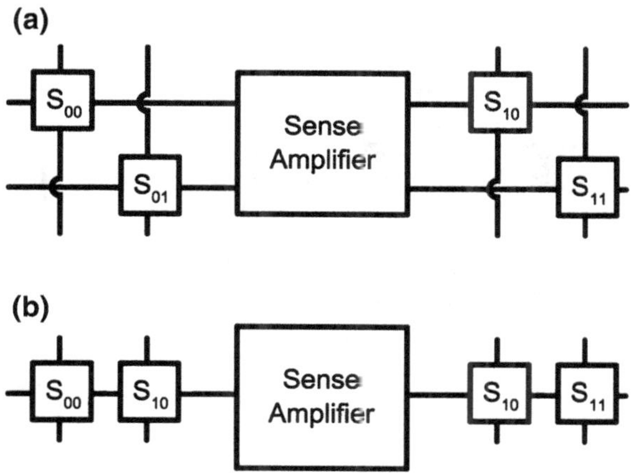

Fig. 4.2 Schematic diagrams for cell arrangement in **a** Folded and **b** Open B/L architectures

Now, we can understand why the folded and open B/L architectures have 8 and 6 F^2 cell sizes, respectively, from Fig. 4.3. Here, the arrangements of cell transistors are projected on a wafer surface according to the two B/L architectures. Cell capacitors are placed on top of a capacitor contact (here, represented by "D", the drain contact), so they do not consume additional area in principle if its lateral size is small enough. However, actual capacitor geometry requires a bit more space than the ideal, so the actual cell size is slightly larger than the ideal values. Thanks to the sequential operation of W/Ls, one source region can be shared by the neighboring two MOSFETs. If one capacitor contact is shared by two neighboring cells, data cannot be stored when the neighboring cell is read, i.e., independent data reading is impossible so this type of geometry cannot be built. An ideal MOSFET must have 3 F^2 area (3 F for source + gate + drain in one direction and F for their width). However, as just mentioned, one F can be saved for a set of two MOSFETs in DRAM, allowing two cells to have 5 F^2 of area instead of 6 F^2. However, in order to separate them from the other cells, an additional F is necessary in both directions, and thus two cells take up 12 F^2, meaning that each cell has 6 F^2 in an open B/L architecture, as shown in Fig. 4.3a. The cell arrangement in a folded B/L architecture can be understood in a similar fashion. The only difference is that cells must be omitted in alternating B/Ls. However, this leaves a large space between the B/Ls that have memory cells attached to them, so we can place memory cells along alternating W/Ls as shown in Fig. 4.3b. This type of arrangement results in a cell size of 8 F^2.

From the discussions given above, we can understand clearly why a high capacitance is necessary for stable DRAM operation even with a shrunk cell size. The margin for the sense amplifier to work appropriately is ΔV, which is proportional to $C_c V_{cc}$ and inversely proportional to C_b. C_b is mainly determined by the number of cells attached to one B/L, which is usually 1,024–2,048. A smaller value gives a better reading margin for the given $C_c V_{cc}$ but it is unfavorable for higher density because a larger number of sense amplifiers are necessary. V_{cc}

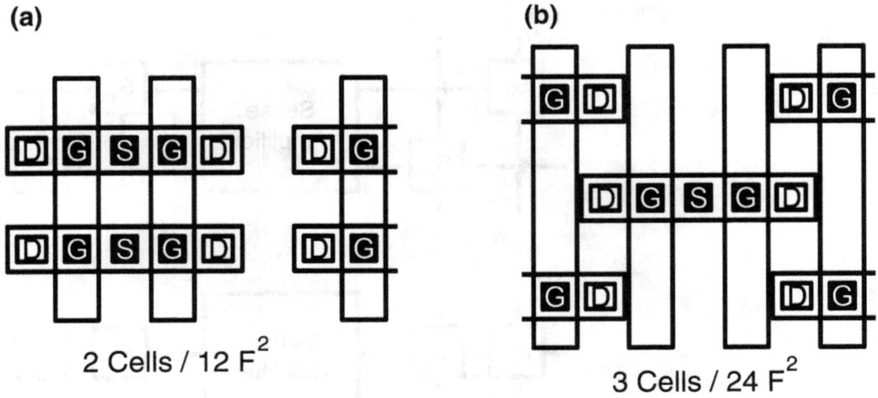

Fig. 4.3 Arrangements of cell transistors in **a** Open and **b** Folded B/L architectures

usually decreases with increasing density due to the excessive power consumption that is required to maintain it at a constant level. Therefore, C_c must be maintained as high as possible or even increased as scaling progresses. Fortunately, the smaller circuits are, they are less susceptible to the noise and the necessary minimum ΔV also decreases slightly. Therefore, an approximately constant cell capacitance (25–30 fF) is necessary irrespective of DRAM generation. This trend is well summarized in Fig.4.9 in p. 536 of ref. [3]. Actually, decreasing V_{cc} is beneficial in terms of sensing margin since it can be more accurately represented by $(V_{cc} + \Delta V)/V_{cc}$. However, the decrease in ΔV with decreasing V_{cc} is much more dominant; ΔV actually does not decrease linearly with V_{cc} but it decreases with a much faster trend. This is mainly due to the relatively constant V_{th} (0.7–0.8 V) of cell MOSFET irrespective of the DRAM generations, but this is over the scope of this chapter. Overall, a high C_c is crucial in every generation of DRAM.

A high C_c can be achieved in three ways; increasing electrode surface area, decreasing dielectric film thickness, and increasing the dielectric constant (k). Among them, increasing electrode area has severe limitations due to the difficulty in etching and maintaining the mechanical strength of the tall capacitor structure. This is discussed in more detail below. Decreasing dielectric thickness and increasing dielectric constant can be represented by decreasing the single term, equivalent oxide thickness (EOT) (t_{ox}), which is given as $t_{phys} \times (3.9/k)$ where t_{phys} is the physical thickness of the dielectric film. The smaller the t_{ox} is, the higher the capacitance.

For a given V_{cc}, stored charge (Q_s) in a cell capacitor is given as $\alpha \times C_c \times V_{cc}$, where α is a constant <1 determined by the circuit design. When system power is turned off, Q_s is removed, so DRAM is a volatile memory. Even when the power is kept on and the cell transistor remains in the off state, Q_s decreases with time due to dielectric leakage and transistor leakage. Transistor leakage originates from several reasons such as p–n junction leakage, source-drain leakage due to the drain-induced barrier lowering effect, which is serious for very short channel lengths, etc. The actual V_{th} of the cell transistor deviates from its nominal value (0.7–0.8 V) during circuit operation under the various bias conditions applied to the gate, source and drain, which also contributes to the transistor leakage. Again, this is over the scope of this chapter. Considering these issues in design, an approximately 10 % loss of Q_s can be tolerated before the stored data must be read out. This corresponds to a 2–3 fC loss for a cell capacitance of 20–30 fF and a capacitor write voltage of 1 V. This means that the specifications for dielectric leakage and transistor leakage are 1–2 fA. Assuming the capacitor area of $\sim$1–2 μm^2, this leakage current corresponds to a leakage current density of $\sim 10^{-7}$ A/cm^2 at the capacitor voltage. Mass productive capacitors usually have leakage current densities smaller than this specification by $\sim$10 times to ensure the uniformity of numerous cells. Due to the finite dissipation of Q_s in a certain time period, the stored data must constantly be read out and rewritten back dynamically, typically 5–10 times/s, which is why this is called dynamic memory.

4.1.3 How is the Data in NAND Flash Written and Read Out?

As mentioned above, the basic building element of flash memory is the double gate transistor of which the schematic diagram and its transfer characteristics (drain current (I_d) vs. control gate (CG) voltage (V_{cg})) are shown in Fig. 4.4. When the V_{cg} increases to a high enough value ($\sim$20 V) relative to the channel potential, electrons are injected from the channel (or from source and drain since the channel was already turned on by the high V_{cg}) into the floating gate (FG). This is called programming, which increases the V_{th} of the transistor as shown in Fig. 4.4b. On the other hand, a low enough V_{cg} (~ -20 V) repels electrons from the FG and injects hole from the channel (here, the channel is under the accumulation condition due to its p-type nature and low gate potential) into the FG. This is called erasing, which decreases the V_{th} of the double gate MOSFET. Both program and erase utilize the Fowler-Nordheim tunneling mechanism. One can assign the 0 and 1 data states to the programmed and erased states, which corresponds to the basic mechanism of NAND flash memory operation. NOR flash adopts a slightly different program method, which was discussed in Chap. 2, while the same erase method is used. While the program is done in a cell-by-cell manner, as described in detail below, erase is performed simultaneously on many cells (usually contained in a memory sector, called block), which results in the nonrandom access of the cell. In addition, program and erase take a much longer time (μs – ms) than the necessary writing time in DRAM (ca. < $\sim$50 ns), although data reading is as speedy as DRAM. Therefore, NAND flash memory cannot function as the main memory in computers even though its density is higher than DRAM in every generation.

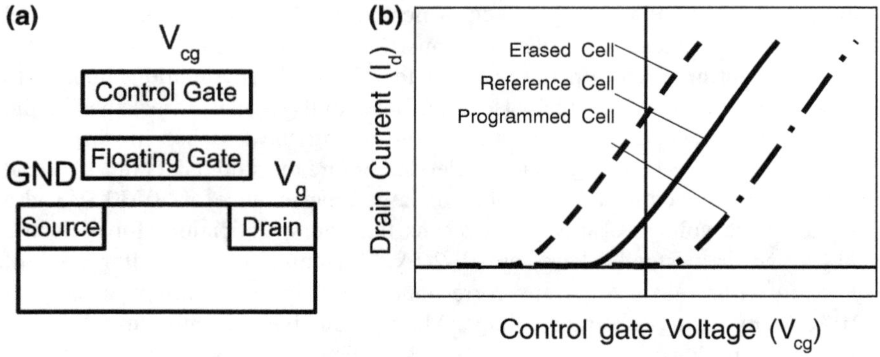

Fig. 4.4 **a** Schematic diagram of double gate transistor for flash memory. **b** Erasing and programming operations of double gate MOSFET

While the NOR flash and DRAM utilize a matrix-type cell arrangement, NAND flash uses a characteristic "string structure" where 32–64 memory cells are connected in series by sharing their source and drains. This is in a stark contrast to the structures of matrix-type memories, where each source and drain has an individual contact which increases the overall cell size due to the contact and cell separation areas. Because of its characteristic string structure, the cell size of NAND flash memory is the smallest of all memory cells, being $\sim 5\ F^2$, which enables the highest density to be achieved in each technology node. In addition, the simpler cell structure compared to DRAM, due to the lack of any capacitors, further contributes to the higher density, even though the FG height is quite high (see Fig. 4.20 of Sect. 4.2.2.1). As of year 2011, 16–32 Mb NAND flash is under active mass production while main production DRAM remains at 8–16 Gb density. The various resolution enhancement techniques, such as double pattering, contributed to such a higher integration density for given lithographic capabilities. Double patterning is discussed in Sect. 4.2.2.2 in detail.

Despite these positive aspects, the serially connected structure of the NAND string requires a rather complicated circuit operation, because accessing each cell requires the turn on of all connected cells, which might disturb the written data in the connected cells. Therefore, the writing and reading in these devices require some special operation schemes as described below.

Figure 4.5 shows the schematic circuit diagram of a NAND flash cell where two B/Ls are drawn. This will help readers envision the disturbance issues caused by the high W/L voltage. In Fig. 4.5, the memory cell marked by "P" is about to be programmed. Here, all memory cells are assumed to be initially erased by the previous block erase operation. In order to achieve this goal, first, the DSL, which is connected to the normal NMOSFET located at one end of the string, is boosted up to the V_{cc} level to let the B/L voltages (0 V in the string containing cell P, and V_{cc} in the next string) be transferred to each string. In the meantime, SSL connected to the other normal NMOSFET located at the opposite end of the string

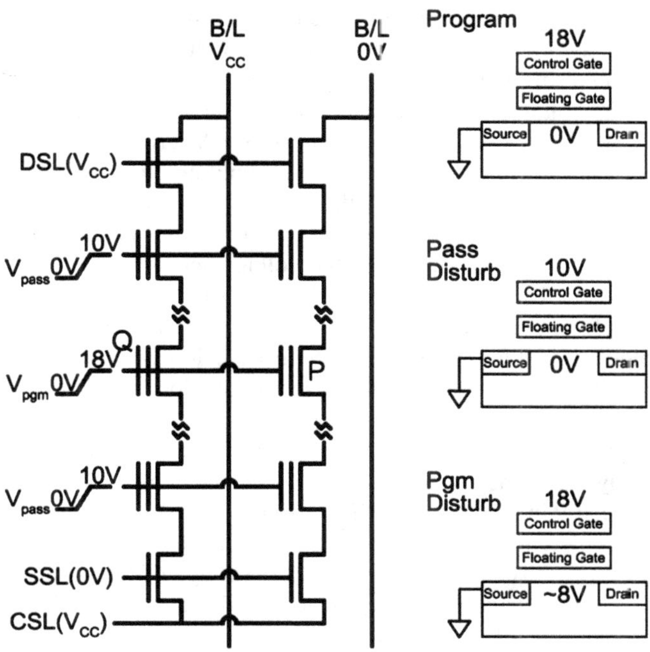

Fig. 4.5 Schematic circuit diagram of a NAND flash cell

remained at 0 V to fill up the channel of the string with the intended voltage. Then, a high enough voltage (called program voltage, V_{pgm} $\sim$18 V) is applied to the corresponding CG line while the channel is connected to the ground potential of the corresponding B/L. This type of operation requires the turning on of all connected cells in the same string, which is accomplished by applying another high enough voltage (called V_{pass} $\sim$10 V) to all the CG lines. It must be noted here that the V_{pass} of $\sim$10 V is high enough to turn on the cells irrespective of their memory states (the V_{th} of a programmed cell is $\sim$1 V and that of an erased cell is ~-3 V) but low enough to prevent the programming of non-targeted cells. It should be recognized that any of the cells in the same string could be programmed in the previous steps, although all the cells were initially erased. Now, the cell P is programmed by electron injection to the FG of that memory cell, while other cells in the same string remained in their previous states since the voltage of those cells is only $\sim$10 V. Actually, the V_{pass} of $\sim$10 V is far higher than the necessary voltage to turn on the programmed cells ($\sim$2 V) because they have V_{th} of $\sim$1 V, being high enough to induce a slight programming of the non-targeted cells, which is called program disturb. However, the reason for such a high V_{pass} can be understood from the following.

Let us now look at what is happening in the next string which is connected to a B/L of which voltage is V_{cc}. First, we can immediately understand that the cell marked "Q" is about to be programmed due to the simultaneously applied V_{pgm}, which must be avoided to not disturb the data written in cell Q. This can be achieved by increasing the voltage of the channel region of the cell Q to ~8 V meaning that the voltage across the cell is only ~10 V. Now, the question is why the channel potential of cell Q is as high as ~8 V given the much lower V_{cc} (~3 V) of that B/L. This requires some understanding on the rather complicated operation of NAND strings with different V_{cg}, which is only qualitatively explained below. When the DSL is boosted up, the V_{cc} is transferred into the string containing the cell Q. At this moment, the voltage across the DSL NMOSFET is only $V_{cc} - (V_{cc} - V_{th}) = V_{th}$, where $(V_{cc} - V_{th})$ is approximately channel potential. Therefore, the NMOSFET is just turned on. Then, the V_{pass} is applied to the nearby memory cell to turn it on. The high V_{pass} drags channel electrons out of the channel region of the DSL NMOSFET to the junction region between the DSL NMOSFET and memory cell due to the proximity effect, eventually turning the DSL NMOSFET off even though the DSL still remains at V_{cc}. Then, an immediate question is why the same did not happen in the DSL NMOSFET of the string containing the cell P in previous case. This can be understood from the fact that the initial voltage difference between the CG and channel of this NMOSFET was $\sim V_{cc} + V_{th}$ $(= V_{cc} - (0 - V_{th}))$ since the B/L voltage of that string remained at 0 V. Therefore, even though the V_{pass} of the nearby memory cell was high, this NMOSFET remained in the on state.

Now, it can be understood that the memory cell string containing cell Q is separated from outside circuits because the DSL and SSL NMOSFETs are turned off. Therefore, when high V_{pass} and V_{pgm} are applied to the passing cells and target cell, respectively, the channel potentials of the memory cells increase according to the voltage partitioning between the insulating layers and channel region. According to this circuit operation, the channel potential of cell Q was increased to ~8 V, and thereby the programming of the cell Q was avoided. Now, it can be understood why such a high V_{pass} is necessary; the DSL NMOSFET of the string containing cell Q must be turned off to boost up the channel potential to a high enough level, otherwise the channel potential remains at $\sim(V_{cc} - V_{th})$, which is too low to prevent the programming of cell Q. Programming of other cells can be proceeded in a similar manner, and each cell can be programmed independently by the sequential operation of W/Ls.

Unfortunately, the erasing cannot be proceeded in a similar one-by-one manner. The erasing is performed in units of blocks, where ~128 kilobytes are contained, by applying 0 V to all CG and 20 V to the p-well containing whole channel regions of all strings. When a sufficiently long time elapses (~ms), all FGs are filled with holes (erased state) irrespective of their former states. Then, the programming starts again as described above. Therefore, random access, which requires the independent erase and program of each cell, is not possible in NAND flash memory.

The reading action of a NAND flash cell is easier to understand as shown in Fig. 4.6. Let us assume again that the cell P is read out; it could be either in the erased or programmed state. First, the B/L is set to ~1 V, and the DSL and SSL

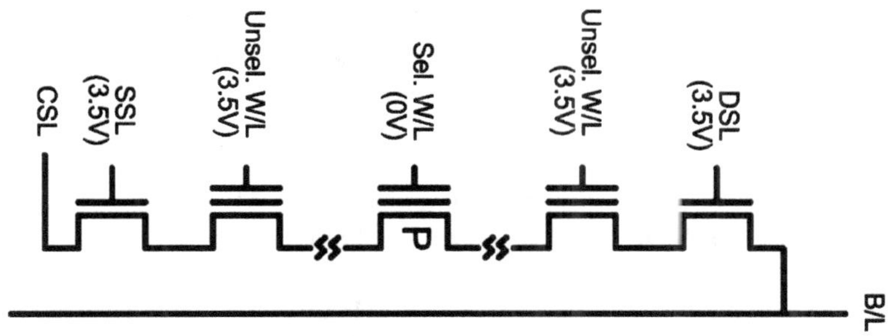

Fig. 4.6 Schematic diagram of NAND flash memory for data reading operation

are boosted up to ~3–4 V to turn on the normal MOSFETs, so that the B/L potential can or cannot be transferred to the sense amplifier through CSL depending on the state of cell P. The V_{cg} of cell P is set to 0 V while the V_{cg} of other cells in the same string is set to ~3–4 V to turn them on. Here, 3–4 V is high enough to turn them on because of the relatively low B/L voltage (~1 V), and the channel voltage is <1 V. If the cell P was in the programmed state, the V_{cg} of 0 V is not enough to turn the memory cell on so the B/L voltage cannot be transferred to the sense amplifier through CSL because the V_{th} of the programmed cell is ~1 V. On the other hand, if the cell P was in the erased state, the V_{cg} of 0 V is high enough to turn on the memory cell and the B/L voltage can be transferred to the sense amplifier because the V_{th} of the erased cell is ~−3 V. Therefore, we can read out the data state of cell P in this way.

Although the descriptions given above for understanding the operation of NAND flash memory are much more complicated compared to the more straightforward cells in DRAM, it is believed that these principles would not be too difficult for readers who did not study semiconductor circuitry, and that knowledge in this field will be beneficial for developing novel materials and processes for such an important device.

There is another important NAND-type flash memory, called charge trap flash (CTF), where the FG is replaced with a charge trap (CT) layer which is typically made of a low-pressure chemical vapor deposited Si_3N_4 layer containing many deep and uniform traps for electrons and holes (See Fig. 4.20 of Sect. 4.2.2.1). In this case, the injected charges are stored in the traps of the CT layer instead of the FG, which provides a relatively simple solution to several problems in highly integrated conventional FG-NAND flash memory. Two of the merits are described below briefly. In flash memory, the V_{cg} is transferred to the Si channel surface via capacitive coupling, which is necessary for the sufficient band bending of Si. The presence of a floated conducting layer (FG) is detrimental to this action, since the mobile carriers in the FG effectively shield the V_{cg}. Therefore, this problem has been overcome by increasing the surface area of the FG, making the FG height very high (see Fig. 4.5), which improves the capacitive coupling. However, with this high FG structure, cross-coupling effect between neighboring cells, where the charging and discharging of a certain cell influences the charge state of neighboring cells by the capacitive coupling between them, also increases, meaning that the circuit could malfunction. The space between FGs is filled with dielectric materials along the B/L direction while that along the W/L is taken by a gate, so coupling in the same string is serious when the surface area of the FG is large. In CTF, the CT layer is a dielectric layer (Si_3N_4) so such field shielding effects are negligible, and thus a lower V_{cg} can be used. More notably, a high FG is not necessary, and the fabrication process becomes much simpler. Actually, the structure of the FG in conventional NAND almost looks like a DRAM capacitor when it is seen in B/L direction. Therefore, this is a big advantage. Another merit also contributes to fluent fabrication; since Si_3N_4 is an insulator and carriers are trapped in the discrete trap levels within it, the layer in each cell does not necessarily need to be separated from each other. Therefore, one critical photo-etching step, which is necessary for FG formation in conventional NAND flash, can be omitted. Currently, the total thickness of the tunneling oxide/CT layer/blocking oxide stack is $< \sim 20$ nm which is almost negligible compared to the height of a conventional FG.

The operation principles and cell layout of CTF are almost identical to that of the FG-NAND memory with generally lower operation voltages ($V_{cg} < \sim 15$ V). Therefore, additional explanation for the circuitry of CTF is not necessary.

4.2 ALD Process and Materials for Mass-Production Memories

4.2.1 Structure and ALD Layers for DRAM

As mentioned in the introduction chapter, DRAM is a typical active matrix-type memory in which the switches and storage cells are cell metal-oxide-semiconductor field-effect transistors (MOSFETs; mostly n-type) and capacitors, respectively. To write the data within a short enough time (<50 ns) and keep the stored data stable, the cell MOSFET must show a high on/off current (I_{on}/I_{off}) ratio (as high as 10^8), which normally requires a high threshold voltage (V_{th} $\sim$0.7–0.8 V) and a high operating voltage (V_{cc}). This high V_{th} is generally problematic for writing data using the lower drive voltage of modern DRAMs ($V_{cc} < \sim 2$ V) because the overvoltage that can be written to the capacitor, which is approximately proportional to V_{cc}–V_{th}, decreases rapidly with decreasing V_{cc}. Therefore, innovations in select transistors and cell capacitors are necessary to maintain DRAM function.

As the requirement for faster on–off action in a DRAM select transistor is less severe than logic transistors, the deleterious short channel effects by the drain voltage have been overcome by the adoption of a 3D transistor structure such as a recessed channel array transistor (RCAT) [4, 5]. In the RCAT structure, the channel length is increased by etching of the Si channel (see Fig. 4.7) to form a

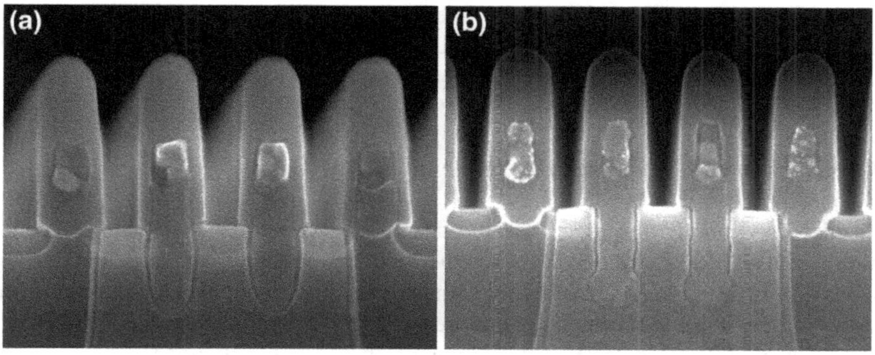

Fig. 4.7 Vertical SEM images for **a** RCAT DRAM cell and **b** S-RCAT DRAM cell [5]

trench shape. The physical gate length was further increased by adoption of the so-called S-RCAT structure (Fig. 4.7b) for the given projection area of the cell MOSFET.

The physically longer channel length decreases the drain-induced barrier lowering and local electric field enhancement effect by lowering the channel doping concentration. The latter is also important for achieving a lower junction leakage current. It should be noted that this strategy cannot be applied to logic transistors, because it will certainly reduce the on current and the operating speed. Due to this physically longer dimension, it is not expected that high-k gate dielectrics will be used in the select DRAM transistor at this moment. However, the nonuniform growth of gate SiO_2 on the various crystallographic planes of an etched Si channel surface (see Fig. 4.7) is becoming more problematic as gate insulator scaling-down. Thin, extremely high quality SiO_2 layer grown by ALD might be necessary in future DRAMs for use as gate insulators of the select transistor for conformal formation of the gate dielectric layer. This will also be the case for the vertical type 3D flash memory (Sect. 4.2.2.3). If the use of metal gate technology is necessary in cell MOSFET, the highly developed TiN ALD process can be well adopted with some tuning of work function control since the cell MOSFETs are n-type only. Metal-organic (MO) ALD would be beneficial compared to $TiCl_4$-based ALD on account of its higher flexibility of composition control, which will allow more room for work function control.

On the other hand, the transistors in the periphery area require a higher operating speed, so that high-k gate dielectric/metal gate and related technologies are necessary as in the case of logic devices. As the transistors in the periphery area are composed of n- and p-type transistors, similar complications to those of high-k logic transistors exist. The required t_{ox} is slightly less stringent than that of logic devices. There are many reports of high-k gate dielectrics and related integration issues. Recent reviews by Kuesters et al. [6] and Niinistö et al. [7] summarize these issues quite well. Several chapters in this monograph are also devoted to this issue.

In the meantime, dense W/L (gate line) and bit-line patterning are other key technologies for DRAM fabrication. Double patterning technology (DPT) is now indispensable in this area, which will be discussed in detail in Sect. 4.2.2.2.

4.2.1.1 ALD Layers for the Capacitor in DRAM Cells

As discussed above, the electrical and structural environments for capacitors in highly scaled DRAM are becoming harsher, which does not allow sufficient design space for maintaining performance with smaller cell capacitance. The sluggish improvement in the performance of the sense amplifier requires an almost constant cell capacitance of ~25 fF irrespective of the technology node. The smaller footprint of the shrunken cell capacitor has a negligible effect on the cell capacitance, because >95 % of the surface area of the cell capacitor is provided by the sidewall area of the bottom electrode. Figure 4.8 shows the change in the shape of DRAM capacitors according to the design rule shrinkage. Along with material technology innovations, the major contribution to keeping the capacitance has been achieved from the structural innovation of the storage node. It can be understood that complicated storage node structures such as cylinder or hemispherical grains can be fabricated when the design rule is >50 nm. However, as the feature size decreases extremely (~30 nm), these surface area-enhancing technologies can no longer be allowed. For example, if a cylindrical structure is still attempted at the design rule of ca. 30 nm, the wall thicknesses of the bottom electrode and the space between them are only 10 nm. To make the capacitor structure, three layers (two dielectric layers and one top electrode layer) must be placed within the narrow gap, meaning that each layer must be as thin as 3.3 nm. No dielectric layer at this thickness can withstand the capacitor voltage. Therefore, the structure must return to its simple stud structure but have extreme height, which makes the aspect ratio of the features extremely high (~100).

The larger influence of the smaller design rule comes from the weaker mechanical strength of the tall storage node (bottom electrode), which has a smaller footprint. This generally results in the leaning or even the breakage of the storage node during capacitor fabrication (see Fig. 4.9 [5]).

Therefore, the storage node height should be decreased as the design rule shrinks further. The shift in the bottom electrode material from heavily doped polycrystalline Si (poly-Si) to TiN greatly improved the mechanical strength of the capacitor structure and decreased the detrimental interfacial low-k dielectric (mostly SiO_2) effect. To date, in a traditional Si-based capacitor, target cell capacitance has been achieved by increasing the surface area of the capacitor (semiconductor-insulator-semiconductor [SIS]; Fig. 4.8) while the dielectric thickness is scaled down according to the design rules [8]. Innovations have recently been made in the component materials, which now use a higher dielectric constant (k) material in conjunction with the metal electrode in a capacitor to obey Moore's law. A metal electrode, TiN or Ru, and a dielectric material with a higher

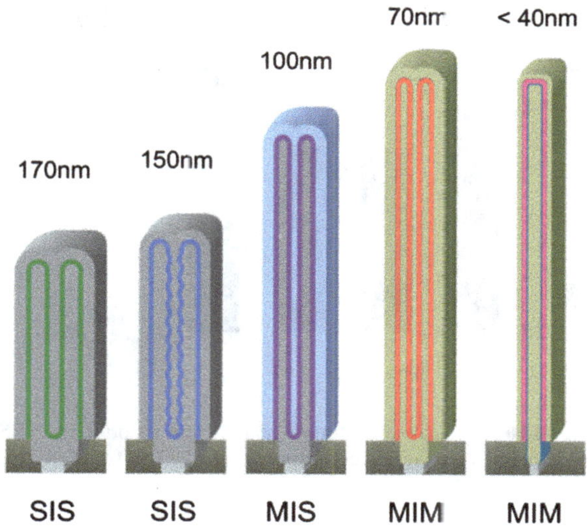

Fig. 4.8 Evolution of DRAM capacitor structures [8]

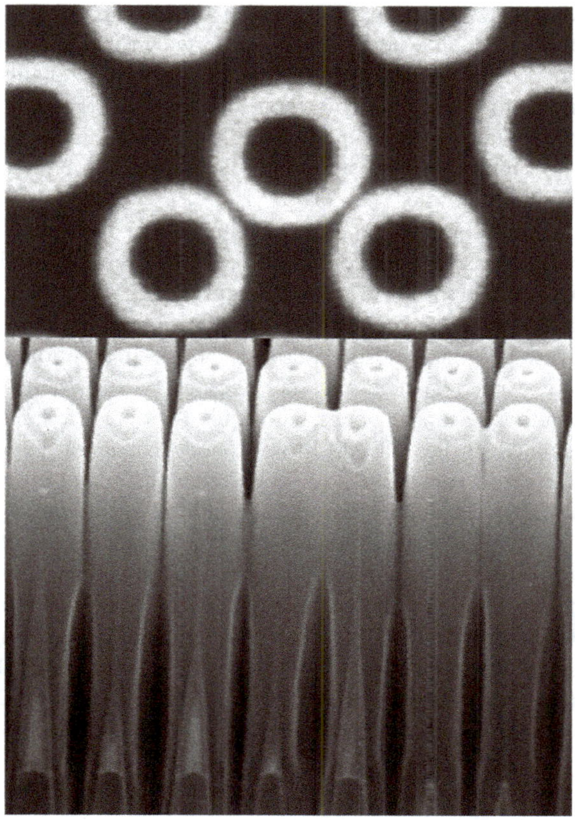

Fig. 4.9 Top view and vertical SEM images of leaning storage nodes [5]

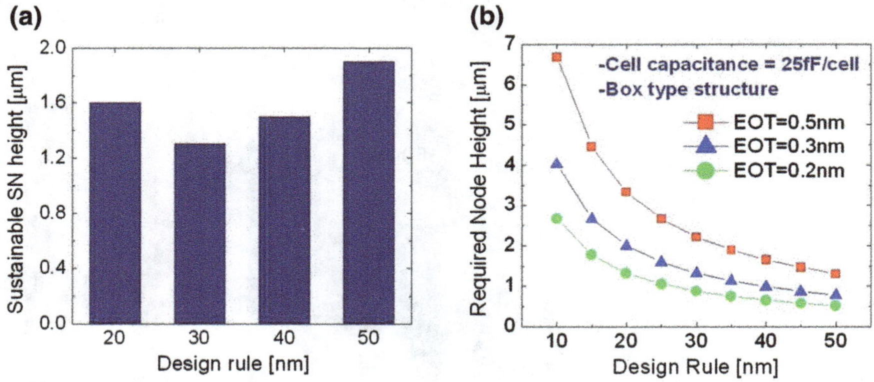

Fig. 4.10 **a** Sustainable storage node height reported in ITRS 2009 [9] and **b** required storage node height to achieve a cell capacitance of 25 fF for the various EOT as a function of the design rule

k value are being explored in gigabit-scale DRAMs (metal-insulator-semiconductor [MIS] and metal-insulator-metal [MIM]; Fig. 4.8). Figure 4.10a, b shows the sustainable height that is reported in the International Technology Roadmap for Semiconductors (ITRS) [9] and necessary heights, respectively, of the storage node to achieve a cell capacitance of 25 fF for the various EOT (sometimes abbreviated as t_{ox}) values of the capacitor dielectric as a function of the design rule <50 nm. Here, the structure was assumed to be a simple box type rather than a cylindrical type.

From Figs. 4.10a, b, it can be understood that EOT values of 0.4 nm and 0.3 nm are essential for the DRAM with a design rule of sub-30 nm and -20 nm, respectively, which are quite aggressive values viewed from the presently available technology. Up until now, there have been no mass-production compatible ways to obtain an EOT value of 0.4 nm. In addition, the equivalent electric field of the capacitor dielectric increases to >18 MV/cm for the <30 nm design rule, another serious concern. The equivalent electric field in the capacitor dielectric is defined as the DRAM storage node capacitor voltage divided by EOT. If the dielectric comprises high-k material, the actual electric field is [equivalent field/(k/3.9)]. Therefore, the leakage current specification becomes even tighter.

Another critical factor is that the low thickness of the dielectric as well as the top electrode required to fit them into the narrow gap between neighboring storage nodes. For the design rule of 20 nm (pitch 40 nm), only 6–8 nm-thick films (dielectric and top electrode) are allowed. Considering the height of the storage node, this corresponds to an aspect ratio of 80 and 200 for the capacitor dielectric and top electrode, respectively. Therefore, it is evident that all of the very thin dielectric and electrode layers should be conformally deposited by ALD.

Dielectrics in DRAM Capacitors

Binary and Doped Oxides

Al_2O_3

Al_2O_3 thin films have been extensively studied as a gate dielectric material in high performance logic chips as well as a capacitor dielectric in DRAM devices. Although Al_2O_3 alone has a relatively small dielectric constant (<10), the Al_2O_3 thin films have an excellent leakage property due to both a wide band-gap (8.8 eV) comparable to that of SiO_2 and its amorphous nature. The good insulating properties of Al_2O_3 thin films have attracted interest as a capacitor dielectric material in DRAM devices. See Chap. 5 for more information about its application in logic chips.

Al_2O_3 is one of the most studied oxide materials grown by ALD. Although there are several available Al precursors for the growth of Al_2O_3 thin films by ALD including $AlCl_3$ [10], $Al(mmp)_3$ [11], $Al(OEt)_3$ [12], and $Al(O^iPr)_3$ [13], $Al(CH_3)_3$ (TMA, trimethylaluminum) is the most commonly used as an Al precursor. The ALD reaction with TMA and H_2O is well-known as one of the ideal ALD reactions [14]. The ALD process with TMA shows a wide ALD temperature window (R.T.–400 °C) and a moderate growth rate (0.08–0.1 nm/cycle) [15]. Due to its excellent thermal stability and self-limiting property, TMA can provide high uniformity on 300-mm wafers (or even larger glass substrates) and allows very conformal ALD reaction over 3D capacitor structures with high aspect ratios.

Studies on the capacitors with Al_2O_3 thin films as the capacitor dielectric have been based on the use of electrodes of heavily doped poly-Si and TiN. On heavily doped poly-Si substrates, Kim and co-workers showed that an EOT value as low as 3.5 nm for the SIS capacitors using an Al_2O_3 layer was achieved with a very low leakage current (<0.1 fA/cell) at an applied voltage of 1.5 V when the Al_2O_3 thin films were grown on heavily doped poly-Si electrodes with TMA and H_2O [16]. Seidl et al. [17] fabricated capacitors using Al_2O_3 grown by ALD on hemispherical grain silicon exhibiting significant capacitance enhancement (>50 %) at a low leakage current compared to the capacitor with conventional SiO_xN_y. The use of TiN as the electrode rather than heavily doped poly-Si contributed significantly to improving the DRAM capacitors due to the negligible depletion thickness caused by its high conductivity and lack of interfacial layer with low dielectric constant. Kim et al. reported that a gain of 0.9 nm EOT value was obtained by employing a TiN top electrode instead of the heavily doped poly-Si [16]. Additionally, TiN/Al_2O_3/TiN capacitors achieved a minimum EOT of 2.2 nm and showed excellent thermal stability [18]. The TiN/Al_2O_3/TiN capacitors have been primarily investigated for deep trench DRAM technology due to its superior thermal stability of the structure to other high-k materials. In the deep trench technology, the capacitors are fabricated prior to dopant activation of the cell MOSFETs; as such, the capacitors have to withstand a high thermal budget (~1,000 °C, 30 s) [18].

Capacitors using Al_2O_3 thin films as the dielectric material were primarily developed for the 100–150 nm design rule. The EOT values achieved from the

Al_2O_3 thin films (>2 nm) are too large for the <100 nm design rule. Therefore, the Al_2O_3 single layer has not been further studied as the capacitor dielectric for gigabit DRAM devices due to its low dielectric constant. However, its good insulating properties continue to make Al_2O_3 useful as a blocking layer or doping material in the MIM capacitor for the sub-100 nm design rule.

Ta_2O_5

Ta_2O_5 has two polymorphic phases of orthorhombic and hexagonal structures. Ta_2O_5 is known to have a phase-dependent dielectric constant. The amorphous and orthorhombic structures possess a relatively low dielectric constant of 25, whereas the hexagonal Ta_2O_5 structures have an anisotropic dielectric constant of $\sim$65 along the c-axis. Thus, the most important factor for the use of Ta_2O_5 as the capacitor dielectric is stabilization of the c-axis preferred hexagonal Ta_2O_5 phase. However, Ta_2O_5 thin films deposited by ALD are known to have an amorphous or an orthorhombic structure on conventional electrodes such as Si and TiN [19, 20]. Ta_2O_5 thin films on (001)-oriented Ru electrodes were recently found to have a hexagonal structure with c-axis preferred orientation due to structural compatibility between the (002) Ru and (001) hexagonal Ta_2O_5 planes [21].

The formation of hexagonal Ta_2O_5 film structures formed by ALD could reduce an EOT value to <1 nm while maintaining a low leakage current. However, the Ta_2O_5 films grown by ALD have an amorphous phase, even on Ru electrodes, due to the low temperature of the ALD process. Therefore, a post-annealing process >650 °C was necessary to crystallize the films. Although this high temperature annealing resulted in a high dielectric constant >60, post-annealing easily deformed the Ru electrode by thermal stress [22]. In addition, high-temperature annealing can induce oxidation of the TiN diffusion barrier, creating a too-high contact resistance between the capacitor and the cell MOSFET. Therefore, decreasing the post-annealing temperature of Ta_2O_5/Ru is indispensable. It is known that Nb_2O_5 has an identical crystalline structure to that of hexagonal Ta_2O_5 and has a lower crystallization temperature. To reduce the post-annealing temperature, the use of very thin Nb_2O_5 films as a seed layer or the growth of $Ta_{2-x}Nb_xO_5$ films was attempted. Ma et al. reported growth of Ta_2O_5/Nb_2O_5 bilayer films on Ru electrodes by ALD using $Ta(OEt)_5$ and $Nb(OEt)_5$ precursors [23]. Using 6 nm-thick Nb_2O_5 film as the seed layer, the annealing temperature for the crystallization of the films could be decreased to 575 °C. A high dielectric film constant ($\sim$66) was achieved and the EOT value decreased to 0.75 nm with quite low leakage currents due to formation of the hexagonal Ta_2O_5 films.

However, the annealing temperature for crystallization of the Ta_2O_5 films is still too high for the fabrication of the capacitor with a design rule of deep sub-100 nm. Since use of Ta_2O_5 with the Ru electrodes does not have a distinct advantage over ZrO_2, which is able to use a conventional TiN electrode, the material currently attracts little interest as a capacitor dielectric.

HfO_2 and ZrO_2

HfO_2 and ZrO_2 are known polymorphism materials that have several crystalline structures—monoclinic, tetragonal, and cubic structures—each of which has a different dielectric constant. Their energy band-gap in thin film form is 5.5–5.8 eV, which is much lower than that of SiO_2 or Al_2O_3 but still high enough to induce electron and hole barriers using most electrode materials. The dielectric constant of their amorphous and monoclinic phases processes a relatively low dielectric constant (<20), whereas high-temperature stable phases such as the cubic or tetragonal phase have higher dielectric constants of $\geq$30 [24]. Therefore, it is important to stabilize the high dielectric constant cubic or tetragonal phase within a reasonable thermal budget (<500 °C). The desired phase could be obtained from either the as-deposited film or the post-annealed sample. In some cases, doping is also an effective method to stabilize the high temperature phases at typical ALD temperatures. As-grown thin ZrO_2 films deposited by ALD adopt the cubic/tetragonal phases even without doping, while the as-grown HfO_2 films generally have amorphous or monoclinic phase.

Kim et al. recently reported that ZrO_2 thin films grown on TiN electrodes by ALD only at 275 °C has a high dielectric constant of approximately 40 due to the formation of cubic or tetragonal ZrO_2 [25]. Minimization of the surface and grain boundary energies of the polycrystalline ZrO_2 thin films resulted in the formation of cubic or tetragonal ZrO_2 [25]. Contrary to ZrO_2, as-grown HfO_2 thin films have an amorphous or monoclinic phase that is not changed even after the post-annealing process at temperatures as high as 1,000 °C. Incorporation of small amounts of other oxides induced the transformation of HfO_2 films into the cubic/tetragonal phase, resulting in a high dielectric constant >40 [26]. Incorporation of SiO_2, Al_2O_3, or Y_2O_3 into HfO_2 films is known to be able to transform the HfO_2 phase to a cubic or tetragonal phase [27]. The reason for such a phase transition by doping has been understood from the first principles calculation [28]. However, the stabilization of cubic or tetragonal HfO_2 films by doping still required a post-annealing process >700 °C, which can induce thermal deformation of the capacitor in DRAM devices. The easy formation of the cubic/tetragonal ZrO_2 phases makes ZrO_2 a more attractive capacitor dielectric than HfO_2.

ALD growth of HfO_2 and ZrO_2 thin films has been extensively studied since both materials have attracted great interest as most promising candidates for gate dielectric material in high-performance MOSFET as well as capacitor dielectrics in DRAM. There are several available Hf and Zr precursors. Hf and Zr chlorides are known common precursors for Hf and Zr. The chloride precursors have a very wide temperature window for ALD up to 600 °C due to their very high thermal decomposition temperature. Carbon contamination is not an issue with these precursors because they do not contain carbon. Thanks to these beneficial features of the chloride precursors, these precursors have been widely used for the growth of HfO_2 and ZrO_2 thin films. Despite the merits of the chloride precursors, the ALD processes using these precursors have struggled at mass-production scale due to their drawbacks of relatively low growth rate, formation of corrosive by-product (HCl), and possible chlorine film contamination. In particular, the conformality of the films grown from the chloride precursors and H_2O is reported to be inadequate in the complex structure [29, 30], which might be related with the insufficient precursor supply. However, these precursors are generally in powder form at typical vaporization temperature, so increasing the precursor supply is not a trivial task.

As a result, it appears that use of the chloride precursors in capacitor applications is not appropriate, even though they can be used in high-k gate dielectrics of high-performance MOSFETs.

Alkoxides have also been used extensively for HfO_2 and ZrO_2 film ALD. The limited thermal stability of the alkoxide precursors, however, results in generally inadequate film qualities. For example, the $Zr(O^iBu)_4$ and $Hf(O^iBu)_4$ precursors fail to show self-limiting reactions, one of the unique properties of ALD, at temperatures >200 °C due to their facile thermal decomposition, which is usually accompanied with a significant level of carbon contamination (5–8 atomic % [at%]) in the films [31–33]. β-diketonate precursors such as $Zr(thd)_4$ have been used to grow ZrO_2 films [34]; however, the growth rate is quite low due to the bulky ligand structure of the precursor [34]. The low reactivity of the precursor required the use of ozone as the oxygen source and left a large carbon residue in the films.

Alkylamides are the most commonly used precursors for HfO_2 and ZrO_2 film ALD. Such films from alkylamides exhibited a moderate growth rate of ~0.1 nm/cycle and showed a facile reactivity with various oxygen sources such as H_2O, ozone, and O_2 plasma [25, 35–37]. The films grown from alkylamides were dense and had rather low impurity contamination; as such, they exhibited desirable dielectric performances. Due to the superior properties of the films grown from the alkylamides, alkylamides ($Hf(NEtMe)_4$ [TEMAH] and $Zr(NEtMe)_4$ [TEMAZ] in particular) have received great attention as promising precursors for HfO_2 and ZrO_2 ALD in memory applications.

ALD growth of HfO_2 and ZrO_2 thin films using a different precursor is still under investigation. ALD of HfO_2 thin films using a $HfO^tBu(NEtMe)_3$ (BTEMAH, *tert*-butoxytris(ethylmethylamido)hafnium) precursor and O_3 was recently reported [38]. The growth rate of the HfO_2 films from BTEMAH and O_3 was 0.16 nm/cycle at 300 °C, the highest value for HfO_2 ALD reported to date. In addition, the HfO_2 films grown from BTEMAH exhibited higher film density than that of the HfO_2 films grown from tetrakis(dimethylamido)hafnium. These properties might result from the replacement of one of the ethlymethylamido ligands with a highly reactive *tert*-butoxy ligand in the heteroleptic precursor structure. The use of cyclopentadienyl-based precursors for ZrO_2 ALD was recently introduced. Cyclopentadienyl derivatives such as Cp_2ZrMe_2 [39, 40], $(MeCp)_2ZrMe_2$ [41, 42], and $(MeCp)_2Zr(OMe)Me$ [41, 42] precursors offer the promising feature of high thermal stability up to ~375 °C and low impurity contamination. The growth rate is 0.05–0.06 nm/cycle, which is slightly lower than the value of the films from TEMAZ. Readers who are not familiar with the chemistry of such precursors are encouraged to read Chap. 3, *ALD Precursors and Reaction Mechanisms,* of this monograph.

Studies on capacitors using HfO_2 or ZrO_2 thin films have been carried out based on conventional TiN electrodes. Aoki et al. achieved a low EOT of 1.2 nm from $TiN/HfO_2/TiN$ capacitors [43]. Oh et al. demonstrated that the use of O_2 plasma instead of O_3 improved the leakage current properties of the HfO_2 films and reported an EOT of 1.3 nm from the HfO_2 films on TiN electrodes [44]. However,

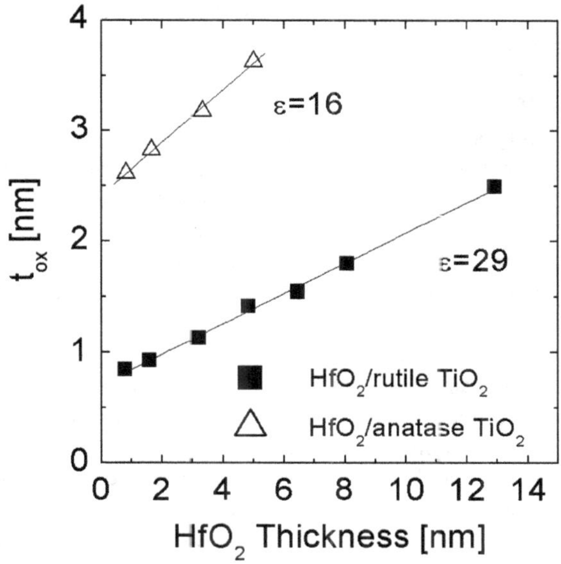

Fig. 4.11 Changes in the t_{ox} of HfO_2/rutile TiO_2 and HfO_2/anatase TiO_2 stacks as a function of the HfO_2 thicknesses [45]

ALD HfO_2 films generally have a monoclinic structure and a relatively low dielectric constant ≤ 20. Because of this problem, it is generally difficult to decrease EOT values to <1.0 nm using HfO_2 films and most studies on HfO_2/TiN capacitors have been performed for DRAMs using the 70–90 nm design rule.

Quite recently, the formation of cubic or tetragonal HfO_2 films grown using ALD without the post-annealing process has been reported. Seo et al. attempted to grow HfO_2 films on anatase and rutile TiO_2 surfaces, respectively [45]. As shown in Fig. 4.11, HfO_2 films on anatase TiO_2 showed a relatively low dielectric constant of 16, whereas the HfO_2 films on rutile TiO_2 exhibited a much higher dielectric constant of 29 due to the cubic or tetragonal HfO_2 structure. The formation of the cubic or tetragonal HfO_2 on rutile TiO_2 resulted in crystallographic compatibility between certain planes of both structures [45]. Since a dielectric constant of rutile TiO_2 is quite high as well, the HfO_2/rutile TiO_2 bi-layer structures are expected to be able to significantly contribute to sub-50 nm DRAM applications [46].

Contrary to HfO_2, ZrO_2 thin films grown by ALD (as-deposited) are usually crystallized into the cubic or tetragonal phase with a high dielectric constant of approximately 40 even on conventional TiN [47]. Hence, ZrO_2 films have recently received attention as a candidate for MIM capacitors of DRAM with small feature size (<60 nm). However, the single ZrO_2 films still have high leakage currents, perhaps due to the presence of grain boundaries. The $ZrO_2/Al_2O_3/ZrO_2$ stack was, therefore, suggested as a dielectric material for DRAMs using the sub-60 nm design rule. The stacked structure/TiN allowed the production of DRAM devices down to the ~45 nm design rule [48]. However, the minimum EOT achieved by the $ZrO_2/Al_2O_3/ZrO_2$ stack is approximately 0.6 nm, which is not low enough for sub-40 nm DRAM technologies. This is closely related with the presence of a low dielectric intervening Al_2O_3 layer, but its presence is inevitable in this MIM structure.

TiO_2

TiO_2 is an eccentric dielectric material among the binary metal oxide in terms of dielectric constant; that of rutile phase TiO_2 is 90 and 170 along the a- and c-axis, respectively [49], which is much higher than the value of the anatase phase of TiO_2 (~40). Hence, rutile structured TiO_2 films have received considerable attention as a strong candidate material for sub-40 nm capacitor applications due to its high dielectric constant. Actually, TiO_2 is one of the most studied materials grown by ALD.

There have been large numbers of studies on the effects of deposition parameters such as precursors, oxidants, and growth temperatures on physical, chemical, and electrical properties of the grown TiO_2 films. Halide and alkoxide precursors have been commonly used as Ti sources for the growth of TiO_2 films. ALD growth using $TiCl_4$ and H_2O had a wide temperature window up to 600 °C and exhibited a moderate growth rate of 0.04–0.07 nm/cycle [50–55]. However, like $HfCl_4$, $TiCl_4$ has a corrosive by-product problem. Therefore, the precursor is not apt for mass production. There are several Ti alkoxide precursors [56–61], such as $Ti(O^iPr)_4$ (TTIP, titanium isopropoxide), $Ti(OEt)_4$ (titanium ethoxide), and $Ti(O_tBu)_4$ (titanium tertiary-buthoxide). Among them, TTIP was extensively used because the precursor showed a facile self-limiting behavior within the ALD temperature window and exhibited a low impurity concentration, although the ALD window of the TiO_2 films from TTIP is limited to 250 °C due to low thermal stability [58]. The growth rate of TiO_2 films varied in the range of 0.015–0.06 nm/cycle when H_2O was used as the oxidant [61]. It has been reported that the growth rate suddenly increased by more than double after crystallization, accompanied by surface roughening. The sudden increase in growth rate is thought to be due to the increase in the number of hydroxyl groups adsorbed on the reaction surface after the crystallization [56]. Indeed, ALD of TiO_2 films from TTIP and O_3, which is based on different reaction mechanisms, did not show a sudden growth rate increase, even after crystallization [62]. $Ti(OEt)_4$ has a narrow ALD window (up to 200 °C) due to its lower thermal stability and can induce high impurities such as carbon and hydrogen [59, 60].

Alkylamides and β-diketonates were also used to grow TiO_2 films. ALD growth of TiO_2 films from tetrakis-dimethyl-amido titanium, one of the alkylamides, showed excellent step coverage [63]. However, the growth rate was generally low (0.028 nm/cycle). The ALD window of the precursor is limited to 250 °C. On the other hand, use of $Ti(O^iPr)_2(tmhd)_2$ (tmhd = 2,2,6,6-tetramethyl-3,5-heptanedione) as a precursor allowed a high ALD temperature up to ~400 °C due to its better thermal stability [64]. The films grown from the precursor showed much smoother surface morphology than the films grown from TTIP. Despite these merits of the precursor, the growth rate is quite low (~0.02 nm/cycle) due to the bulky structure of the ligands in the precursor.

TiO_2 films grown via ALD can be formed with amorphous or crystalline structures such as anatase and rutile depending on the process conditions including growth temperature, precursor, and substrate. Rutile-structured TiO_2 is desirable for DRAM applications as mentioned above. However, most ALD-grown TiO_2 films are formed with the amorphous or anatase phase because formation of the high-temperature phase, rutile, requires high-temperature processes (>500 °C or post-annealed at >800 °C [65–67]), whereas most Ti precursors decompose thermally at high temperatures (>400 °C). Although halide precursors such as $TiCl_4$ or TiI_4 achieved rutile TiO_2 formation at temperatures >500 °C [55], those precursors are less compatible with the mass-production worthy ALD tools.

It was recently reported that rutile TiO_2 film is achieved even at temperatures as low as 250 °C by the interposing of an interfacial conducting layer with a reasonable lattice match with rutile TiO_2. RuO_2, or IrO_2 grown on bottom metal electrodes (Ru or Ir) either by in situ or ex situ methods, which has also rutile structure and small lattice mismatch with rutile TiO_2 allowing for the formation of rutile TiO_2 films on the top at low temperatures [62, 68]. TiO_2 films grown on Si or Pt substrates by ALD using TTIP and O_3 were crystallized into the anatase phase, resulting in a relatively low dielectric constant of ~36 as shown in Fig. 4.12a. However, rutile TiO_2 films were grown on Ru substrates when an oxygen source with a strong oxidation potential such as O_3, O_2-, or N_2O plasma was used [62, 68, 69]. The oxygen sources oxidized the Ru surface to RuO_2 at the initial growth stage and induced rutile TiO_2 formation [62, 68]. Figure 4.12b shows Ru 3d X-ray photoelectron spectra of the Ru electrode coated with 1 nm-thick TiO_2 films grown using O_3 or H_2O as the oxygen source, respectively. The Ru substrate with the TiO_2 grown using O_3 clearly showed the strong signal of the RuO_2 phase, whereas the Ru substrate with TiO_2 grown using H_2O was less oxidized to induce rutile growth. Most importantly, the dielectric constant of the TiO_2 films grown on the Ru electrode was as high as 80–100 due to the crystalline phase as shown in Fig. 4.12a. An even higher dielectric constant of ~155 was also obtained from the TiO_2 grown on RuO_2 electrodes due to the different crystallographic orientation [70].

Although these are quite promising results for sub-40 nm DRAM technologies, the rutile TiO_2 films have high leakage current properties due to their small band-gap (~3.1 eV) and n-type nature that induce a small Schottky barrier height for electron transport. This problem was overcome by doping the TiO_2 film with Al_2O_3 (called ATO film). An EOT value of 0.48 nm with quite low leakage currents was achieved by this approach as shown in Fig. 4.12c [71]. Doping with Al ions was accomplished by replacing one of the TiO_2 cycles (60–120) with one Al_2O_3 cycle [71, 72]. Despite discrete feeding of the Al_2O_3 cycle, Al ions are uniformly distributed across the TiO_2 films, and some of them piled up at the interface with the electrode [71, 72]. This interfacial Al positions the Fermi level at the middle point of the TiO_2 band-gap by an appropriate charge transfer that results in significant improvement in the leakage current properties [8, 71]. ATO films were still crystallized into the rutile phase even after Al doping, resulting in a higher dielectric constant (50–80) than anatase phase. Therefore, ATO films have received considerable attention for sub-40 nm DRAM technologies.

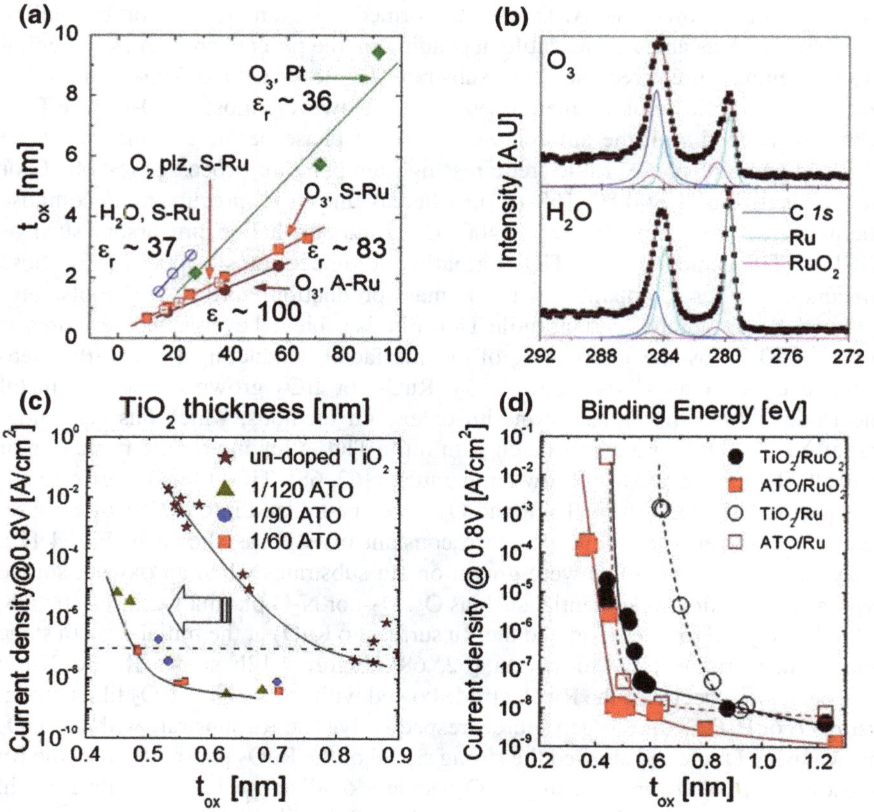

Fig. 4.12 **a** Changes in the t_{ox} of TiO_2 films grown on various substrates as a function of TiO_2 film thickness (A-Ru: ALD Ru, S-Ru: Sputtered Ru) [8]. **b** Ru 3d X-ray photoelectron spectra of Ru substrate coated with 1 nm-thick ALD TiO_2 film which are grown using O_3 and H_2O as the oxygen source, respectively [8]. **c** Leakage current density at 0.8 V versus t_{ox} value curves of un-doped TiO_2 and ATO films [71]. **d** Leakage current density at 0.8 V versus t_{ox} value curves of TiO_2 and ATO films on Ru and RuO_2 bottom electrodes, respectively [73]

Further, improved results were reported very recently [73] in which a pulse-CVD (p-CVD)–grown bulk RuO_2 (~30 nm-thick) was used as the electrode rather than the ~1–2 nm RuO_2/(30 nm) Ru electrodes that were used previously [71]. The higher work function of RuO_2 and smoother surface of the p-CVD RuO_2 electrode further reduced the leakage current for the given ATO and TiO_2 film. The electrical properties of the ultra-thin RuO_2 layer that formed on the thick Ru layer were influenced by the Ru layer, of which the work function is lower than that of RuO_2. These results are well summarized in Fig. 4.12d, where the minimum EOT of 0.46 nm with a leakage current density of ~1×10^{-8} A/cm^2 is reported from the ATO film on a p-CVD RuO_2 electrode. Here, the physical thickness of the ATO film was ~6.9 nm. One even more crucial merit of the ATO material is that its leakage current versus EOT performance does not noticeably vary with Al concentration in the 5–10 at% range, which is very important considering the process variation in mass production and the possible compositional variation along the 3D structure.

Ternary oxides

Even though the dielectric performance of the TiO_2-based films shown above is impressive, the limited bulk dielectric constant of <100 will eventually restrict the achievable EOT to no lower than ~0.4 nm. Therefore, dielectric materials with an even higher dielectric constant must be pursued. The most promising candidate in this regard is perovskite-based material such as $SrTiO_3$ (STO) or $(Ba,Sr)TiO_3$ (BSTO). There have been extensive studies on the perovskite-structured dielectrics (ABO_3 structure, space group of P_{m3m}). The representative material is STO, a cubic perovskite crystal that has a lattice parameter of 0.3905 nm and a bulk dielectric constant of 300 at room temperature due to its ionic polarization (displacement of Ti^{4+} ion from the center of the O_6 octahedron along the applied bias) [74, 75]. Along the (001) direction, STO can be considered an alternating stack of charge neutral TiO_2 and SrO atomic layers as shown in Fig. 4.13. The energy band-gap of STO is ~3.2 eV [76], while the electronic band edges at the top of the valence band and the bottom of the conduction band mostly consist of the O 2p and Ti 3d orbitals, respectively [77–79]. Unlike several other perovskite-structured materials, such as $(Pb,Zr)TiO_3$, STO does not show ferroelectricity [80] at device operating temperatures.

Despite their merits as dielectric materials, perovskite oxides have serious demerits. First of all, the k value decreases severely with decreasing physical thickness in the range of interest. This is due to the degraded dielectric

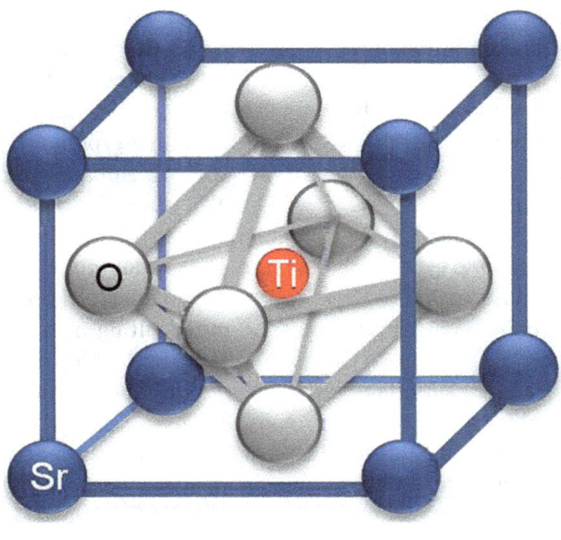

Fig. 4.13 A unit cell of $SrTiO_3$; alternating stack of charge neutral TiO_2 and SrO atomic layers along the (001) direction

performance at the electrode (metal or conducting oxides)–dielectric interfaces. [81–84]. The second is the much more complicated ALD processes compared to binary oxides due to their multi-cation composition. Crystallization of the perovskite oxides is necessary to realize a higher k, because the amorphous phase shows a *k* value of ~20, which might be due to the lack of collective ionic polarization effect. The optimum *k* value of crystalline STO is acquired only in a very narrow stoichiometric range [85–88], thus the crystallization issue becomes more severe when it comes to extreme 3D structure of the DRAM capacitors, because the composition step coverage over a complicated 3D structure is difficult to achieve. A low thermal budget of the deposition and post-deposition annealing (PDA) processes (<500–600 °C) and a low growth rate of perovskite oxide ALD are another key concerns against adopting perovskite oxides.

ALD of multi-cation perovskite oxides has quite interesting aspects compared to ALD of typical binary oxides. In an ideal point of view, the alternating growth of each AO and BO_2, with their fully saturated surface coverage, would result in an invariant A/B ratio in the film, which is totally determined by the chemistry used. However, steric hindrance from volumetric precursor molecules usually results in partial coverage of the surface in a single growth cycle. Thus, the appropriate cycle ratio (*m*/*n*, where *m* and *n* are the cycle numbers of AO and BO_2, respectively, or SrO and TiO_2 in this case) should be selected for the precise composition control of the film. It should be noted that the chemisorption behavior of the metal precursor is strongly dependent on the pre-existing surface state; thus, there is a chance that the growth rate of AO and BO_2 when they are grown alternatively can differ compared to the deposition when AO or BO_2 grow separately as a binary oxide since the metal precursor encounters different surface at the precursor pulse step. This is one reason why the multi-component ALD process is much more complicated than binary oxide ALD processes [89, 90].

In the facile STO ALD process, selection of the proper metal precursor is very important. The maturity of the ALD TiO_2 film using several Ti precursors was discussed in detail in the previous section. However, the development of the SrO layer (and the STO) ALD has been hindered by the complicated ALD chemistry of Sr precursor, which originates from low volatility, insufficient thermal stability, and weak reactivity to surface functional groups [91]. In practice, selection of the Sr precursor has been restricted to β-diketonate and cyclopentadienyl compounds. In the early stages of ALD STO research, the most viable Sr precursor was $Sr(tmhd)_2$ ($Sr(C_{11}H_{19}O_2)_2$), which was reported to have limited ALD reactivity to common oxygen sources like H_2O. Therefore, an oxygen source with higher reactivity (such as O_3) was used to induce a more active ALD reaction with the $Sr(tmhd)_2$ precursor. However, the resulting film was $SrCO_3$ instead of SrO since the $SrCO_3$ phase is more thermally favorable than the SrO phase when CO_2 is generated simultaneously as a by-product, which was not alleviated even by combining TiO_2 layer growth [89].

Ahn et al. [92] used O_2 plasma for the same $Sr(tmhd)_2$, while Langereis et al. also introduced O_2 plasma using bis(trisisopropylcyclopentadienyl)-strontium with the 1,2-dimethoxyethane adduct [$Sr(^iPr_3Cp)_2DME$] [90]. They obtained reasonable

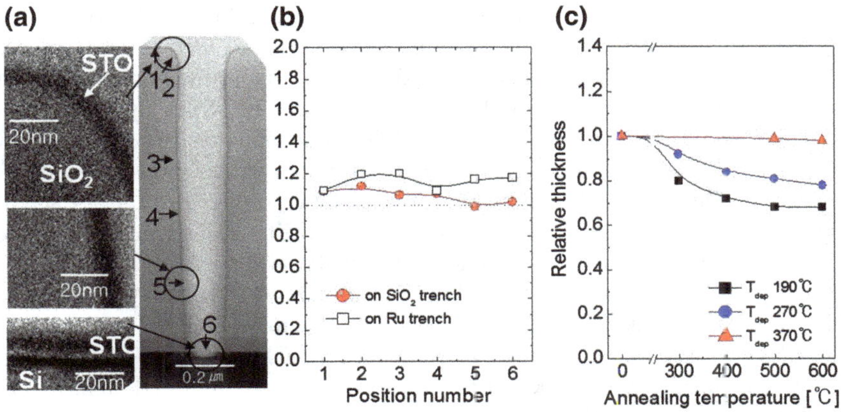

Fig. 4.14 **a** Cross-sectional TEM image of ALD STO film on a contact hole [8], **b** Sr/Ti atomic ratio at the various points shown in **a** [8], and **c** the variations in the normalized thickness of the STO films as a function of a PDA temperature [8]

STO films in the ALD mode as well as promising electrical properties of the films after the appropriate PDA was used to crystallize them. However, there is a significant concern with plasma-based ALD processes in achieving excellent step coverage over a severe 3D geometry with a high aspect ratio.

Kwon et al. reported that a SrO layer can be grown on a TiO_2 surface by ALD even with $Sr(tmhd)_2$ and H_2O with the help of the catalytic effect of the TiO_2 surface when the Sr precursor was vaporized at temperatures <200 °C, and stoichiometric STO films were grown by appropriate control of the SrO/TiO_2 cycle ratio [93, 94]. A highly conformal thickness and cation-composition step coverage over an extreme 3D geometry was also confirmed (see Fig. 4.14a, b) [8, 94].

The as-grown films were amorphous, thus the PDA at temperatures >600 °C was necessary to achieve the crystallization of ALD STO films. However, the as-grown films have lower density than the theoretical values, which caused a serious problem during crystallization annealing. The thickness of the films decreased significantly ($\sim$70–80 %) than the original thickness after PDA (Fig. 4.14c) [8]. This degraded the leakage current density severely by making huge grains and micro-cracks, so that the crystallization of amorphous STO film by means of PDA is incompatible with the DRAM application. These discouraging properties were attributed mainly to the low ALD temperature (<280 °C), which resulted in an amorphous film due to the limited thermal stability of the Ti-precursor used, TTIP.

It was recently reported that the in situ crystallization of STO at the as-deposited state could be achieved with promising electrical properties by increasing the deposition temperature as much as possible within the ALD regime, and adopting a crystalline seed layer [95]. In an attempt to increase the deposition temperature, $Ti(O^iPr)_2(tmhd)_2$ molecules were used as the Ti precursor, which showed a thermal decomposition behavior at ALD temperatures $\sim$400 °C. The $Ti(O^iPr)_2(tmhd)_2$ and $Sr(tmhd)_2$ precursors allowed for the growth of highly dense TiO_2 and STO films on

$Ru/SiO_2/Si$ substrates at a growth temperature as high as 390 °C in the ALD mode (Fig. 4.14c). In situ crystallization was achieved using a so-called two-step process in which a thin crystalline STO layer (~3–5 nm) was first formed, followed by the growth of main STO layer [8, 93, 94]. The thin crystallized STO layer (called the seed layer) can be grown at 370 °C by ALD under the same growth condition, where the film is still amorphous, followed by crystallization annealing at 650–700 °C. Due to local lattice match between the crystallized seed layer and the main layer, the overall film was crystallized in situ with an average grain size of ≪50 nm (Fig. 4.15a, where no boundary between the seed layer and the main layer is observed). Using a Pt/20 nm-thick STO/Ru (bottom) capacitor, an EOT of 0.72 nm and a low leakage current density ($\sim 10^{-7}$ Acm^{-2} at 0.8 V) were achieved (Fig. 4.15b). However, the acquired growth per cycle (GPC) of 0.015 nm/cycle was too low to be used in mass production. This low GPC was attributed mainly to the very low chemical reactivity between the $Sr(tmhd)_2$ and H_2O. Therefore, an alternative Sr precursor with better chemical reactivity and thermal stability is needed for the development of a mass-production compatible STO ALD process.

There was an earlier report of the ALD of STO and $BaTiO_3$ films using modified cyclopentadienyl (Cp)-based group II precursors and H_2O as the oxygen source [91, 97]. These precursors react with H_2O and generally have a higher vapor pressure than $Sr(tmhd)_2$, which facilitates fluent delivery of the precursor to the ALD chamber. The chemical bond energy between the Sr ion and the Cp ligand is much weaker than that between the Sr ion and the tmhd ligand which would be better for high GPC [98]. However, their limited thermal stability during vaporization and thermal decomposition behavior at the typical ALD temperature window (<350 °C) has hindered the further improvement of this process [91]. However, this problem was largely mitigated by better control of the precursor molecule synthesis process.

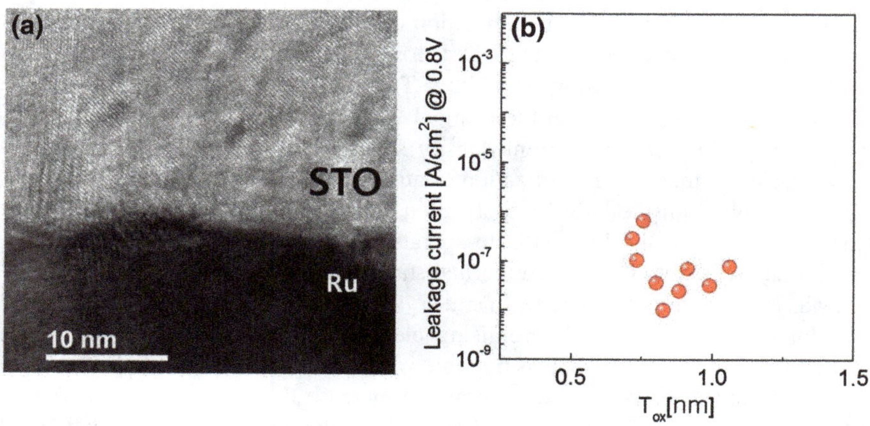

Fig. 4.15 **a** The cross-sectional TEM image and **b** Electrical properties of the in situ crystallized STO film grown on Ru substrate [6]

Several results have recently been published using a Cp-based Sr precursor [90, 99–104]. Lee [100] et al. used $Sr(^iPr_3Cp)_2$ and $Ti(O^iPr)_2(tmhd)_2$ as the Sr and Ti precursors, respectively, to achieve a high GPC of STO ALD. Using this ALD process, highly enhanced growth rates were achieved (0.107 nm/cycle when the cycle ratio (Ti:Sr) was 3:1) that were ~7 times higher than the previous GPC from $Sr(tmhd)_2$. The in situ crystallized STO film showed an EOT of 0.57 nm ($k \sim 146$) with a stably low leakage current (3×10^{-8} A/cm^2 at 0.8 V) using Ru/TiO_2 (3.5 nm)/STO seed layer/STO main layer/Pt capacitor (Fig. 4.16a, c, d). The in situ crystallized STO shows a higher density than the amorphous film (X-ray reflection data in Fig. 4.16b). In addition to electrical performance, excellent step coverage of 95 % was achieved inside a deep capacitor hole with an aspect ratio of 10. Figure 4.16e shows the highly conformal thickness as well as the composition step coverage (>95 %) of the STO film inside a capacitor hole structure deposited at 370 °C.

It should also be noted that a thin TiO_2 layer was deposited prior to deposition of the STO seed layer on the Ru substrate (capacitor stack—Ru/TiO_2/STO seed layer/STO main layer/Pt) since the Sr incorporation in the film was exceptionally high, suggesting that the Sr precursors interact strongly with the Ru substrate during the initial 1–2 super-cycles when $Sr(^iPr_3Cp)_2$ precursor is used. This exaggerated incorporation of Sr on the Ru substrate can be effectively blocked by the addition of a thin (3–5 nm-thick) TiO_2 layer deposited by ALD on the Ru substrate; thus, a thin TiO_2 layer was necessary for normal ALD growth using an

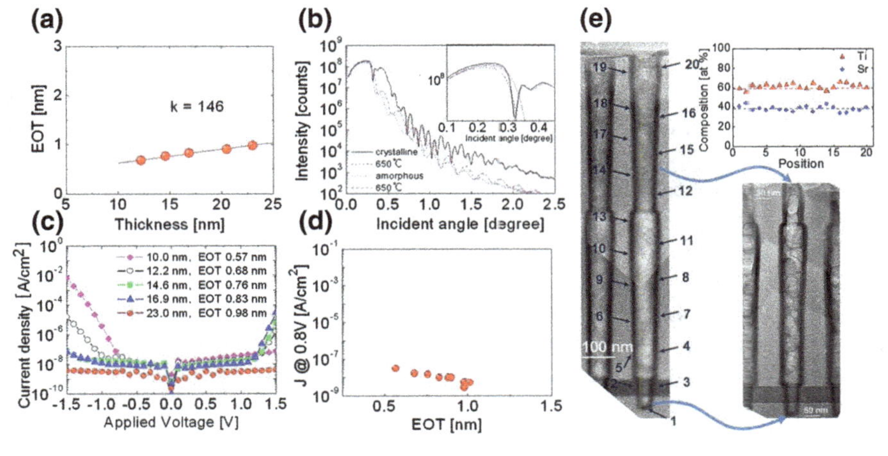

Fig. 4.16 **a** Variation in EOT as a function of physical thickness of STO films ($k \sim 146$) [100]. **b** X-ray reflectivity spectra of the in situ crystallized and amorphous STO films before and after PDA at 650 °C. (*Inset*) The enlargement of the vicinity of critical angles for the clearness [100]. **c** *J–V* plot of the in situ crystallized STO films with different thicknesses [100]. **d** Summary of the dielectric performance of the in situ crystallized STO films [100]. **e** Cross-section TEM image and energy dispersive spectroscopy results that show a highly conformal thickness as well as composition step coverage (>95 %) of the STO film inside a capacitor hole structure deposited at 370 °C [100]

$Sr(iPr_3Cp)_2$ precursor. Thinner (2–3 nm) Al_2O_3 layer also plays a role as the blocking layer over the excessive Sr incorporation, but the lower dielectric constant of Al_2O_3 layer results in undesirable electrical performance of the overall capacitor structure.

Development of a STO ALD process using a TiN electrode was also attempted, because TiN is a much more mature electrode material in the DRAM industry [87, 101–103]. The ALD of a STO film using $Sr(^tBu_3Cp)_2$ ($^iBu_3Cp$ = trisisobuthylcyclopentadienyl) and $Ti(O\text{-}iMe)_4$ as the Sr and Ti precursors, respectively, and H_2O as the oxygen source at an ALD temperature of 250 °C was reported. In those works, the as-grown films were still amorphous due to the low growth temperature; thus, the PDA step was necessary [93, 95]. However, the approach to achieve a lower leakage current was slightly different from the reports mentioned above. The authors showed that the Sr-rich composition, which is generally not favorable compared to the stoichiometric composition due to the decreased dielectric constant, was essential to achieve the optimal electrical properties for TiN/STO/TiN capacitors (EOT = 0.6 nm and $J = \sim 10^{-6}$ A/cm^2 at 1 V) when the maximum thermal budget was limited to 550 °C [101, 102]. Sr-rich films [(Sr/Sr + Ti) ~ 62 at%] showed a significantly improved leakage current at similar EOT values compared to the stoichiometric films [(Sr/Sr + Ti) ~ 50 at%] after the crystallization annealing. In fact, only a slight increase in band-gap was observed with increasing Sr content from the photoconductivity experiment, from 3.0 (stoichiometric STO) to 3.2 eV [(Sr/Sr + Ti) ~ 66 at%]. It was assigned the origin of the different leakage behavior between stoichiometric and Sr-rich films to be correlated with different microstructures developed by crystallization annealing. Sr-rich films showed promising leakage behavior with lower EOT values than stoichiometric films; however, there is a potential uppermost limitation of this process due to its lower dielectric constant (~74). Therefore, Sr-rich film is not suitable for the further aggressive scaling. A more promising capacitor fabrication scheme was recently proposed using optimized Ru/RuO_x/TiO_x/STO/TiN stack to obtain better electrical performances [103]. Sr-rich STO films with compositions in the range of Sr/(Sr + Ti) > 54–64 at% were grown by ALD on a pre-deposited thin interfacial TiO_x layer, followed by crystallization annealing at 600 °C in an N_2 atmosphere. Prior to the TiO_x layer deposition, the RuO_x bottom electrode was formed by low pressure oxidation of pre-existing Ru at 250 °C, which resulted a thin (1 nm) RuO_2 layer on top of the Ru electrode. The significant decrease of the EOT value down to 0.4 nm was demonstrated with a leakage current of 10^{-7} A/cm^2 at 0.8 V for the optimized capacitor stack (Fig. 4.17a, b). These excellent electrical performances are quite promising for the aggressive scaling of the DRAM capacitor. However, this TiN/Ru/RuO_x/TiO_x/STO/TiN capacitor stack still requires a Ru layer that can cancel out the merits of the low-cost TiN process.

The thickness of the dielectric layer should be <6–8 nm for adoption in DRAM using the ~20 nm design rule due to the physical limitation. At such a thin STO thickness, the leakage current might be very serious, so perhaps a doping approach, as for the TiO_2 case with Al-doping, would be necessary. Although there have been several attempts to grow Al-doped STO or BSTO films using several

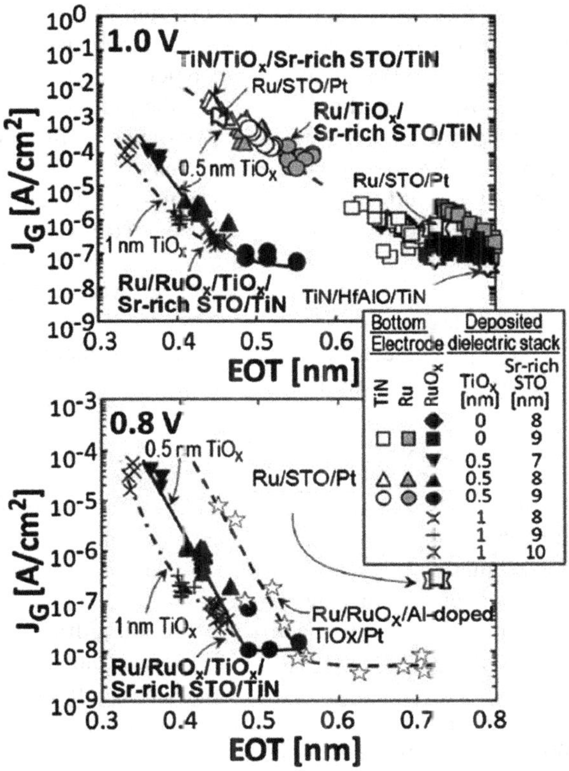

Fig. 4.17 J_G versus EOT plots at **a** 1.0 and **b** 0.8 V which show the significant decrease of the EOT value down to 0.4 nm with a leakage current of 10^{-7} A/cm^2 at 0.8 V for the optimized capacitor stack [103]

methods, other than ALD, which resulted in a large leakage current decrease, the appreciable decrease in the dielectric constant overcompensates for the leakage gain [105, 106]. Therefore, further doping optimization is necessary.

Electrodes for DRAM Capacitors

Sub-40-nm DRAM technologies require the development of new electrode material that is compatible with the newly developed dielectric material, such as ATO and STO. Heavily doped poly-Si electrodes, which were traditionally used as the electrode for the $SiO_2/Si_3N_4/SiO_2$ dielectric, cause severe problems for sub-100-nm DRAM technologies. Nonnegligible depletion thickness caused by the low conductivity of the poly-Si electrode reduces the effective capacitance. In addition, the SiO_2 interfacial layer that is inevitably formed at the high-k oxide/poly-Si interface largely decreases the capacitance density. The TiN electrodes are now used because it is metallic, so it has negligible depletion layer thickness and a negligible interfacial oxidation problem, because the dielectrics Al_2O_3, HfO_2, and ZrO_2 have higher oxidation potentials than does TiO_2.

New electrode materials for the sub-40-nm DRAM capacitor have to fulfill stringent requirements [107]. The electrode has to remain conductive after the back-end process and should not react with the dielectric material. Additionally, the electrode should adhere well to the plug material and the interlevel dielectric and must maintain low contact resistance to the underlying plug. A sufficiently high work function of the electrode material is favorable for a low leakage current. TiN may not fulfill such stringent requirements; thus, noble metals and conducting oxides have been extensively investigated. In addition, the conformal growth of the electrode materials on 3D structures is indispensably required. Therefore, ALD must be the process of choice for growth of the electrode materials and the dielectric layers.

ALD processes of noble metals and conducting oxides for the electrode in the DRAM capacitor are discussed below.

Noble Metals

Noble metals such as Pt, Ir, and Ru are expected to be feasible electrode materials for sub-40-nm DRAM applications due to their outstanding properties such as high work function, sufficient chemical inertness over the reaction with the dielectrics, and excellent thermal stability. Among those noble metals, Ru has been considered the most promising electrode material. Although the work function of Ru ($\sim$4.8 eV) is lower than those of Pt and Ir (5.6 and 5.3 eV, respectively), Ru could be easily etched due to the formation of gas phase by-products such as RuO_3 and RuO_4 gas under O_2 plasma. This ease of the reactive dry etching of Ru is favorable for patterning the electrodes.

There are large numbers of studies on the ALD growth of Ru metallic films. Most of the ALD processes of Ru thin films use oxygen as the reactant. Molecular oxygen gas is dissociatively adsorbed on the Ru surface, which has a catalytic effect on oxygen dissociation, and some of the adsorbed oxygen diffuses into the subsurface region [108]. The subsequent injection of the Ru precursor consumes the adsorbed oxygen by a reaction between the ligands in the precursor and the adsorbed oxygen atoms, finally producing Ru metallic film.

The use of several Ru precursors has been attempted to grow Ru thin films using the oxidative ALD reaction. The β-diketonate precursor $Ru(C_{11}H_{19}O_2)_3$ ($Ru(tmhd)_3$) showed quite a wide ALD window because it is thermally stable up to 450 °C [109, 110]. The growth rate slightly increased with growth temperature from 0.033 to 0.045 nm/cycle at the temperature range of 325–450 °C. However, the grown films from $Ru(tmhd)_3$ have relatively high impurity contents (3 at% H, 2 at% C, and 6 at% O) and long incubation cycles due to the low reactivity of the precursor [109]. Cyclopentadienyl derivatives such as $Ru(Cp)_2$ and $Ru(EtCp)_2$ have also been used as the precursor [111, 112]. Those precursors exhibited growth rates of Ru thin films range (0.03–0.04 nm/cycle). The high reactivity of the Cp ligands with oxygen resulted in low impurity contents <1 at%, producing Ru films with very low resistivity ($\sim$13 μΩ cm). In addition, relatively small

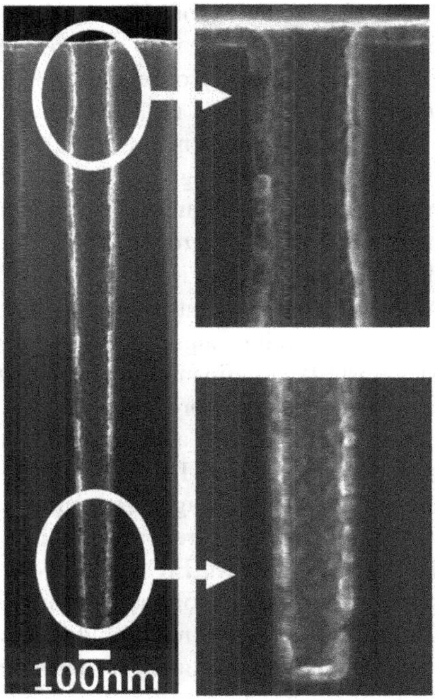

Fig. 4.18 Vertical SEM images of Ru film grown on the SiO_2 contact hole with an aspect ratio of 17 [113]

incubation cycles were obtained on an Al_2O_3 surface from the ALD process using $Ru(Cp)_2$ and O_2 [111].

The 2,4-(dimethylpentadienyl)(ethylcyclopentadienyl)Ru (DER) precursor with an asymmetric ligand structure was recently employed to grow Ru thin films using ALD [113]. The ALD process of DER and O_2 produced excellent quality Ru films at a growth temperature <300 °C. The grown Ru films showed negligible impurity contents and short incubation cycles due to the high reactivity of the ligands toward oxygen. In addition, very promising step coverage of the Ru films on the contact-hole structure with an aspect ratio of ~17 and an opening diameter of 150 nm were confirmed from the precursor as shown in Fig. 4.18.

Although there have been significant developments in the ALD process of Ru thin films, two critical issues in Ru ALD for DRAM applications remain: non-negligible incubation time and substrate-dependent growth. The long incubation cycles in the ALD suggest inadequate nucleation behavior resulting in a roughened surface of the grown films. The rough surface of the Ru films can deteriorate the capacitor quality due to the high leakage currents caused by the field concentration. Moreover, the long incubation cycles are also quite detrimental to the growth of nanometer-thick scale uniform and continuous films. In particular, the film thickness of the electrode in the sub-40-nm DRAM capacitor should be <7 nm. Hence, the formation of very thin and continuous metal films will become more crucial as scaling down continues. In addition, the bottom electrode in sub-40-nm

DRAM capacitors should be formed on both TiN diffusion barriers on plug and interlevel dielectrics with different surface properties. Insufficient nucleation behavior is generally observed on TiN and SiO_2 with weak ionic characteristics, whereas incubation is relatively small on transition metal oxide surfaces.

The substrate-dependent growth of Ru films can result in severe problems for the capacitor. The long incubation cycle and the substrate-dependent growth of Ru films have a close relationship to the initial growth behavior. An interesting study on the initial ALD growth of Ru films has recently been reported. Kim et al. demonstrated that the initial growth of Ru films was determined by the adsorption of the Ru precursor rather than by the catalytic dissociation of molecular oxygen [114]. This finding suggests that enhanced reactivity between the Ru precursor and the substrate significantly contributes to fluent nucleation.

There are two approaches for the fluent nucleation of Ru ALD: use of NH_3 plasma rather than O_2 as the reactant, and the development of a novel Ru precursor with better reactivity. The high reactivity of NH_3 plasma can activate the surface at the initial growth stage, generating multiple available nucleation sites on the surface. Kwon et al. reported the negligible incubation cycle of Ru ALD on the TiN surface from $Ru(EtCp)_2$ and NH_3 plasma [115]. Park et al. also showed very smooth morphology of the Ru films from $Ru(EtCp)_2$ and NH_3 plasma on TaN substrates due to the negligible incubation cycle [116]. Yim et al. demonstrated that NH_3 plasma treatment on a SiO_2 surface significantly reduced the incubation cycle of Ru ALD with $Ru(EtCp)_2$ and O_2 [117]. However, plasma-enhanced ALD might be concerning in the area of step coverage, which is critical in the 3D-structured capacitor in sub-40-nm DRAM devices.

Several other Ru precursors were developed for fluent nucleation. ALD of Ru films from *N,N*-dimethyl-1-ruthenocenylethylamine achieved the formation of very thin ($\sim$4 nm) continuous Ru films on an Al_2O_3 surface [118]. ALD of Ru films from (ethylcyclopentadienyl)(pyrrolyl)ruthenium showed smooth surface morphology on a TiN surface [119]. However, the films grown from the precursor still exhibited improper nucleation behavior on SiO_2. Although there are many report on the ALD of Ru films from various Ru precursors such as bis(2,6,6-trimethyl-cyclohexadienyl)ruthenium [120], $CpRu(CO)_2Et$ [121], $C_{16}H_{22}Ru$ [122], and bis(*N,N*′-di-tert-butylacetamidinato)ruthenium(II) dicarbonyl [123], the slow nucleation and substrate-dependent growth issues in Ru ALD are not entirely addressed by those precursors.

There is another approach for producing Ru thin films in which an inorganic RuO_4 precursor was adopted and reduced with the assistance of H_2 gas, which is in contrast to the oxidative deposition of MO Ru precursors mentioned above [124, 125]. It was recently reported that Ru film growth using this precursor was governed mainly by a CVD-type deposition mechanism, where the ALD-like source pulse-purge-reactant pulse-purge steps corresponding to p-CVD were repeated [124].

When the Ru films were grown at 230 °C, a process in which RuO_4 was dissolved in a methyl-ethyl fluorinated solvent, vaporized at 3 °C, and pulsed for 3 s and 5 % H_2/95 % N_2 gas was used, a high growth rate (0.18 nm/cycle) and a negligible incubation cycle were observed on both the TiN and Si substrates. The nucleation was mainly governed by the thermal decomposition of the RuO_4 molecules on the substrate surface rather than by charge transfer, which is facilitated by the (partial) oxidation of the TiN or Si substrates. The lower surface roughness of the as-deposited film by the p-CVD process using RuO_4 compared to the DER ALD process corroborates the idea of more fluent nucleation, which was further confirmed by the AFM images [124]. Furthermore, the p-CVD Ru films showed higher thermal stability due to their very low impurity concentrations [124]. Interestingly, their oxygen concentrations were even lower than that of the ALD films, suggesting the complete removal of oxygen by the H_2 reduction gas [124]. It was confirmed by high-resolution TEM that the highly uniform Ru film was formed on the TiN substrate even at a film thickness as low as 2 nm. High conformity of the Ru film over the severe 3D contact hole was also confirmed. The RuO_4 precursor can be used to grow RuO_2 conducting oxide films as discussed in next section.

ALD processes of Pt and Ir thin films were examined as well. The ALD reaction for those materials is based on the oxidative reaction of the Pt and Ir precursors, like the case of Ru ALD, because those metals also have a catalytic effect on oxygen dissociation. $MeCpPtMe_3$ [126] and $Ir(acac)_3$ [127, 128] are mainly used as precursors for Pt ALD and Ir ALD, respectively. However, those materials are not seriously considered as candidate electrode materials in next-generation DRAM capacitors due to patterning difficulties and cost issues.

Conducting Oxides

Although perovskite materials such as STO and BSTO have a very high dielectric constant (>>100), it is well-known that the dielectric constant of the materials decreases largely with decreasing film thickness due to the effect of several interfacial effects, including the intrinsic dead layer at the interface between the dielectric and the metal electrode [82–84]. Adoption of the conducting oxide as an electrode mitigates the decrease in the dielectric constant of the thin STO and BSTO films; as a result, RuO_2, IrO_2, and $SrRuO_3$ films have been investigated as potential conducting oxide electrodes. Those oxide films have quite low resistivity and maintain the high dielectric constant of the STO and BSTO thin films at very thin film thickness (<10 nm).

There are a few reports on the ALD of RuO_2 and IrO_2 films [129–131]. The ALD mechanism of RuO_2 and IrO_2 thin films is almost identical to the reaction mechanism of Ru and Ir metallic films except for the amount of injected oxygen. A larger flow rate or a longer injection time of O_2 gas is required for oxide film growth. During the injection of oxygen, the molecular oxygen is dissociatively adsorbed on the reaction surface like in the case of metallic film ALD. The subsequently injected metal precursor reacts with the adsorbed oxygen. However, the previously adsorbed oxygen is not completely consumed during the reaction with the metal precursor injection, and it eventually produces oxide films [129]. The ALD of RuO_2 and IrO_2 films exhibits a higher growth rate compared to the

ALD of their metallic films [129]. However, the oxide films grown using the oxidative reaction tend to be oxygen deficient. Therefore, the modified ALD process in which the O_2 gas was continuously flown during the growth or the use of ozone was suggested [130, 131].

RuO_2 thin films were also grown by p-CVD using RuO_4 and 95 % N_2/5 % H_2 mixed gas as the Ru precursor and the reducing reactant, respectively, at a growth temperature of 230 °C [132]. In contrast to the previous attempts to grow RuO_2 films using MO precursors and an oxidative reactant, this study adopted an alternative route to synthesize RuO_2 via the kinetically limited reduction of RuO_4 by H_2. In addition to the generic thermal decomposition of RuO_4 at the growth temperature, surface OH group-mediated chemical adsorption of the precursor on the RuO_2 surface made a significant contribution to RuO_2 film growth at relatively low N_2/H_2 gas pulse times (3–10 s). This surface OH-mediated growth also induced a saturated growth behavior with the N_2/H_2 gas pulse time. When the N_2/H_2 gas pulse time was increased to more than a critical value (10–15 s), the RuO_2 abruptly reduced to Ru by the auto-catalytic reduction effect of the Ru nuclei. On the Ru layer, film deposition was enhanced further by the disproportionation reaction between RuO_4 and Ru. The resistivity of the RuO_2 thin film was ~250 μΩ cm. On the capacitor hole structure of TiO_2 (4 nm)/SiO_2 with an aspect ratio of 10 (opening diameter is 100 nm), the RuO_2 film showed excellent step coverage [132]. Longer N_2/H_2 gas pulse times were beneficial in obtaining a smoother surface morphology on the capacitor hole structure. The better performance of the p-CVD RuO_2 films as the electrode for the ATO dielectric film was mentioned above [73].

On the contrary to the rutile structures of RuO_2 and IrO_2, which are desirable for achieving the high dielectric rutile phases of TiO_2-based materials, $SrRuO_3$ is another type of oxide electrode with a pseudocubic perovskite structure. Its pseudocubic lattice parameter (0.393 nm) is very close to the value of STO, so that $SrRuO_3$ provides a suitable base for heteroepitaxial growth of STO and BSTO films. Very thin STO or BSTO films have very high dielectric constants at even <10 nm thicknesses thank to this structural compatibility between STO (or BSTO) and $SrRuO_3$ [82]. As described above, $SrRuO_3$ is a very promising electrode material for sub-20-nm DRAM capacitors due to its perovskite-like structure, which can potentially omit the use of seed layers for achieving the in situ crystallization. However, no successful ALD growth of $SrRuO_3$ films has yet been reported. This is mainly due to the lack of ALD mechanism of the Ru(O) layer on top of the previously formed SrO layer.

As mentioned above, all the ALD mechanism of Ru(O) relies upon the oxidative adsorption reaction of MO precursors on the growing surface. However, the much higher bonding energy between Sr and O atoms generally does not allow the oxidative adsorption reaction of MO precursors of Ru, so the Ru(O) layer does not grow on the SrO layer. Therefore, the combination of the ALD of SrO and the p-CVD of the RuO_2 layers could be a viable solution to solve this problem, because RuO_2 can be deposited on the ALD SrO surface via thermal decomposition or the hydrogen reduction-mediated CVD reaction.

4.2.1.2 Sacrificial Layer

As mentioned in the introduction, ALD can be used to grow several layers that will be eventually removed during the chip fabrication process but subsidize other critical process steps. The use of low-temperature SiO_2 for the double patterning is a typical example that will be addressed in Sect. 4.2.2.2.

Another example of sacrificial ALD for DRAM capacitor fabrication is shown below. The fabrication of DRAM capacitors depends heavily on the development of bottom and top electrodes composed of noble metals (or noble metal oxides) and higher-k dielectric layers. Figure 4.19a–d shows the process flow for the fabrication of $TiN/ZrO_2/Al_2O_3/ZrO_2/TiN$ stacked capacitors with a cylindrical cell structure. After the TiN layer is deposited on the capacitor hole pattern as the bottom electrode (called mold oxide; Fig. 4.19a), it must be separated to form each storage node (Fig. 4.19c). This can be achieved by either chemical mechanical polishing or an etch-back process of the deposited bottom TiN layer. During this storage node separation process, the bottom area of the TiN electrode in the capacitor hole can be damaged and the electrical contact between the TiN bottom electrode and the drain contact stud becomes problematic. To protect the contact region (indicated by arrow in Fig. 4.19c) from damage, the inside space of the TiN bottom electrode is filled with SiO_2 (Fig. 4.19b). ALD SiO_2 at low temperatures is indispensable due to the extremely high aspect ratio and potential oxidation of the TiN bottom electrode. After storage node separation, the remaining SiO_2 inside the storage node hole is removed concurrently with the mold oxide by wet etching to create the cylinder structure of the TiN electrode. This ALD SiO_2 layer is a sacrificial layer and does not require stringent electrical performance.

4.2.2 *Structure and ALD Layers for Flash Memory*

4.2.2.1 Floating Gate and Charge Trap Flash Memories

The structure and operation principles of NAND flash memory were explained in detail in Sect. 4.1.3. Figure 4.20 shows the cross-sections of two types of NAND flash memories, FG flash and CTF, along the bit-line and word-line directions [134]. It is evident that the tall height of the FG makes the fabrication process complicated in the case of FG flash. However, the low-pressure CVD process of the SiO_2, Si_3N_4, and poly-Si gate processes provides sufficient step coverage and electrical performance, so there is little need to use ALD for this type of device. The new functional materials in CTF include the Al_2O_3 blocking oxide, the TaN control gate, and the nonstoichiometric SiN_x CT layer [135, 134]. The blocking oxide should have a negligible leakage current ($\sim 10^{-9}$ A/cm^2) to ensure the safe operation of flash memory since charge carriers must be exchanged only between the channel and the CT layer, not via the blocking oxide. Therefore, ALD at high temperature may be a viable option for the growth of the blocking oxide and the

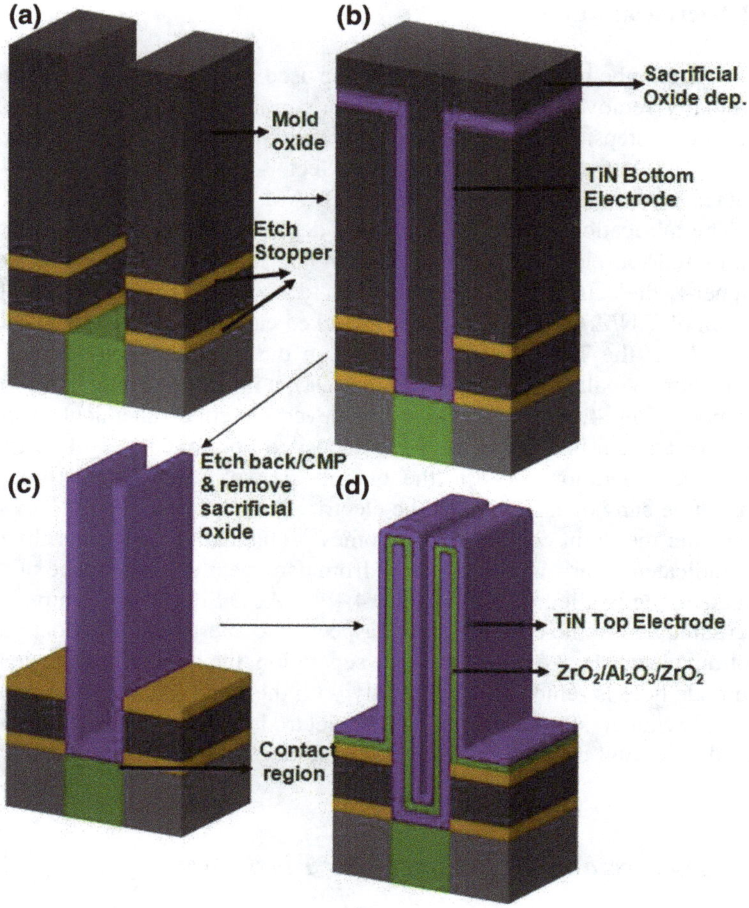

Fig. 4.19 **a–d** Schematic process flow diagram for the fabrication of $TiN/ZrO_2/Al_2O_3/ZrO_2/TiN$ stacked capacitors with a cylindrical cell structure [133]

control metal gate. To ensure a low enough leakage current, the blocking oxide layer normally undergoes high temperature annealing (~950–1,000 °C), so exceptional thermal stability is needed for this ALD layer.

The higher capacitance density of the blocking oxide layer is another requirement for achieving better flash cell operation. Due to the very strict requirement for a low leakage current and thermal stability, other high-k dielectrics (HfO_2 or silicates) may be less likely to be used as the blocking oxide layer in the near term. However, there have been several attempts to adopt high-k dielectrics by further improving the ALD process conditions or combining different high-k layers [137, 138]. This application does not have the severe restriction of the ALD process temperature limit. Overall, these applications correspond to the active function of the ALD layers.

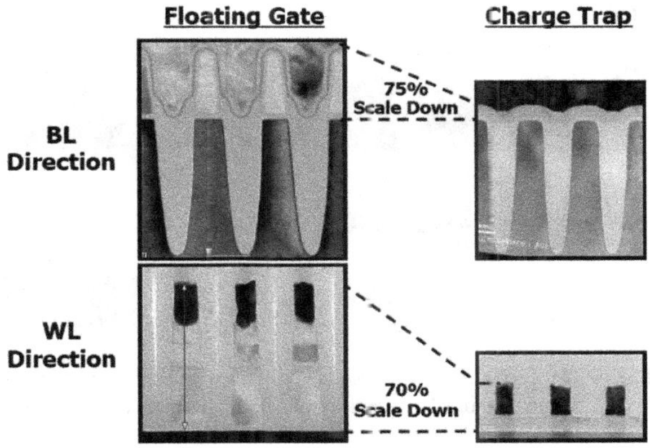

Fig. 4.20 Cross-section images of two types of NAND flash memories: floating gate flash and charge trap flash [134]

4.2.2.2 Double Patterning

On the other hand, ALD plays a crucial role in fabricating the extremely narrow gate and bit-line patterns in flash memories via DPT. To achieve the smallest flash feature size, which is generally smaller than that of DRAM for the given technology node, particularly the gate and bit-line pitch, the lithographic process requires revolutionary changes. The extensive use of DPT subsidizes the limited capability of fine pattering with the existing 193-nm immersion lithography. The oxide spacer method of DPT for making fine line/space patterns, which is the typical structure of control gates and B/Ls in NAND flash memory, uses ALD of low-temperature SiO_2. Figure 4.21a, b shows the two process sequences of DPT using the ALD SiO_2 layer, which are used mainly as the shallow trench isolation/ gate and bit-line patterning, respectively.

In either case, the final pitch is half of the first patterning pitch but the minimum feature size was determined to be the thickness of the deposited ALD SiO_2 layer (side wall thickness). In Fig. 4.21a, the final mask line dimension (called the bar dimension) is the ALD thickness, whereas in Fig. 4.21b, the distance between the final mask lines (called space) is the ALD thickness. Therefore, the constant critical dimension of the bar in Fig. 4.21a and the space in Fig. 4.21b is achieved. These techniques are suitable to apply to the dense and regularly arranged patterns, whereas complicated shape patterns require further process optimization.

For both cases, the final mask layer is made of a bottom hard mask layer that could be SiO_2, Si_3N_4, or TiN depending on the layer to be patterned. Therefore, the ALD layer works as a sacrificial layer for this application. Very tight thickness control ($\sim$1 % variation within wafer and wafer to wafer), a step coverage of $\sim$100 % and low deposition temperature (<200 °C) so as not to deform the finely patterned PR are required for this type of application. Low stress is another key factor. This is a very stringent requirement, even for ALD. ALD of SiO_2 has been quite challenging, but it is now widely developed, mainly due to the improvement in Si precursors [139, 140]. Some other materials with higher dry etching

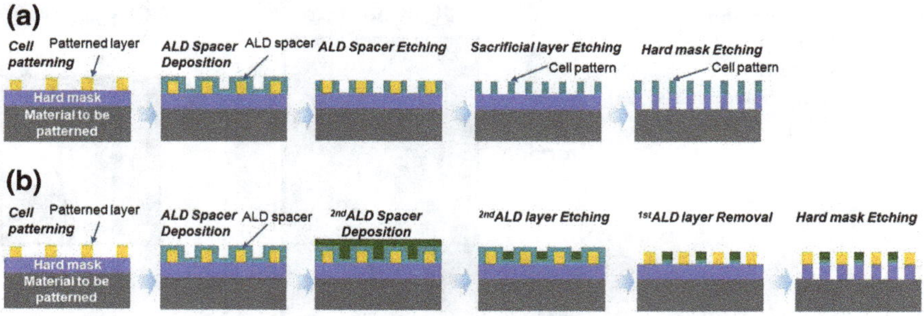

Fig. 4.21 Two different process sequences of double patterning technology using the ALD SiO_2 layer, which are used mainly as (**a**) Shallow trench isolation/(**b**) Gate and B/L, respectively

selectivity than SiO_2 might be necessary in the future depending on the patterning process. These DPT was originally developed to pattern the extremely fine pitches in flash memory, but it is currently used in many other devices requiring extremely fine patterns, such as DRAMs.

4.2.2.3 Vertically Integrated 3D Flash Memory

A very interesting application of the CTF cell is constructing a 3D stacked flash memory cell [141–143]. Vertical integration of memory cells is considered as the ultimate structure of the highest density memory cells. However, it has been hindered by the relatively low performance and large scatter of threshold voltage in poly-Si channel materials. The stacking of laterally integrated poly-Si-based memory cell layers does not offer an optimal solution for the most economical integration schemes, because each memory cell layer requires expensive photolithography steps. Figure 4.22a, b shows schematic diagrams of the two types of vertically integrated 3D stacked NAND-type flash cells. In Fig. 4.22a, a hole was etched after alternative deposition of the control gate and insulating layers was performed. A blocking oxide, CT layer, and tunneling oxide layers were deposited inside the hole. Finally, the semiconductor layer was deposited. In some cases, the semiconductor layer was made very thin to make the threshold voltage spread smaller, while the remaining core volume was filled with an insulator. In this case, the electronic structure of the semiconductor resembles the silicon-on-insulator structure. Hence, the electrical characteristics also resemble it.

When a standard 32-cell transistor string is formed along one vertical structure with a hole dimension of 50 nm, the aspect ratio of the hole easily exceeds 30, suggesting that ALD should play a key role in depositing various layers inside the hole. It is possible that the initial stage of these vertical cells could be fabricated using standard materials: ONO blocking oxide, SiN_x CT layer, and SiO_2 tunneling oxide. Among these materials, the tunneling SiO_2 layer can be formed by oxidation of a certain portion of the SiN_x CT layer or by ALD. In this regard, ALD

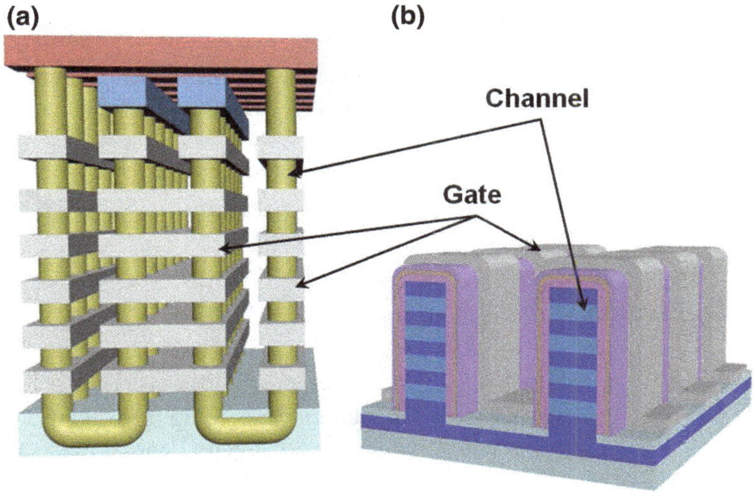

Fig. 4.22 Schematic diagrams of the two types of vertically integrated 3D stacked NAND type flash cells

SiO_2 must overcome the range of challenges from the very serious electrical requirements for the tunneling oxide layer as well as the blocking oxide layer. If other materials such as Al_2O_3 are considered, they must also be of supreme quality. The poly-Si deposition process may not require an ALD process, because the present low-pressure CVD still offers sufficient conformity.

For the other structure in Fig. 4.22b, the semiconductor and insulating layers are stacked first. They were then etched to form a vertical stud structure composed of alternating layers of semiconducting and insulating layers. The tunneling oxide, CT layer, and blocking oxide layers are subsequently formed, which is the reverse order of deposition shown in the previous case. Again, the narrow width and high aspect ratio of the trench between the active studs highlights the need for the ALD processes to deposit the layers. Here, the ALD layers are in direct contact with the etched side area of the semiconductor layers that will eventually form the channel. Therefore, extreme care of the ALD process is important to prevent damage to the sensitive semiconductor channel region.

For these applications, the ALD layers, i.e., the tunneling oxide, CT layer, blocking oxide, and hopefully the control gate, work as active materials. In general, CTF memory suffers from insufficient retention of trapped charges in the CT layer [144, 145]. This has been attributed mainly to the insufficiently deep trap level and its distribution over a certain energy range in the energy gap of the CT material. Therefore, there remains a need for the development of new CT materials using a better controlled ALD process and precursor chemistry. Oxide semiconductors, such as crystalline ZnO [146] or amorphous $InGaZnO_x$ [147], are attractive channel material candidates in vertically or laterally stacked 3D flash memory.

There could be several other device fabrication and integration schemes with a single crystalline Si-nano wire structure [148, 149]. In any case, the device structure becomes more and more complicated, and severe structures require conformal coating. Therefore, use of the ALD technique becomes even more crucial in these devices.

4.2.3 3D Stacked Memories

The future of highest memory density integration is 3D stacking. Chip level stacking is already used in mass production (multi-chip packaging [MCP]). However, MCP is not the optimum structure for the highest density memory, nor is it most cost-effective fabrication method. Wafer-level stacking using through-silicon-via (TSV) technology has been highlighted [150]. In TSV technology, the back-side grounded wafers with the memory patterns are diced and the fabricated chips are stacked to form multi-level memory. Each stacked layer is connected electrically through the TSV, where large and deep via holes were dry-etched and filled with interconnection metals. In TSV technology, the side wall area of the via holes are insulated from the interconnection metal by the adoption of a SiO_2 insulating layer. ALD SiO_2 with a higher growth rate is necessary due to the high aspect ratio of the TSV hole and the necessary thickness of the SiO_2 layer. Processing of these steps on such small stacked chips is technically challenging. Meanwhile, the first stacking process may suffer from low yield. In addition, expensive photolithography steps for the fabrication of each wafer layer remain necessary. Therefore, the wafer-level stacking technology via TSV is not the ideal method for reducing the cost of ultra-high-density memory chips.

On the other hand, film- or cell-level stacking must be the ultimate structure for the highest memory devices without increasing the fabrication cost, which is related to the extremely fine patterning, to an uneconomically high level. This structure and the related necessity for ALD was discussed in Sect. 4.2.2.3 for vertical NAND flash fabrication. In general, the usefulness of ALD increases as the memory structure becomes increasingly complex in the 3D structures to achieve a higher integration density.

Acknowledgments The author acknowledges support of the IT R&D program of MKE/IITA, "Capacitor technology for next generation DRAMs having mass-production compatibility" 2009-F-013-01, and the Converging Research Center Program through the National Research Foundation of Korea (2012K001299) funded by the Ministry of Education, Science and Technology, Korea. CSH greatly appreciates the crucial help from postdoctoral fellows and graduate students in his group at Seoul National University.

References

1. Ranganathan P (2011) Computer 44:39
2. Mayer M (2009) The Physics of Data. Xerox PARC Forum Distinguished Lecture. www.parc.com/event/936/innovation-at-google.html
3. Waser R (2008) Nanotechnology. Wiley-VCH Verlag GmbH & Co. KGaA, Weinheim
4. Kim JY, Lee CS, Kim SE, Chung IB, Choi YM, Park BJ, Lee JW, Kim DI, Hwang YS, Hwang DS, Hwang HK, Park JM, Kim DH, Kang NJ, Cho MH, Jeong MY, Kim HJ, Han JN, Kim SY, Nam BY, Park HS, Chung SH, Lee JH, Park JS, Kim HS, Park YJ, Kim K (2003), Symposium on VLSI technology, Digest of Technical Papers, p 11
5. Kim K (2005) International electron devices meeting, IEDM Technical Digest, p 323
6. Kuesters KH, Beug MF, Schroeder U, Nagel N, Bewersdorff U, Dallmann G, Jakschik S, Knoefler R, Kudelka S, Ludwig C, Manger D, Mueller W, Tilke A (2009) Adv Eng Mater 11:241
7. Niinistö J, Kukli K, Heikkilä M, Ritala M, Leskelä M (2009) Adv Eng Mater 11:223
8. Kim SK, Lee SW, Han JH, Lee B, Han S, Hwang CS (2010) Adv Funct Mater 20:2989
9. International Technology Roadmap for Semiconductors (2009)
10. Räisänen PI, Ritala M, Leskelä M (2002) J Mater Chem 12:1415
11. Min Y-S, Cho Y-J, Hwang CS (2005) Chem Mater 17:626
12. Hiltunen L, Kattelus H, Leskelä M, Mäkelä M, Niinistö L, Nykänen E, Soininen P, Tiitta M (1991) Mater Chem Phys 28:379
13. Tiitta M, Nykänen E, Soininen P, Niinistö L, Leskelä M, Lappalainen R (1998) Mater Res Bull 33:1315
14. Puurunen RL (2005) J Appl Phys 97:121301
15. Kim SK, Lee SW, Hwang CS, Min Y-S, Won JY, Jeong J (2006) J Electrochem Soc 153:F69
16. Kim YK, Lee SH, Choi SJ, Park HB, Seo YD, Chin KH, Kim D, Lim JS, Kim WD, Nam KJ, Cho M-H, Hwang KH, Kim YS, Kim SS, Park YW, Moon JT, Lee SI, Lee MY (2000) International Electron Devices Meeting, IEDM Technical Digest, p 369
17. Seidl H, Gutsche M, Schroeder U, Bimer A, Hecht T, Jakschik S, Luetzen J, Kerber M, Kudelka S, Popp T, Orth A, Rekinger H, Saenger A, Schupke K, Sell B (2002) International Electron Devices Meeting, IEDM Technical Digest, p 839
18. Böscke T, Kudelka S, Sänger A, Müller J, Krautschneider W (2006) European solid-state device research conference, p 391
19. Kukli K, Ritala M, Leskelä M (1995) J Electrochem Soc 142:1570
20. Aarik J, Kukli K, Aidla A (1996) Appl Surf Sci 103:331
21. Hiratani M, Hamada T, Iijima S, Ohji Y, Asano I, Naknishi N, Kimura S (2001) Symposium on VLSI Technology, Digest of Technical Papers, p 41
22. Sun H-J, Kim K-M, Kim Y, Cho K-J, Park K-S, Lee J-M, Roh J-S (2003) Jpn J Appl Phys 42:582
23. Ma D, Park S, Seo B-S, Choi S, Lee N, Lee J-H (2005) J Vac Sci Technol B 23:80
24. Zhao X, Vanderbilt D (2002) Phys Rev B 65:075105
25. Kim SK, Hwang CS (2008) Electrochem Solid-State Lett 11:G9
26. Park PK, Kang S-W (2006) Appl Phys Lett 89:192905
27. Govindarajan S, Böscke TS, Sivasubramani P, Kirsch PD, Lee BH, Tseng H-H, Jammy R, Schröder U, Ramanathan S, Gnade BE (2007) Appl Phys Lett 91:062906
28. Lee C-K, Cho E, Lee H-S, Hwang CS, Han S (2008) Phys Rev B 78:012102
29. Kukli K, Ritala M, Leskelä M, Sundqvist J, Oberbeck L, Heitmann J, Schröder U, Aarik J, Aidla A (2007) Thin Solid Films 515:6447
30. Schroeder U, Jakschik S, Erben E, Avellan A, Kudelka SP, Kerber M, Link A, Kersch A (2006) ECS Trans 1:125
31. Kim Y, Koo J, Han J, Choi S, Jeon H, Park C-G (2002) J Appl Phys 92:5443
32. Kukli K, Ritala M, Leskelä M (2000) Chem Vap Deposition 6:297

33. Endo K, Tatsumi T (2003) Jpn J Appl Phys 42:L685
34. Putkonen M, Niinistö L (2001) J Mater Chem 11:3141
35. Hausmann DM, Kim E, Becker J, Gordon RG (2002) Chem Mater 14:4350
36. Niinistö J, Kukli K, Kariniemi M, Ritala M, Leskelä M, Blasco N, Pinchart A, Lachaud C, Laaroussi N, Wang Z, Dussarrat C (2008) J Mater Chem 18:5243
37. Granneman E, Fischer P, Pierreux D, Terhorst H, Zagwijn P (2007) Surf Coat Technol 201:8899
38. Seo M, Min Y-S, Kim SK, Park TJ, Kim JH, Na KD, Hwang CS (2008) J Mater Chem 18:4324
39. Putkonen M, Niinistö J, Kukli K, Sajavaara T, Karppinen M, Yamauchi H, Niinistö L (2003) Chem Vap Deposition 9:207
40. Putkonen M, Niinistö L (2001) J Mater Chem 11:3141
41. Dezelah CL, Niinistö J, Kukli K, Munnik F, Lu J, Ritala M, Leskelä M, Niinistö L (2008) Chem Vap Deposition 14:358
42. Niinistö J, Kukli K, Tamm A, Putkonen M, Dezelah CL, Niinistö L, Lu J, Song F, Williams PA, Heys PN, Ritala M, Leskelä M (2008) J Mater Chem 28:3385
43. Aoki Y, Ueda T, Shirai H, Sakoh T, Kitamura T, Arai S (2002) International electron devices meeting, IEDM technical digest, p 831
44. Oh S-H, Chung J-H, Choi J-H, Yoo C-Y, Kim YS, Kim ST, Chung U-I, Moon JT (2003) Symposium on VLSI technology, Digest of technical papers, p 73
45. Seo M, Kim SK, Han JH, Hwang CS (2010) Chem Mater 22:4419
46. Seo M, Rha SH, Kim SK, Han JH, Lee W, Han S, Hwang CS (2011) J Appl Phys 110:024105
47. Yoon KR et al (2005) European solid-state device research conference, p 188
48. Cho HJ, Kim YD, Park DS, Lee E, Park CH, Jang JS, Lee KB, Kim HW, Chae SJ, Ki YJ, Han IK, Song YW (2006) European solid-state device research conference, p 146
49. Parker RA (1961) Phys Rev 124:1719
50. Sammelselg V, Rosental A, Tarre A, Niinistö L, Heiskanen K, Ilmonen K, Johansson L-S, Uustare T (1998) Appl Surf Sci 134:78
51. King JS, Graugnard E, Summmers CJ (2005) Adv Mater 17:1010
52. Mitchell DRG, Attard DJ, Triani G (2003) Thin Solid Films 441:85
53. Aarik J, Aidla A, Kiisler A-A, Uustare T, Sammelselg V (1997) Thin Solid Films 305:270
54. Aarik J, Aidla A, Sammelselg V, Siimon H, Uustare T (1996) J Cryst Growth 169:496
55. Ritala M, Leskelä M, Nykänen E, Soininen P, Niinistö L (1993) Thin Solid Films 225:288
56. Kim WD, Hwang GW, Kwon OS, Kim SK, Cho M, Jeong DS, Lee SW, Seo MH, Hwang CS, Min Y, Cho YJ (2005) J Electrochem Soc 152:C552
57. Ritala M, Leskelä M, Niinistö L, Haussalo P (1993) Chem Mater 5:1174
58. Aarik J, Aidla A, Uustare T, Ritala M, Leskelä M (2000) Appl Surf Sci 161:385
59. Aarik J, Karlis J, Mändar H, Uustare T, Sammelselg V (2001) Appl Surf Sci 181:339
60. Ritala M, Leskelä M, Rauhala E (1994) Chem Mater 6:556
61. Rahtu A, Ritala M (2002) Chem Vap Deposition 8:21
62. Kim SK, Kim W-D, Kim K-M, Hwang CS, Jeong J (2004) Appl Phys Lett 85:4112
63. Pheamhom R, Sunwoo C, Kim D-H (2006) J Vac Sci Technol A 24:1535
64. Kim SK, Hoffmann-Eifert S, Mi S, Waser R (2009) J Electrochem Soc 156:D296
65. Byun C, Jang JW, Kim IT, Hong KS, Lee BW (1997) Mater Res Bull 32:431
66. Kadoshima M, Hiratani M, Shimamoto Y, Torii K, Miki H, Kimura S, Nabatame T (2003) Thin Solid Films 424:224
67. Hiratani M, Kadoshima M, Hirano T, Shimamoto Y, Matsui Y, Nabatame T, Torii K, Kimura S (2003) Appl Surf Sci 207:13
68. Kim SK, Han S, Han JH, Lee W, Hwang CS (2011) Phys Status Solidi RRL 5:262
69. Choi G-J, Kim SK, Won S-J, Kim HJ, Hwang CS (2009) J Electrochem Soc 156:G138
70. Fröhlich K, Ťapajna M, Rosová A, Dobročka E, Hušeková K, Aarik J, Aidla A (2008) Electrochem Solid-State Lett 11:G19

71. Kim SK, Choi G-J, Lee SY, Seo M, Lee SW, Han JH, Ahn H-S, Han S, Hwang CS (2008) Adv Mater 20:1429
72. Kim SK, Choi GJ, Kim JH, Hwang CS (2008) Chem Mater 20:3723
73. Han JH, Han S, Lee W, Lee SW, Kim SK, Gatineau J, Dussarrat C, Hwang CS (2011) Appl Phys Lett 99:022901
74. R. Wyckoff, Crystal structures, 2nd ed, Vol. 2, Interscience, New York, 1964
75. Deak DS (2007) Mater Sci Technol 23:127
76. Cardona M (1965) Phys Rev 140:A651
77. Henrich V, Dresselhaus G, Zeiger H (1978) Phys Rev B 17B:4908
78. Reihl B, Bednorz JG, Müller KA, Jugnet Y, Landgren G, Morar JF (1984) Phys Rev B 30B:803
79. Møller P, Komolov S, Lazneva E (2000) J Phys: Condens Matter 12:7705
80. Lines ME, Glass AM (1977) Principles and application of ferroelectrics and related materials. Clarendon Press, Oxford
81. Hwang CS, Park SO, Cho H-J, Kang CS, Kang H-K, Lee SI, Lee MY (1995) Appl Phys Lett 67:2821
82. Hwang CS (2002) J Appl Phys 92:432
83. Black CT, Welser JJ (1999) IEEE Trans Electron Devices 46:776
84. Natori K, Otani D, Sano N (1998) Appl Phys Lett 73:632
85. No SY, Oh JH, Jeon CB, Schindler M, Hwang CS, Kim HJ (2005) J Electrochem Soc 152:C435
86. Lesaicherre PY, Yamaguchi H, Miyasaka Y, Watanabe H, Ono H, Yoshida M (1995) Integr Ferroelectr 8:201
87. Menou N, Popovici M, Clima S, Opsomer K, Polspoel W, Kaczer B, Rampelberg G, Tomida K, Pawlak MA, Detavernier C, Pierreux D, Swerts J, Maes JW, Manger D, Badylevich M, Afanasiev V, Conard T, Favia P, Bender H, Brijs B, Vandervorst W, Van Elshocht S, Pourtois G, Wouters DJ, Biesemans S, Kittl JA (2009) J Appl Phys 106:094101
88. Popovici M, Van Elshocht S, Menou N, Swerts J, Pierreux D, Delabie A, Brijs B, Conard T, Opsomer K, Maes JW, Wouters DJ, Kittl JA (2010) J Electrochem Soc 157:G1
89. Kosola A, Putkonen M, Johansson L-S, Niinistö L (2003) Appl Surf Sci 11:102
90. Langereis E, Roijmans R, Roozeboom F, van de Sanden MCM, Kessels WMM (2011) J Electrochem Soc 158:G34
91. Vehkamäki M, Hänninen T, Ritala M, Leskelä M, Sajavaara T, Rauhala E, Keinonen J (2001) Chem Vap Deposition 7:75
92. Ahn JH, Kim JY, Kang SW, Kim JH, Roh JS (2007) Appl Phys Lett 91:062910
93. Kwon OS, Lee SW, Han JH, Hwang CS (2007) J Electrochem Soc 154:G127
94. Kwon OS, Kim SK, Cho M, Hwang CS, Jeong J (2005) J Electrochem Soc 152:C229
95. Lee SW, Kwon OS, Han JH, Hwang CS (2008) Appl Phys Lett 92:222903
96. Lee SW, Han JH, Kwon OS, Hwang CS (2008) J Electrochem Soc 155:G253
97. Vehkamäki M, Hatanpää T, Hänninen T, Ritala M, Leskelä M (1999) Electrochem Solid-State Lett 2:504
98. Holme TP, Prinz FB (2007) J Phys Chem A 111:8147
99. García H, Castán H, Gómez A, Dueñas S, Bailón L, Kukli K, Kariniemi M, Kemell M, Niinistö J, Ritala M, Leskelä M (2011) J Vac Sci Technol B 29:01AC04
100. Lee SW, Han JH, Han S, Lee W, Jang JH, Seo M, Kim SK, Dussarrat C, Gatineau J, Min Y-S, Hwang CS (2011) Chem Mater 23:2227
101. Menou N, Wang XP, Kaczer B, Polspoel W, Popovici M, Opsomer K, Pawlak MA, Knaepen W, Detavernier C, Blomberg T, Pierreux D, Swerts J, Maes JW, Favia P, Bender H, Brijs B, Vandervorst W, Van Elshocht S, Wouters DJ, Biesemans S, Kittl JA (2008) International electron devices meeting, IEDM technical digest, p 929
102. Pawlak MA, Kaczer B, Kim M-S, Popovici M, Tomida K, Swerts J, Opsomer K, Polspoel W, Favia P, Vrancken C, Demeurisse C, Wang W-C, Afanas'ev VV, Vandervorst W, Bender H, Debusschere I, Altimime L, Kittl JA (2010) Appl Phys Lett 97:162906

103. Pawlak MA, Kaczer B, Kim M-S, Popovici M, Swerts J, Wang W-C, Opsomer K, Favia P, Tomida K, Belmonte A, Govoreanu B, Vrancken C, Demeurisse C, Bender H, Afanas'ev VV, Debusschere I, Altimime L, Kittl JA (2011) Appl Phys Lett 98:182902
104. Riedel S, Neidhardt J, Jansen S, Wilde L, Sundqvist J, Erben E, Teichert S, Michaelis A (2011) J Appl Phys 109:094101
105. Huang CC, Cheng CH, Lin CW, Chang LM (2010) J Electrochem Soc 157:H624
106. Chen TL, Li XM, Wu WB (2005) J Appl Phys 98:064109
107. Kang SY, Hwang CS, Kim HJ (2005) J Electrochem Soc 152:C15
108. Aaltonen T, Rahtu A, Ritala M, Leskelä M (2003) Electrochem Solid-State Lett 6:C130
109. Aaltonen T, Ritala M, Arstila K, Keinonen J, Leskelä M (2004) Chem Vap Deposition 10:215
110. Kim SK, Hoffmann-Eifert S, Waser R (2009) J Phys Chem C 113:11329
111. Aaltonen T, Alen P, Ritala M, Leskelä M (2003) Chem Vap Deposition 9:45
112. Kim W-H, Park S-J, Son J-Y, Kim H (2008) Nanotechology 19:045302
113. Kim SK, Lee SY, Lee SW, Hwang GW, Hwang CS, Lee JW, Jeong J (2007) J Electrochem Soc 154:D95
114. Kim SK, Han JH, Kim GH, Hwang CS (2010) Chem Mater 22:2850
115. Kwon O-K, Kwon S-H, Park H-S, Kang S-W (2004) J Electrochem Soc 151:C753
116. Park S-J, Kim W-H, Lee H-B-R, Maeng WJ, Kim H (2008) Microelectron Eng 85:39
117. Yim S-S, Lee D-J, Kim K-S, Kim S-H, Yoon T-S, Kim K-B (2008) J Appl Phys 103:113509
118. Kukli K, Ritala M, Kemell M, Leskelä M (2010) J Electrochem Soc 157:D35
119. Kukli K, Kemell M, Puukilainen E, Aarik J, Aidla A, Sajavaara T, Laitinen M, Tallarida M, Sundqvist J, Ritala M, Leskelä M (2011) J Electrochem Soc 158:D158
120. Gregorczyk K, Henn-Lecordier L, Gatineau J, Dussarrat C, Rubloff G (2011) Chem Mater 23:2650
121. Leick N, Verkuijlen ROF, Lamagna L, Langereis E, Rushworth S, Roozeboom F, van de Sanden MCM, Kessels WMM (2011) J Vac Sci Technol A 29:021016
122. Eom T-K, Sari W, Choi K-J, Shin W-C, Kim JH, Lee D-J, Kim K-B, Sohn H, Kim S-H (2009) Electrochem Solid-State Lett 12:D85
123. Wang H, Gordon RG, Alvis R, Ulfig RM (2009) Chem Vap Deposition 15:312
124. Han JH, Lee SW, Choi GJ, Lee SY, Hwang CS, Dussarrat C, Gatineau J (2009) Chem Mater 21:207
125. Gatineau J, Yanagita K, Dussarrat C (2006) Microelectron Eng 83:2248
126. Aaltonen T, Ritala M, Sajavaara T, Keinonen J, Leskelä M (2003) Chem Mater 15:1924
127. Aaltonen T, Ritala M, Sammelselg V, Leskelä M (2004) J Electrochem Soc 151:G489
128. Hämäläinen J, Puukilainen E, Kemell M, Costelle L, Ritala M, Leskelä M (2009) Chem Mater 21:4868
129. Kwon S-H, Kwon O-K, Kim J-H, Jeong S-J, Kim S-W, Kang S-W (2007) J Electrochem Soc 154:H773
130. Kim J-H, Kil D-S, Yeom S-J, Roh J-S, Kwak N-J, Kim J-W (2007) Appl Phys Lett 91:052908
131. Hämäläinen J, Kemell M, Munnik F, Kreissig U, Ritala M, Leskelä M (2008) Chem Mater 20:2903
132. Han JH, Lee SW, Kim SK, Han S, Hwang CS, Dussarrat C, Gatineau J (2010) Chem Mater 22:5700
133. Pinna N, Knez M (2012) Atomic layer deposition of nanostructured materials. Wiley-VCH Verlag GmbH & Co. KGaA, Weinheim
134. Unpublished; courtesy of Samsung
135. Lee CH, Choi KI, Cho MK, Song YH, Park KC, Kim K (2003) International electron devices meeting, IEDM technical digest, p 26
136. Jeon S, Han JH, Lee JH, Choi S, Hwang H, Kim C (2005) IEEE Trans Electron Devices 52:2654

137. Sung S-K, Lee S-H, Choi BY, Lee JJ, Choe J-D, Cho ES, Ahn YJ, Choi D, Lee C-H, Kim DH, Lee Y-S, Kim SB, Park D, Ryu B-I (2006) Symposium on VLSI technology digest of technical papers, p 86
138. Lue H-T, Lai S-C, Hsu T-H, Du P-Y, Wang S-Y, Hsieh K-Y, Liu R, Lu C-Y (2009) IEEE International reliability physics symposium, p 847
139. Kamiyama S, Miura T, Nara Y (2006) Thin Solid Films 515:1517
140. Katamreddy R, Feist B, Takoudisb C (2008) J Electrochem Soc 155:G163
141. Tanaka H, Kido M, Yahashi K, Oomura M, Katsumata R, Kito M, Fukuzumi Y, Sato M, Nagata Y, Matsuoka Y, Iwata Y, Aochi H, Nitayama A (2007) Symposium on VLSI technology digest of technical papers, p 14
142. Kim W, Choi S, Sung J, Lee T, Park C, Ko H, Jung J, Yoo I, Park Y (2009) Symposium on VLSI technology digest of technical papers, p 188
143. Kim J, Hong AJ, Kim SM, Song EB, Park JH, Han J, Choi S, Jang D, Moon J-T, Wang KL (2009) Symposium on VLSI technology digest of technical papers, p 186
144. Huo ZL, Yang J, Lim S, Baik S, Lee J, Han JH, Yeo I-S, Chung U-I, Moon JT, Ryu B-I (2007) Symposium on VLSI technology digest of technical papers, p 138
145. Hong SH, Jang JH, Park TJ, Jeong DS, Kim M, Won JY, Hwang CS (2005) Appl Phys Lett 87:152106
146. Hoffman RL, Norris BJ, Wager JF (2003) Appl Phys Lett 82:733
147. Nomura K, Ohta H, Takagi A, Kamiya T, Hirano M, Hosono H (2004) Nature 432:488
148. Hubert A, Nowak E, Tachi K, Maffini-Alvaro V, Vizioz C, Arvet C, Colonna J-P, Hartmann J-M, Loup V, Baud L, Pauliac S, Delaye V, Carabasse C, Molas G, Ghibaudo G, De Salvo B, Faynot O, Ernst T (2009) International electron devices meeting, IEDM technical digest, p 637
149. Yun J-G, Kim G, Lee J-E, Kim Y, Shim WB, Lee J-H, Shin H, Lee JD, Park B-G (2011) IEEE Trans Electron Devices 58:1006
150. Motoyoshi M (2009) Proceedings of the IEEE, p 43

Chapter 5
PCRAM

Simone Raoux and Mikko Ritala

5.1 Introduction to Phase Change Materials and a Brief History

Phase change materials are a unique class of materials with properties that make them useful for data storage. A first description of their memory switching capabilities was published [1] and variations of phase change memory cell designs were patented [2] by Ovshisky in the 1960s. Early phase change materials based on the Te–As–Si–Ge system, however, switched too slowly for a viable memory technology. The breakthrough discovery of fast-switching phase change materials based on the compositions on the pseudo-binary line between GeTe and Sb_2Te_3 by Yamada et al. [3] in 1987 led to a revival of phase change technology. The first technological application of phase change materials was realized in rewritable optical storage disks. First products included rewritable CDs (CD-RW introduced in 1996) followed by rewritable DVDs such as DVD-RW up to the latest Blu-ray disks technology (e.g., BD-RE). Overviews of materials development for optical storage and optical storage technology can be found elsewhere [4–7]. The rapid and successful development of phase change optical storage technology triggered a revival of interest in phase change random access memory (PCRAM) technology. Intense research and development efforts have led to a rapid progress in this field and first PCRAM products have just entered the market [8].

Phase change thin films for optical storage are typically deposited by sputter process. In this chapter we will concentrate on the PCRAM application and material parameters that are relevant for this technology because atomic layer deposition (ALD) offers the opportunity for novel and improved PCRAM cell

S. Raoux (✉)
IBM T.J. Watson Research Center, Yorktown Heights, New York, USA
e-mail: simoner@us.ibm.com

M. Ritala
Department of Chemistry, University of Helsinki, Helsinki, Finland

C. S. Hwang (ed.), *Atomic Layer Deposition for Semiconductors*,
DOI: 10.1007/978-1-4614-8054-9_5,

designs that require conformal and very well controlled deposition of phase change materials that cannot be achieved by sputter deposition.

5.2 Principle of Phase Change Technology

Phase change materials exist in two phases—an amorphous phase and a crystalline phase (some materials have two crystalline phases, a metastable and a stable phase). Phase change memory technology is based on the large difference between the optical and electronic properties between the phases and the possibility to switch these materials repeatedly between the phases. The switch from the amorphous to the crystalline phase is performed by heating the material above its crystallization temperature T_x for a sufficiently long time, so that the amorphous-crystalline phase transition can occur. This operation is called "set" in PCRAM technology. The reverse operation (called "reset") is achieved by melting the crystalline phase change material and cooling it rapidly (quenching), so that it solidifies in the amorphous phase. In optical storage, the switching is performed with a laser pulse, the large difference in reflectivity between the phases is used as contrast mechanism, and the reading is done with a laser pulse as well. In PCRAM electrical pulses are used for switching and reading (high, short pulse for reset, longer, lower pulse for set, and even lower pulse for reading). An access device connected to each memory cell provides the current pulses. In most cases this is a transistor (bipolar junction transistor [9] or field effect transistor [10]), but PCRAM cells addressed with diodes have been reported also [11]. The very large difference of several orders of magnitude between the electrical resistances of the two phases is used as the storage mechanism. A special switching mechanism occurs in phase change materials that is called threshold switching. It allows the set operation at small voltages. Without this effect PCRAM would not be possible because very large voltages would be required to heat the amorphous, highly resistive material above the crystallization temperature. However, if a voltage applied to a PCRAM cell surpasses a certain threshold voltage, the amorphous, originally highly resistive material becomes suddenly highly conductive. Now a large current can flow through the cell and Joule heating can occur to heat the material above its crystallization temperature. Physically, it is a certain material-dependent threshold field [12, 13] that translates into a voltage depending on the memory cell design and the dimensions of the amorphous phase change material. Typical threshold fields are on the order of 10–100 V/μm [13].

As mentioned before, some of the most commonly applied phase change materials can be found on the pseudo-binary line between GeTe and Sb_2Te_3, notably $Ge_2Sb_2Te_5$, which is the most studied among the phase change materials and is also the base material used in most PCRAM applications. Figure 5.1 shows a ternary phase diagram of different phase change alloys, their year of discovery, and their use in different optical storage products.

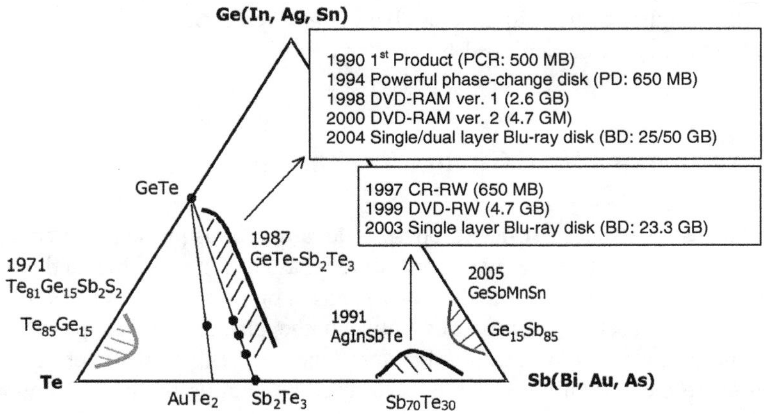

Fig. 5.1 Ternary phase diagram of different phase change alloys, their year of discovery, and their use in different optical storage products. Reprinted from [6] by permission from Macmillan Publishers Ltd: Nature Materials, copyright (2007)

Besides the Ge–Sb–Te materials, also materials close to Sb_2Te with a number of dopants such as Ag and In (AIST) have been applied in optical storage. The term dopant in the phase change materials community is used somewhat misleading because it is not a doping comparable to conventional semiconductor doping but rather an alloying with concentrations of the additive element in the several atomic percent range. Even though AIST has been used for prototype PCRAM device fabrication in academic studies [13] it is unlikely that it will be found in PCRAM products because of the incompatibility of Ag and In containing alloys with CMOS technology. However, using another dopant, successful array level integration in an advanced 65 nm CMOS process flow of a PCRAM memory concept based on a narrow line of a doped-Sb_2Te phase has been demonstrated [14, 15].

Other alloys found in prototype devices include materials based on the eutectic $Ge_{15}Sb_{85}$ composition [13–16]. They show very short switching times and high crystallization temperature but it was observed later that these materials tend to elemental segregation [17, 18].

In phase change optical storage, higher data storage densities have been achieved mainly by moving to shorter wavelengths and larger numerical apertures (780 nm, 0.45 for CDs, 650 nm, 0.6 for DVDs, and 405 nm, 0.85 for Blu-ray disks, respectively [5]). With the move to shorter wavelengths new alloys were explored because $Ge_2Sb_2Te_5$ shows weak contrast at shorter wavelength. It was found that alloys with higher GeTe and lower Sb concentrations are better suited for optical storage at blue wavelengths [19, 20]. Similarly, GeTe and doped GeTe PCRAM devices have been reported recently [21–23] and were shown to have very fast switching speed and improved data retention because the GeTe rich alloys are characterized by higher crystallization temperature and better stability of the amorphous phase.

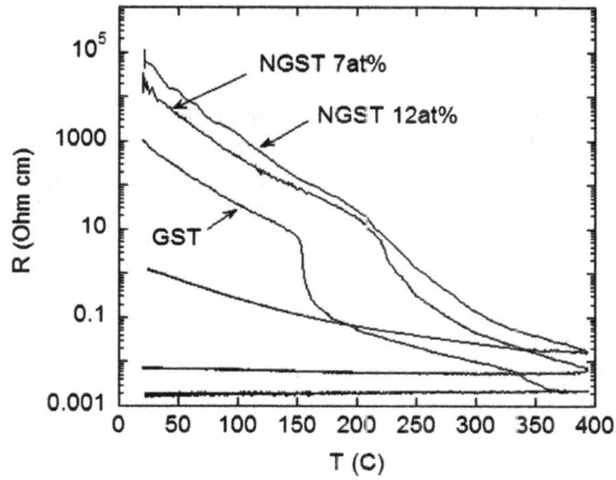

Fig. 5.2 Resistivity of $Ge_2Sb_2Te_5$ and N-$Ge_2Sb_2Te_5$ films is shown while heating the film at a rate of 1 °C/s, followed by cooling back to room temperature. Undoped $Ge_2Sb_2Te_5$ shows a sharp drop in resistivity around 150 °C corresponding to the transition from the amorphous to the rocksalt phase, and a second smaller drop around 350 °C caused by the rocksalt-hexagonal transition. Both N-$Ge_2Sb_2Te_5$ films show more gradual amorphous-rocksalt transitions at higher temperatures and no formation of the hexagonal phase for temperatures of up to 400 °C. The resistivities increase in both the amorphous and crystalline phases by nitrogen doping. Reprinted with permission from [24]

$Ge_2Sb_2Te_5$ is still the most commonly used material for PCRAM applications, often doped with various materials such as N, O, Sn, SiC, SiO_x [24–28]. These dopants improve cyclability, increase crystallization temperature and resistances, and reduce grain size thus leading to better device performance. Figure 5.2 shows as an example the resistance as a function of temperature for undoped and nitrogen doped $Ge_2Sb_2Te_5$.

The increased resistance in the crystalline phase leads to a reduction in the reset current and thus improves device performance [28]. The reset current is a very critical device parameter because it determines the size of the access device which in current PCRAM technology determines the cell size. The actual phase change memory element is typically much smaller than the access device, in particular if transistors are applied [29].

Phase change materials are very unique in the combination of their properties. To be of practical use in storage technology they must fulfill a number of criteria that appear sometimes even contradictory. Some of the most important material properties that are dictated by the storage application include:

(1) High optical or electrical contrast for optical storage or PCRAM, respectively.
(2) Short crystallization time. It determines the write speed and is the data rate limiting factor.

(3) High cyclability. The cycle number depends strongly on the application and is 10^5 cycles for Flash replacement, 10^{12} cycles for embedded memory, and even 10^{16} cycles for DRAM replacement.
(4) Threshold switching to allow for the set operation in PCRAM.
(5) High stability of the amorphous phase at operating temperature. This depends also on the application and is 10 years at 80 °C for embedded memory, 10 years at 150 °C for automotive applications, and even 250 °C for 20–30 s if the memory chip needs to be pre-coded and solder-bonded.
(6) Small reset current. Currently, the size of the access device determines the PCRAM cell size and thus the storage density. Reduction in reset current can be achieved by improved PCRAM cell design, improved phase change material properties, or a combination of both.
(7) Good scalability for the development of future technology nodes.

In particular fast crystallization in nanoseconds at high temperature but high stability of the amorphous phase at lower temperature for years are requirements that are difficult to meet simultaneously since they span 10^{17} orders of magnitude difference in the response time [4]! Material optimization is therefore a crucial part of the technology development and much research is directed toward understanding the basic mechanisms that underlie the unique property combination of phase change materials [4, 30, 31] and the design of new materials based on this understanding [32].

5.3 Structure and Properties of Phase Change Materials

It has been pointed out in a recent review article [33] that the structure of phase change materials determines their unique properties and that in order to understand their function we first need to understand their structure. The structure in both phases, the amorphous and the crystalline, are responsible for the properties that make phase change technology possible. In most cases, phase change materials are deposited by sputter process at room temperature and are amorphous after deposition. Heating will lead to crystallization and polycrystalline films with grains typically in the tens of nanometers are formed. Several phase change materials along the pseudo-binary line between GeTe and Sb_2Te_3, notably also $Ge_2Sb_2Te_5$, form first a metastable crystalline phase. This phase is characterized by a distorted rocksalt-like structure with tellurium atoms occupying the anionic sublattice, and germanium, antimony and about 20 % vacancies on random locations on the cationic sublattice [34]. At higher temperature a stable hexagonal phase is formed [35] but most likely this stable phase is never present in PCRAM cells. Another popular phase change material, GeTe, which lies on one side of the GeTe-Sb_2Te_3 pseudo-binary line, crystallizes in a rhombohedral structure (space group *R*3 *m*). It can be thought of as a distorted rocksalt-like structure stretched along the [111] direction [36]. GeTe undergoes a reversible phase transition to a cubic phase around 350 °C [36, 37]. Locally, each atom is nominally octahedrally coordinated

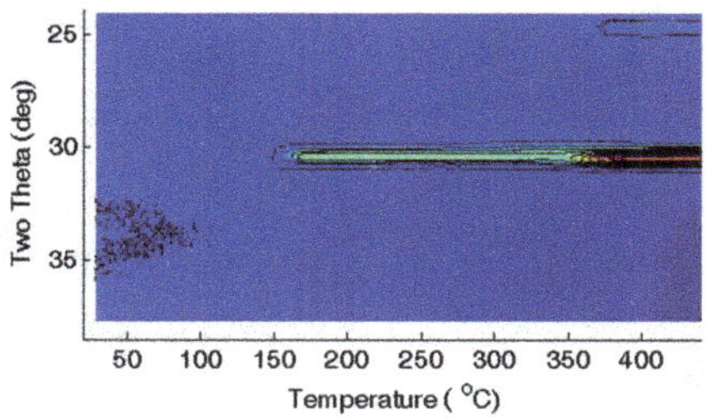

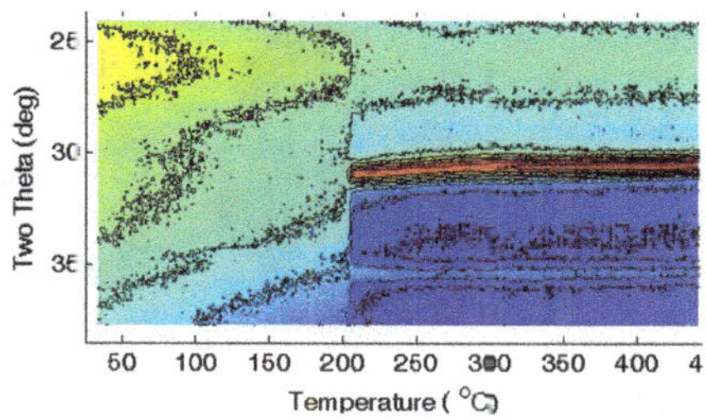

Fig. 5.3 Intensity of diffracted X-ray peaks as a function of temperature during a heating ramp at 1 K/s for 50 nm thick $Ge_2Sb_2Te_5$ (*left*) and GeTe films (*right*). Data were acquired at the National Synchrotron Light Source at an X-ray wavelength of 1.797 Å using a fast linear-diode-array detector that covers a range of 15° in 2θ

with three short (2.843 Å) and three long (3.158 Å) bonds. The crystal structure of Sb_2Te_3, the material on the other end of the pseudo-binary GeTe-Sb_2Te_3 line, consist of an atomic alternation of (Te–Sb–Te-Sb–Te) layers with interlayer bonding between adjacent Te planes being ascribed to Van der Waals interaction. The other stoichiometric Sb–Te alloy, Sb_2Te, also has a rhombohedral symmetry that can be described as a rocksalt-like structure stretched along the [111] direction.

Figure 5.3 shows as an example the crystallization behavior of $Ge_2Sb_2Te_5$ compared to GeTe. Plotted is the intensity of diffracted X-ray peaks as a function of temperature during a heating ramp at 1 K/s for 50 nm thick films.

Both films are amorphous as-deposited and no X-ray diffraction (XRD) peaks are observed. $Ge_2Sb_2Te_5$ transforms first into the metastable rocksalt-like structure around 150 °C and at higher temperature (350 °C) to the stable hexagonal phase, and it remains in this phase upon cooling. GeTe crystallizes at around 200 °C into the rocksalt-like structure and undergoes a reversible rocksalt-cubic phase transition around 350 °C (not very clearly visible because of the limited resolution of the data).

It has been found that resonant bonding is a bonding scheme that can be used to identify phase change materials with property combinations that make them useful for technological applications (high contrast, fast switching, etc.), and it can be applied to identify and design new phase change materials, see Fig. 5.4 [30, 32, 33]. Resonant bonding in phase change materials is characterized by six atomic neighbors but only an average number of three p electrons per atom to facilitate the covalent bond [30]. It can be related to high optical dielectric constants and high electronic polarizability, and rather symmetric, octahedral-like atomic arrangements [33] (materials marked with green circles in Fig. 5.4).

Phase change materials in the amorphous state at low voltages below the threshold voltage show linear current–voltage behavior at very low voltages and an exponential increase of the current at higher voltages. This behavior can be described as thermally activated hopping transport through traps. This model can also describe the threshold switching effect at the threshold field. At the threshold

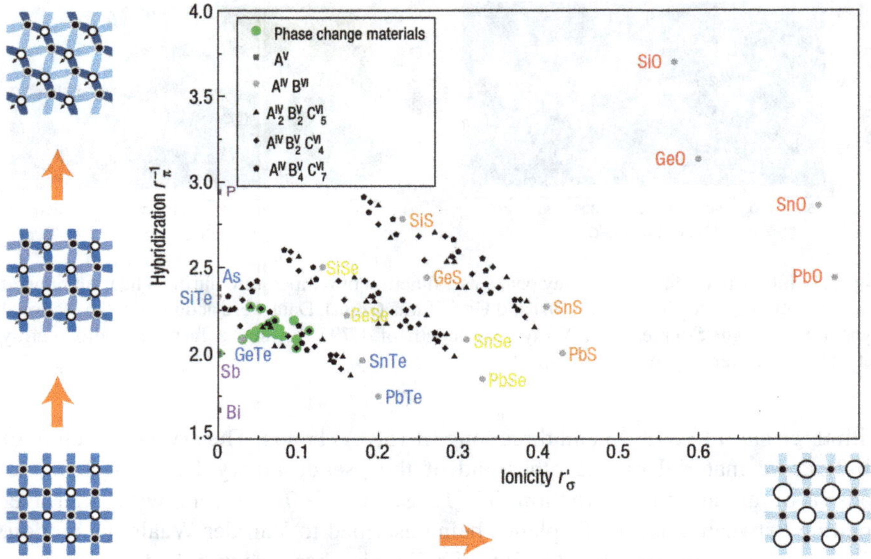

Fig. 5.4 Variety of materials with an average of three *p* electrons per atom and variable hybridization and ionicity. Pure Sb has the most pronounced resonant bonding character. Increase in hybridization weakens the resonant bonding character because it leads to larger distortions favoring a smaller number of more saturated covalent bonds. Increased ionicity also reduces resonant bonding character because it leads to increasingly localized charges. Reprinted with permission from [32]

field electrons gain significant energy leading to a current–voltage instability and a dramatic drop in resistance.

In the crystalline phase, most phase change materials behave like p-type semiconductors. Optical measurements resulted in a bandgap of 0.5 eV [38]. The electrical conductivity in the crystalline phase is orders of magnitude higher than in the amorphous phase. Hall measurements have shown very large hole concentrations of 10^{20}–10^{21} cm^{-3}. Some phase change materials such as GeTe show a metallic behavior of the conductivity with a positive temperature coefficient of resistance. Other phase change materials, however, including $Ge_2Sb_2Te_5$, can show either nonmetallic or metallic behavior with negative or positive temperature coefficient of resistance, respectively, depending on their annealing temperature [39]. This change in temperature coefficient of resistance is connected to a metal–insulator transition observed in these phase change materials.

Scaling properties of phase change materials are among the most important properties since they determine the viability of the PCRAM technology. Scaling studies have been performed on phase change ultra-thin films [40, 41], phase change nanowires [42], phase change nanoparticles [43], and ultra-scaled devices [14–16, 21, 44]. Phase change materials show remarkable scaling behavior. The ultimate limit where they stop being phase change materials, i.e., do not exist

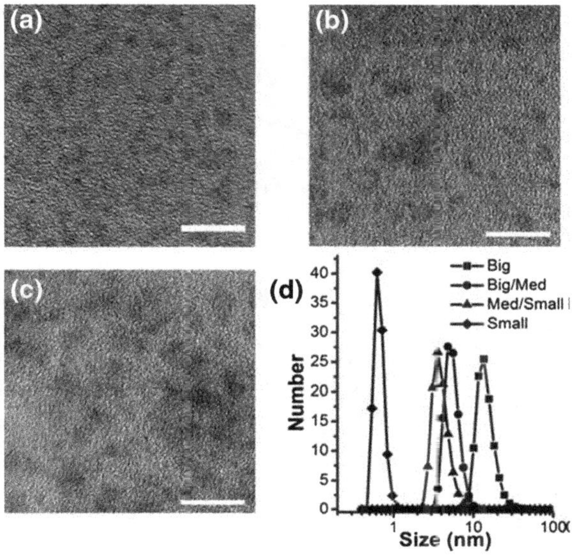

Fig. 5.5 TEM images of size-selected samples. All scale bars are 10 nm. **a** Small nanoparticles of 1.8 ± 0.44 nm in diameter, **b** medium nanoparticles of 2.6 ± 0.39 nm, **c** large nanoparticles of 3.4 ± 0.74 nm and **d** another instance of size-selected nanoparticles. The size distribution was measured by dynamic light scattering. In this case, four samples were obtained using an identical procedure. Reprinted with permission from [43]

anymore in both phases, is on the order of about 1.5–2 nm [40, 41, 43, 44]. This is quite remarkable because such a length scale is just three to four times the crystal lattice constant. Figure 5.5 shows transmission electron microscope (TEM) images of GeTe nanoparticles of various sizes [43]. These particles were amorphous as synthesized and could be crystallized by heating. The crystallization temperature increased when particle size was reduced, this is ideal scaling behavior.

In addition, interfaces play an important role in scaling and phase change crystallization properties depend very strongly on the interface materials [45]. Crystallization times as well as crystallization temperatures can be increased or reduced depending on the substrate and capping layer. Interface tailoring will be increasingly important for ultra-scaled PCRAM devices.

5.4 PCRAM Cell Design and Optimization

PCRAM devices have been fabricated in a large variety of designs. Table 5.1 gives an overview and comparison of the various designs [46]. The phase change memory cell which stores the information needs to be programmed and read by an access device that delivers the required current pulses and is in series with the PCRAM cell. As mentioned before, in most cases this access device is a field effect transistor. The largest current that is required from the access device is for the reset operation. In addition to accessing the PCRAM cell, the access device also isolates the cell when it is not being addressed.

Table 5.1 Characteristics of phase change memory cell device structures. Reprinted with permission from [46]

Reference no.	Structure	Contact area (nm^2)	Technical node (nm)	Ireset (μA)	Schematic view
Pirovano et al. [47][a]	Mushroom	~300–3,000	N/A	~50–1,000	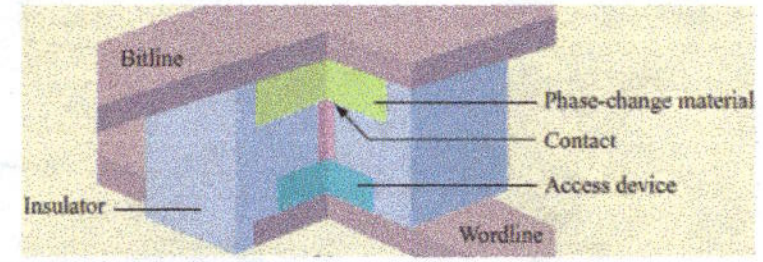 [50]
Sasago et al. [48]	Mushroom	707	80	160	
Pellizzer et al. [49]	Mushroom	3,000	90	700	
Pellizzer et al. [49]	μ-trench	400	90	400	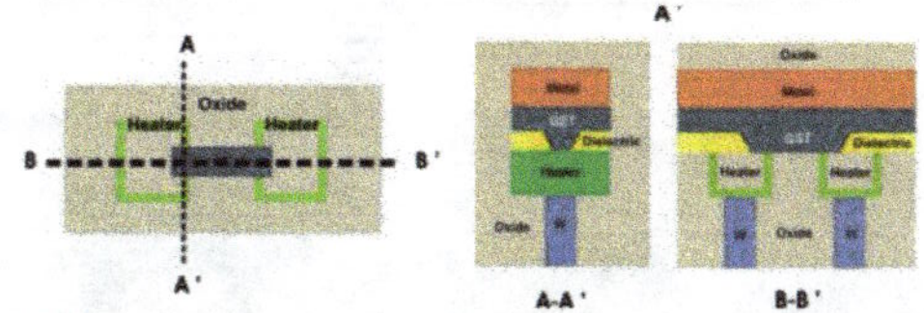 [52]
Pellizzer et al. [51][a]	μ-trench	N/A	180	600	
Oh et al. [53][a]	SA μ-trench (wall)	~400	90	300	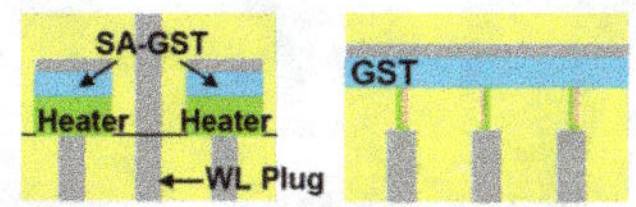 x-direction y-direction [55] *x*-direction *y*-direction
Giraud et al. [54][a]	SA μ-trench (wall)	N/A	45	200	

(continued)

Table 5.1 (continued)

Reference no.	Structure	Contact area (nm^2)	Technical node (nm)	Ireset (μA)	Schematic view
Breitwisch et al. [56]	Pore	1257	180	250	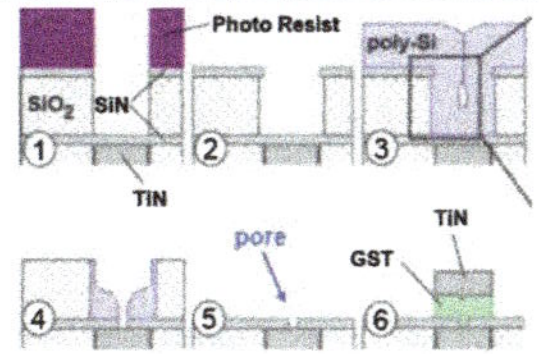 [56]
Chen et al. [57]	Cross-spacer	500	180	80	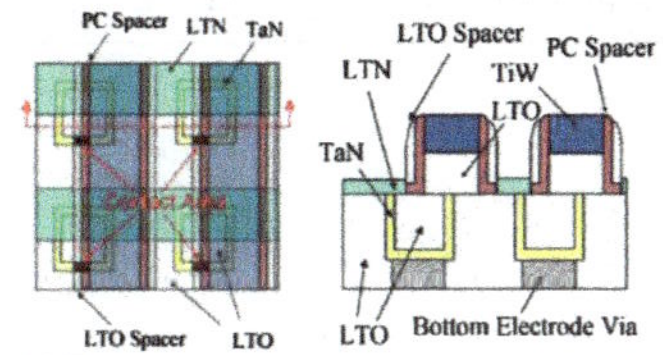 [57]
Chen et al. [57]	Cross-spacer	1,000	180	230	
Ha et al. [58]	Edge	4,000	240	200	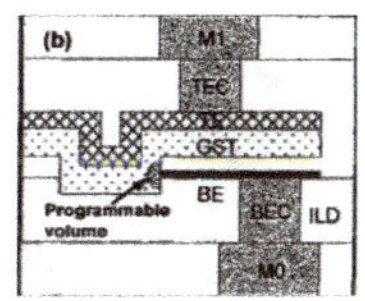 [58]
Oh et al. [59][a]	Ring-shape contact	~1,300	100	600	
Ahn et al. [60][a]	Ring-shaped contact	~1300	90	600	top view 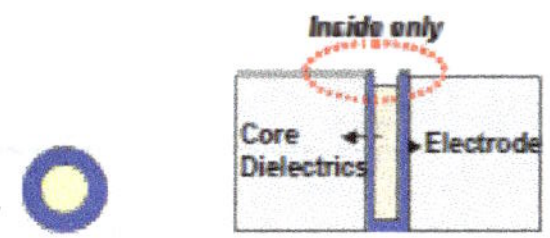 [60]

(continued)

Table 5.1 (continued)

Reference no.	Structure	Contact area (nm^2)	Technical node (nm)	Ireset (μA)	Schematic view
Song et al. [61][a]	Advanced ring-shaped contact	1500	100	450	top view 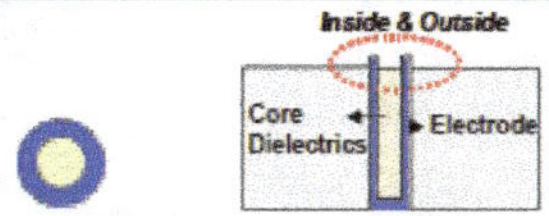 [61]
Hwang et al. [62]	Confined	4418	240	700	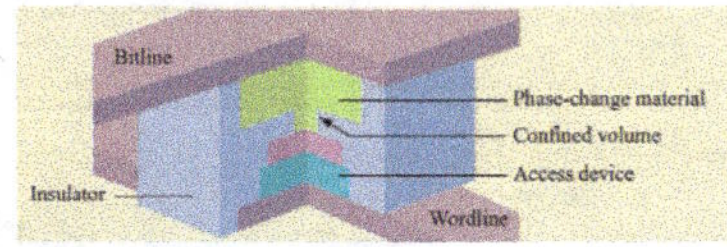 [50]
Lee et al. [63]	Confined	1963	45	260	
Im et al. [64]	Dash-type confined	487.5	<20	160	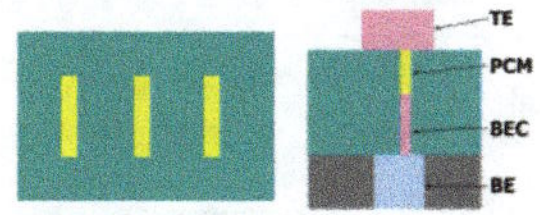 [64]

[a] Indicates the data are either extracted from figures or are estimated from referenced papers, and not directly provided in the literature

One of the simplest and frequently applied designs is the so-called mushroom cell (top line in Table 5.1). A block of phase change material is contacted on the top with a large and on the bottom with a small contact electrode. This leads to a high current density just above the bottom contact. The highest temperature is reached in a mushroom cap shaped area volume above the bottom contact which is switching its phase during set and reset operations giving this design its name. In practical terms one of the most important PCRAM cell parameters is the reset current. Many design aspects and material choices influence the reset current.

Successful and often employed design strategies for reduced reset current include a reduction of the contact area between one of the electrodes and the phase change materials, changes in the thermal properties of the phase change material, better confinement of the heat in the phase change material, or any combination of these principles. Comprehensive overviews on PCRAM prototype designs, cell concepts, and process integration can be found elsewhere [29, 65, 66].

As it can be seen from Table 5.1, a very successful way to reduce the reset current has been the confinement of the phase change material in a small area, e.g., the dash-type confined cell. Typically, this cell fabrication requires deposition of the material in vias. This is a challenge for sputter deposition which is the most common phase change material deposition method. Alternative deposition methods have been developed to address this issue [67]. They include chemical vapor deposition (CVD) [63], electrodeposition [68], solution-phase deposition [69], the application of bottom-up nano-materials [42], and atomic layer deposition [70]. Among those methods atomic layer deposition (ALD) is particularly promising for ultra-scaled devices, because it allows for deposition with atomic layer control of thickness and composition and very high conformality. The following sections of this chapter are devoted to ALD of phase change materials.

5.5 ALD of Phase Change Materials

Like always, also with phase change materials, ALD succeeds or fails with chemistry. The unique benefits of the ALD method, conformality, uniformity, and thickness control in particular, can be explcited only after finding appropriate source compounds that fulfill the specific requirements set to the ALD precursors (Chap. 3, references [71, 72]).

Research on ALD of phase change materials has so far mostly focused on GST and its binary constituents GeTe and Sb_2Te_3. Before the interest in ALD of GST arose from the phase change materials community, all these three elements, Ge, Sb, and Te, had remained sparsely studied in ALD, and even the few existing studies had little relevance to GST [73]. In particular, the lack of a good tellurium precursor had been a major problem because hydrogen telluride, member of the otherwise widely used nonmetal hydrides, is highly toxic and quite unstable. Elemental Te had been used in ALD with elemental Zn and Cd, but besides these two only Hg would be a volatile enough metal for use as a precursor in ALD at reasonably low temperatures. Therefore, the development of ALD processes for GST and related materials needed to start from essentially scratch.

The chemistries explored so far for ALD of GST can be divided into two groups. The first studies exploited precursors that have been typically used in CVD, namely organometallics and alkylamides, the common precursor in all these studies being $^{i}Pr_2Te$ [74–80]. These compounds seem to lack the desired high reactivity toward each other, however, and therefore they have been explored mostly in plasma-enhanced ALD. Even with plasma activation it appears that no true self-limiting ALD was achieved but the processes involved significant CVD contributions. A breakthrough in ALD of GST and related phase change materials, and more broadly metal tellurides and selenides in general, occurred when it was found that alkyl silyl compounds of tellurium and selenium react efficiently with many metal chlorides, including those of germanium and antimony [81], thereby better filling the ALD requirements. In the following, a review will be made of ALD of GST and related materials from alkyl silyl tellurium precursors. This will be followed by a brief summary of studies based on alkyl tellurium precursors.

5.5.1 ALD of Phase Change Materials with Alkyl Silyls as Tellurium Precursors

Film growth. A major step toward ALD of metal tellurides and selenides was made when alkyl silyl compounds with a general formula $(R_3Si)_2M$ (M = Te, Se) were found to meet the key requirements of ALD precursors: Volatility, thermal stability, and high reactivity [81]. The compounds with the smallest alkyl groups, e.g., $(Me_3Si)_2Te$ and $(Et_3Si)_2Te$ are liquids at room temperature, whereas $(^{t}BuMe_2Si)_2Te$ is a low melting point (44 °C) solid (Fig. 5.6). These compounds react efficiently with various metal halides forming the corresponding metal tellurides and selenides. Sb_2Te_3, GeTe, and GST films were deposited by ALD at remarkably low temperature of 90 °C using $(Et_3Si)_2Te$, $SbCl_3$, and $GeCl_2{\cdot}C_4H_8O_2$ as precursors, where $C_4H_8O_2$ is a neutral dioxane adduct ligand.

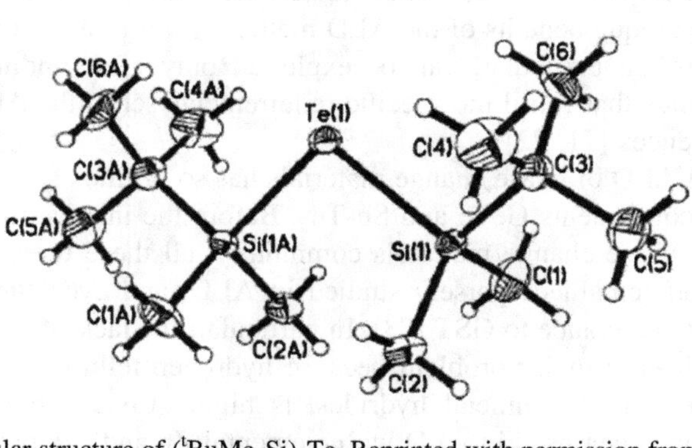

Fig. 5.6 Molecular structure of $(^{t}BuMe_2Si)_2Te$. Reprinted with permission from [81]. Copyright (2009) American Chemical Society

$(R_3Si)_2Te$ offers a straightforward exchange reaction route to the deposition of the metal telluride and efficient elimination of the chloride ligands of the metal precursor, the overall reaction being:

$$(R_3Si)_2\,Te(g) + \; MCl_2(g)\; MTe(s) + \; 2R_3SiCl(g) \qquad (5.1)$$

As the volatile by-product forming in the reaction is alkyl silicon chloride, the reaction can be called as a dechlorosilylation reaction. The facility of this reaction can be understood on basis of the Lewis hard-soft acid–base (HSAB) concept. It may be worth emphasizing that the HSAB concept is just a qualitative concept for assessing reaction feasibilities, and does not imply anything about the reaction mechanism. In the $(R_3Si)_2Te$ molecule silicon, a hard Lewis acid, is bonded to tellurium, a heavy group 16 element which is a soft Lewis base, thus forming an unfavorable Lewis hard-soft acid–base pair. Upon exchange reaction with metal chlorides, silicon becomes bonded more favorably to the harder base of chlorine. Further, when the metal is a soft or borderline Lewis acid, like Ge^{2+} and Sb^{3+}, it serves as a more favorable pair for the soft Lewis base of tellurium in the solid product of the exchange reaction. On the other hand, while also ligands with oxygen donors (e.g., alkoxides, betadiketonates) are classified as hard Lewis bases, little success has been reported so far in using metal precursors with these ligands in place of metal chlorides.

In situ reaction mechanism studies with a quartz crystal microbalance and quadrupole mass spectrometer showed that the ligand exchange and formation of R_3SiCl as a by-product occur stepwise during both metal chloride and alkyl silyl tellurium pulses [82]. For example, in the Sb_2Te_3 process the half reactions were:

$$4{:}6\text{-}Te(SiEt_3)(s) + \; 2\; SbCl_3(g) \rightarrow \text{-}Te_{4.6}Sb_2Cl_{1:4}(s) + \; 4{:}6Et_3SiCl(g) \qquad (5.2a)$$

$$\text{-}Te_{4.6}Sb_2Cl_{1:4}(s) + 3(Et_3Si)_2Te(g) + \text{-}Te_3Sb[Te(SiEt_3)]_{4.6}(s) + 1.4Et_3SiCl(g) \qquad (5.2b)$$

Temperature-wise the Sb_2Te_3 and GeTe processes showed peculiar features compared to the ALD processes in general. The growth of Sb_2Te_3 occurred with a good rate of about 0.6 Å/cycle at a temperature as low as 60 °C but at higher temperatures the growth rate dropped rapidly. Also with GeTe the highest rate was achieved at the lowest possible deposition temperature of 90 °C, being limited by the $GeCl_2{\cdot}C_4H_8O_2$ source temperature that needed to be 70 °C [81, 83]. Above 90 °C, the GeTe deposition rate decreased rapidly reaching a zero level at already 150 °C. Reasons for these temperature dependences are still under discussion but they seem to be related to these specific film materials rather than this chemistry in general, because zinc telluride could be grown at a high temperature of 400 °C using the same tellurium precursor and $ZnCl_2$ [81].

Both Sb_2Te_3 and GeTe binary processes showed the self-limiting behavior as expected in ALD, i.e., the growth rates saturated when the precursor pulses were made longer. Despite the low deposition temperatures, the saturation level was reached with reasonably short pulse times of only 1 s (Fig. 5.7).

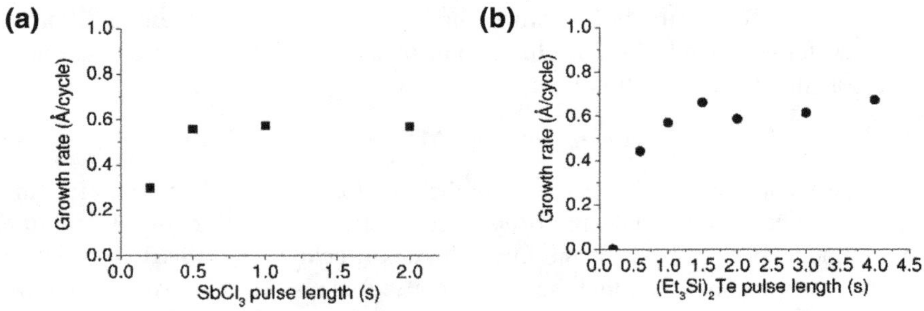

Fig. 5.7 Saturation of growth rate with respect to $SbCl_3$ (**a**) and $(Et_3Si)_2Te$ pulse lengths (**b**). Reprinted with permission from [81]. Copyright (2009) American Chemical Society

Ternary GST films were deposited by mixing the binary cycles of GeTe and Sb_2Te_3. A growth rate of about 0.30 Å/cycle was obtained where the cycle refers to a sum of binary cycles of Ge–Te and Sb–Te [70, 81]. In general, in ALD the control of the ternary composition is much more challenging compared to the binaries which usually adopt automatically the correct composition. A common way for controlling the ternary composition in ALD is to vary the cycle ratio of the binary constituents. Ideally, the film should adopt the stoichiometry corresponding to the composition of the stable ternary compound. However, because GST and related ternary materials are more alloy-like and many stoichiometric compositions exist along the pseudo-binary line between GeTe and Sb_2Te_3 ($Ge_1Sb_2Te_4$, $Ge_2Sb_2Te_5$, $Ge_3Sb_2Te_6$, etc.), there is no strong driving force toward a particular stoichiometry. Indeed, in the ternary process, the antimony content increased and the germanium and tellurium contents decreased with increasing $SbCl_3$ exposure times without approaching any specific composition. Anyhow, by adjusting the ratio of the binary cycles Ge–Te and Sb–Te, the GST composition could be controlled close to the stoichiometric $Ge_2Sb_2Te_5$. If desired, the composition can be adjusted to other stoichiometries on the GeTe–Sb_2Te_3 tie line too by just changing the cycle ratio of the binary constituents.

Subsequently, the dechlorosilylation reaction was extended to elemental antimony and various metal antimonides, also those of germanium [84]. Elemental antimony films were deposited from $SbCl_3$ and $(Et_3Si)_3Sb$, while germanium antimonides were obtained from $GeCl_2{\cdot}C_4H_8O_2$ and $(Et_3Si)_3Sb$. The composition of the latter was quite germanium rich with 82 at. % Ge and 18 at. % Sb, implying that in this case the reactions were more complicated than just the simple ligand exchange reaction, most likely involving also some antimony to germanium substitution. Contrary to the Ge–Sb–Te system, Ge–Sb has no stoichiometric compound and tends to elemental segregation. Anyhow, by mixing ALD cycles of elemental Sb with those of $Ge_{82}Sb_{18}$, the film composition could be tuned in between these two extremes. When also GeTe and/or Sb_2Te_3 ALD cycles were added, the ternary Ge–Sb–Te composition could be tuned over a wide range of compositions in the ternary phase diagram (Fig. 5.8).

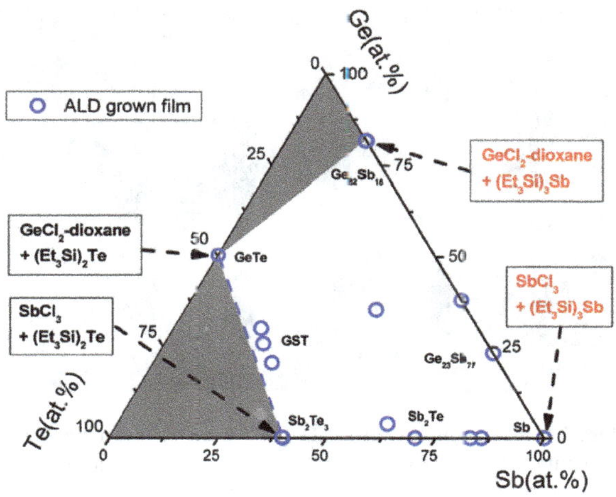

Fig. 5.8 Ge–Sb–Te ternary phase diagram of films deposited by ALD. The white area indicates compositons accessible by ALD and the *blue circles* actual film compositions measured by EDX. Reprinted with permission from [84]. Copyright (2009) American Chemical Society

Film properties. Despite the low deposition temperature, the GST films contained reasonably low impurity contents of about 2.4 at. % oxygen, 1.0 at. % hydrogen, 0.7 at. % carbon, and 0.6 at. % chlorine as measured by time-of-flight energy recoil detection analysis (TOFERDA) [70, 70, 81]. Elemental Sb films had similar amounts of impurities but no chlorine could be detected with ERDA in these films [84]. Because the films analyzed were on silicon, the silicon content was more difficult to measure, but it appeared to be below 1 at. %. Later studies with Secondary Ion Mass Spectrometry (SIMS) verified low levels of silicon and carbon in the GST films [85].

As expected from ALD, the films showed excellent conformality in all structures tested so far. Complete filling of the sub-100 nm holes in actual PCRAM device structures was also demonstrated (Fig. 5.9).

The binary Sb_2Te_3 and elemental Sb films were found to grow in polycrystalline form, while with GeTe both amorphous and polycrystalline films have been obtained; the difference probably arising from film thickness with the thinnest films being completely amorphous. The ternary GST has usually been amorphous in the as-deposited state which is a fortunate consequence of the low deposition temperature, because extensive crystallization during film growth would complicate void-free filling of the three-dimensional PCRAM device structures. In some antimony-rich GST films, a few isolated crystalline grains could be observed, though.

High temperature XRD measurements under a nitrogen flow were exploited for studying the crystallization of the amorphous films. With GST, the crystallization onset was at around 130 °C (Fig. 5.10). First the metastable rock salt (fcc) structure was formed but at 250–300 °C the structure transformed to the stable hexagonal phase [70]. The GeTe films crystallized at 165 °C to the rhombohedral phase which further on transformed to the cubic phase [83]. The crystallization of the ALD GST and GeTe films is essentially similar to the sputtered films, as is

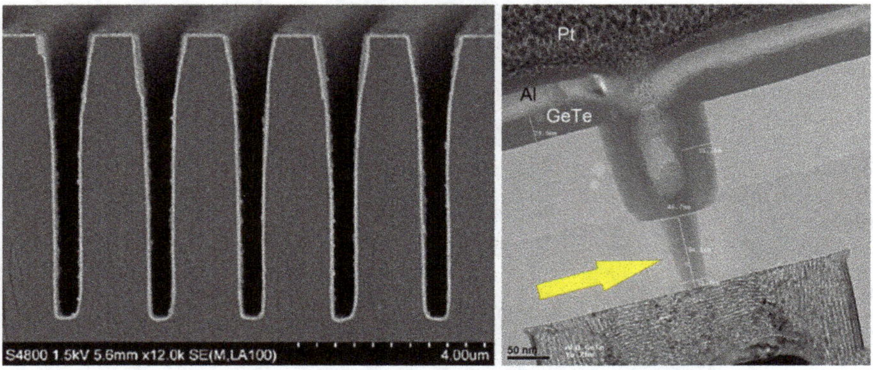

Fig. 5.9 Conformality of ALD GST films in deep trenches and filling of a PCRAM device structure with ALD GeTe. *Left image* reprinted from Ritala et al. [70] Copyright (2009), with permission from Elsevier. *Right image* reproduced from Sarnet et al. [83] by permission of ECS—the Electrochemical Society

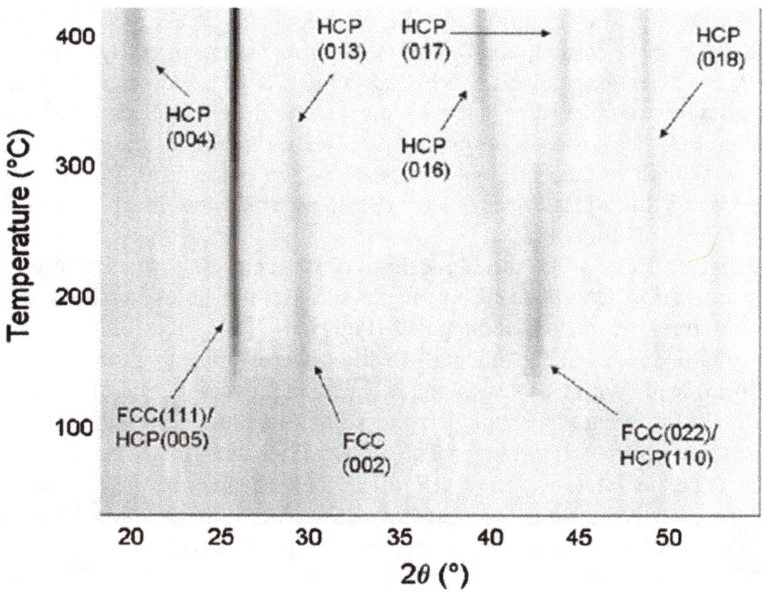

Fig. 5.10 High temperature XRD measurement on an ALD GST film (23 at.% Ge, 28 at.% Sb, 49 at.% Te). Reprinted from Ritala et al. [70] Copyright (2009), with permission from Elsevier

their resistivity versus temperature behavior (Fig. 5.11). The drops in resistivity correlate with phase changes observed with high temperature XRD. More than five orders of magnitude change in resistivity occurs upon crystallization. A second smaller drop in resistivity around 330 °C is correlated to the transition from the

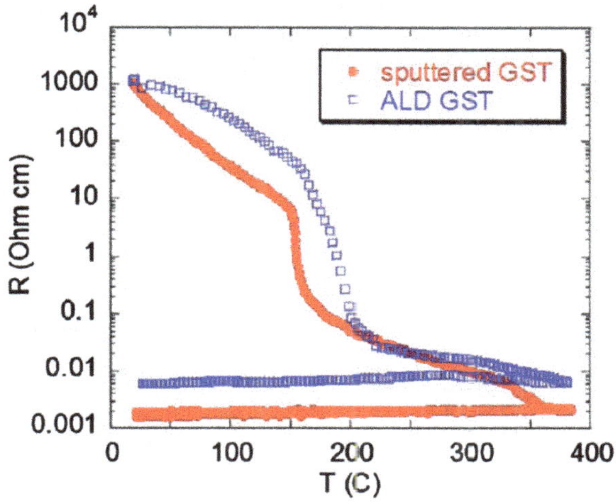

Fig. 5.11 Temperature dependence of resistivity of GST films made by ALD and sputtering. Reprinted from Ritala et al. [70] Copyright (2009), with permission from Elsevier

rocksalt to the hexagonal phase. As compared to the sputtered film, the curve of the ALD GST is shifted to both higher temperature and toward higher resistivities at each state of crystallization. These shifts are most likely due to the impurities or the off-stoichiometric composition of this particular ALD sample. Especially, oxygen impurities are known to increase the crystallization temperature of GST. Fortunately, both shifts may turn out beneficial: The shifts toward higher temperatures increase the stability of the amorphous phase against unintentional crystallization, while the higher resistivities provide more efficient Joule heating in the set and reset processes. An obvious concern related to the impurities is their effect on long-term stability, a topic yet to be studied.

Laser-based studies provide valuable insights into the repeated crystallization of phase change materials without the requirement of fabricating complex devices. One pulsed, high-power laser beam is used to crystallize or amorphize a spot on the film, while another lower power CW laser measures related changes in reflectivity. By varying the laser pulse power and width one can control whether the film attains an amorphous or crystalline state upon cooling, and also allows detailed studies on crystallization kinetics (Fig. 5.12). Full crystallization of the as-deposited GST films occurred within 150–1000 ns depending on the underlayer. Recrystallization after melt-quenching areas in an annealed crystalline film occurred always faster than the crystallization of the original amorphous film. This is due to the fact that crystallization of an amorphous as-deposited film requires crystal nucleation and growth, while recrystallization of an amorphous spot in a crystalline film does not require nucleation (which can be slow) and can proceed by crystal growth from the amorphous-crystalline interface. Also in this respect, the ALD GST films behave similarly to the sputtered GST. ALD GeTe films crystallize from the as-deposited amorphous state in 300 ns, again similar to their sputtered counterparts, but recrystallization from the melt-quenched films is faster (50 ns) with the ALD films [83].

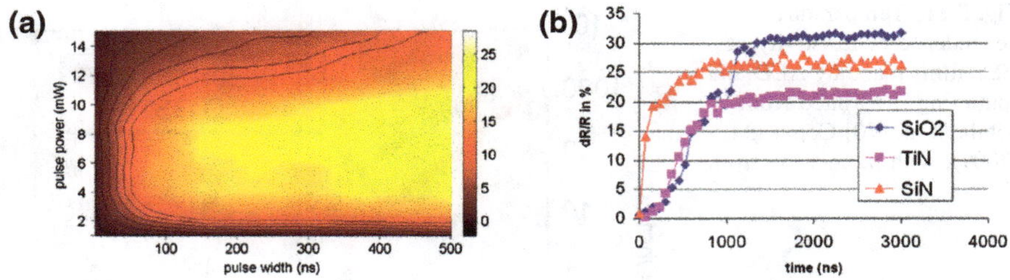

Fig. 5.12 **a** Change of reflectivity (in %) of an as-deposited ALD GST film on SiN_x exposed to laser pulses with different power and pulse widths. **b** Comparison of rate of change of reflectivity R of GST films on SiN_x, SiO_x and TiN underlayers each being measured at the optimum pulse power for fastest crystallization. Reprinted from Ritala et al. [70] Copyright (2009), with permission from Elsevier

An ultimate test for the applicability of the ALD made phase change material films is of course their performance in solid state devices where the crystallization and amorphization are done by electric current rather than laser pulses. While the above cited results were from films made in small research scale reactor on $50 \times 50\ mm^2$ substrates; subsequently, the GST ALD process was scaled up to 200 mm wafers with less than 3 % 1σ thickness nonuniformity, and integrated PCRAM devices were fabricated [85]. With these devices it was demonstrated that the memory cells could be switched between the low and high resistivity states up to 10^6 cycles with a very stable resistance for both states.

5.5.2 ALD of Phase Change Materials with Alkyls as Tellurium Precursors

As noted above, the other group of studies toward ALD of GST materials adopted precursors commonly used in CVD, i.e., organometallics and alkylamides. $^iPr_2Te$ is a precursor used in all these studies. Another common feature of these processes is that the ternary deposition was attempted, at least conceptually, by combining deposition cycles of elementary films, whereas with the alkyl silyl tellurium precursor this was done by combining binary processes.

Lee et al. [74] deposited GST films by PEALD from $Ge(NMe_2)_4$, $Sb(NMe_2)_3$, $^iPr_2Te$, and hydrogen plasma. First proper conditions were identified for depositing each element on its own using a pulsing sequence: precursor pulse–purge–H-plasma–purge. The ternary GST was obtained by mixing the deposition cycles of each element at 250–350 °C. To compensate for the low incorporation rate of Te, seven times higher pulse times were used for $^iPr_2Te$ than for the other two precursors. The as-deposited films were crystalline GST with unspecified phase, but after annealing crystalline elemental Te phase was detected too, most likely as segregates on the surface. Phase change properties were not reported. Also, silylamides were tested as precursors for germanium and antimony, but they exhibited severe decomposition and contaminated the films with silicon, so that the resulting Ge–Si–Sb–Te films did not crystallize even upon annealing at 500 °C.

Choi et al. have produced a series of papers on cyclic PECVD of GST using a precursor combination of $Ge^{i}Bu_4$, $Sb^{i}Pr_3$, and $^{i}Pr_2Te$ with H_2 as a reduction gas [75–80]. After each precursor pulse a reduction gas pulse was supplied, and a radio frequency plasma was applied during both the precursor and reduction gas pulse periods. It is uncommon in plasma-enhanced ALD to apply plasma during a metal precursor pulse, because plasma leads to a fragmentation of the metal precursor into various species, some of which are not volatile and thereby through condensation violate the ALD principle of self-limiting film growth. Indeed, it was observed that Sb and Te were deposited without any saturation [76]. Somewhat surprisingly, Ge deposition did saturate with long enough pulses. This might be due to the fact that even when activated with plasma, germanium does not form condensing species and Ge atoms are deposited only by reacting with Sb and Te atoms on the surface. Anyhow, it was concluded that of the three constituent elements only Ge was deposited in ALD manner while the Sb and Te were deposited by CVD. Consequently, the process is not genuine ALD but a combination of plasma-assisted ALD and cyclic PECVD. Partial ALD cannot provide the benefits of true ALD, however, and therefore this process is not considered further here. The process does exhibit many interesting features like selective growth though [76].

With the same precursors, deposition was also attempted with the plasma on only during the reduction gas (H_2) pulse [76]. Sb and Te growths were possible this way, though less effective than growth with plasma on during also $Sb^{i}Pr_3$ and $^{i}Pr_2Te$ pulses. Germanium deposition, on the other hand, was not possible without plasma activation of $Ge^{i}Bu_4$ up to 450 °C, the highest wafer temperature tested.

5.6 Summary and Outlook

Phase change random access memory is in a very good position to be a contender in the race for the next generation memory technology. The features that make it a promising candidate for Flash memory replacement include very fast switching times, excellent scalability, and high cycle numbers. Other applications such as embedded memory or storage class memory are also being considered. If cycle numbers can be further increased even replacement of DRAM could be possible which would enable very new computer architectures, because it would replace the volatile DRAM by a nonvolatile technology. Key to the development of successful PCRAM technology is the optimization of the phase change materials and their deposition methods, because the phase change material properties determine to a large degree of the functionality and capabilities of PCRAM technology, and the available deposition technique determines which lithographic processes can and need to be applied in the fabrication of the PCRAM chips. The availability of specific deposition techniques dictates many aspects of the memory cell design and can enable certain cell designs that are optimized in terms of switching speed, power consumption, and other performance aspects.

ALD offers to PCRAM manufacturing many advantages compared to other techniques such as very high conformality and atomic precision of film thickness and composition. Compared to CVD, less reactant flux homogeneity is required. The low deposition rate is often named as a disadvantage of ALD but with continued scaling this will become in fact an advantage, because thinner and thinner films are typically required for smaller technology nodes and low rates but excellent control over thickness are beneficial.

ALD processes have been developed for the phase change materials based on the Ge–Sb–Te system. In a search of appropriate ALD chemistry for these materials the finding of alkyl silyl compounds as efficient tellurium precursors when combined with metal chlorides was instrumental. ALD of various thin films such as $Ge_2Sb_2Te_5$, GeTe, Sb_2Te_3, and Sb has been demonstrated. The ALD $Ge_2Sb_2Te_5$ and GeTe films show switching properties that are comparable to sputter deposited phase change materials. The great advantage of ALD deposited phase change materials compared to sputter deposited films is their excellent conformality. This will enable new PCRAM cell designs, in particular those that require a very confined geometry of the phase change material. Such a confined geometry was shown to greatly reduce the switching currents because much less heat is lost to the electrodes. In addition, such a confined geometry increases the cycle numbers, because for very small switching volumes electromigration and void formation is less of a problem. ALD might, therefore, play an important role in future PCRAM fabrication.

References

1. Ovshinsky SR (1968) Reversible electrical switching phenomena in disordered structures. Phys Rev Lett 22:1450–1453
2. Ovshinsky SR (1966) Symmetrical current controlling device. US Patent 3,271,591
3. Yamada N, Ohno E, Akahira N, Nishiuchi K, Nagata K, Takao M (1987) High speed overwritable phase change optical disk material. Jpn J Appl Phys 26(Suppl 26–4):61–66
4. Yamada N (2009) Development of materials for third generation optical storage media. In: Raoux S, Wuttig M (eds) Phase change materials: science and applications. Springer, New York
5. Shi L (2009) Optical memory: from 1st to 3rd generation and its future. In: Raoux S, Wuttig M (eds) Phase change materials: science and applications. Springer, New York
6. Wuttig M, Yamada N (2007) Phase-change materials for rewriteable data storage. Nat Mater 6:824–832
7. Meinders ER, Mijiritskii AV, van Pieterson L, Wuttig M (2006) Optical data storage—phase-change media and recording. Springer, Eindhoven
8. http://www.samsung.com/us/aboutsamsung/news/newsIrRead.do?news_ctgry=irnewsrelease&page=1&news_seq=18828&rdoPeriod=All&from_dt=&search_keyword=

9. Pellizzer F, Benvenuti A, Gleixner B, Kim Y, Johnson B, Magistretti M, Marangon M, Pirovano A, Bez R, Atwood G (2006) A 90 nm phase change memory technology for stand-alone non-volatile memory applications. In: Symposium on VLSI Technology, pp 122–123
10. Kang S, Cho WY, Cho BH, Lee KJ, Lee CS, Oh HR, Choi BG, Wang Q, Kim HJ, Park MH, Ro YH, Kim S, Ha CD, Kim KS, Kim YR, Kim DE, Kwak CK, Byun HG, Jeong G, Jeong H, Kim K, Shin Y (2007) A 0.1 μm 1.8 V 256-Mb phase change random access memory (PRAM) with 66 MHz synchronous burst-read operation. IEEE J Solid-State Circuits 42:210–218
11. Oh JH, Park JH, Lim YS, Lim HS, Oh YT, Kim JS, Shin JM, Park JH, Song YJ, Ryoo KC, Lim DW, Park SS, Kim JI, Kim JH, Yu J, Yeung F, Jeong CW, Kong JH, Kang DH, Koh GH, Jeong GT, Jeong HS, Kim K (2006) Full integration of highly manufacturable 512 Mb PRAM based on 90 nm technology. In: IEDM technical digest, p 2.6
12. Ielmini D (2008) Threshold switching mechanism by high-field energy gain in the hopping transport of chalcogenide glasses. Phys Rev B 78:035308
13. Krebs D, Raoux S, Rettner CT, Burr GW, Salinga M, Wuttig M (2009) Threshold field of phase change memory materials measured using phase change bridge devices. Appl Phys Lett 95:082101
14. Lankhorst MHR, Ketelaars BWSMM, Wolters RAM (2005) Low-cost and nanoscale non-volatile memory concept for future silicon chips. Nat Mater 4:347–352
15. Attenborough K, Hurkx GAM, Delhougne R, Perez J, Wang MT, Ong TC, Tran L, Roy D, Gravesteijn DJ, van Duuren MJ (2011) Phase change memory line concept for embedded memory applications. In: International Electron Devices Meeting
16. Chen YC, Rettner CT, Raoux S, Burr GW, Chen SH, Shelby RM, Salinga M, Risk WP, Happ TD, McCleland GM, Breitwisch M, Schrott A, Philipp JB, Lee MH, Cheek R, Nirschl T, Lamorey M, Chen CF, Joseph E, Zaidi S, Yee B, Lung HL, Bergmann R, Lam C (2006) Ultra-thin phase-change bridge memory device using GeSb. In: IEDM technical digest, pp 777–780
17. Raoux S, Cabral C Jr, Krusin-Elbaum L, Jordan-Sweet JL, Virwani K, Hitzbleck M, Salinga M, Madan A, Pinto TL (2009) Phase transitions in Ge-Sb phase change materials. J Appl Phys 105:064918
18. Krusin-Elbaum L, Shakhvorostov D, Cabral C, Raoux S, Jordan-Sweet JL (2010) Irreversible altering of crystalline phase of phase-change Ge–Sb thin films. Appl Phys Lett 96:121906
19. Yamada N, Kojima R, Uno M, Akiyama T, Kitaura H, Narumi K, Nishiuchi K (2002) Phase-change material for use in rewritable dual-layer optical disk. In: Proceedings of SPIE vol 4342, pp 55–63
20. Krbal M, Matsunaga T, Kolobov AV, Fons P, Kohara S, Honma T, Simpson RE, Tominaga J, Yamada N (2010) Local structure of amorphous $Ge_8Sb_2Te_{11}$. In: Proceedings of the European symposium on phase change and ovonic science, Milano, Italy
21. Bruns G, Merkelbach P, Schlockermann C, Salinga M, Wuttig M, Happ TD, Philipp JB, Kund M (2009) Nanosecond switching in GeTe phase change memory cells. Appl Phys Lett 95:043108
22. Fallica R, Varesi E, Fumagalli L, Grasso S, Erbetta D, Spadoni S, Longo M, Wiemer C, Fanciulli M (2010) Thermal and electrical conductivity of amorphous and crystalline GeTe-N compounds. In: Proceedings of the European symposium on phase change and ovonic science, Milano, Italy
23. Fantini A, Sousa V, Perniola L, Gourvest E, Bastien JC, Maitrejean S, Braga S, Pashkov N, Bastard A, Hyot B, Roule A, Persico A, Feldis H, Jahan C, Nodin JF, Blachier D, Toffoli A, Reimbold G, Fillot F, Pierre F, Annunziata R, Benshael D, Mazoyer P, Vallée C, Billon T, Hazart J, De Salvo B, Boulanger F (2011) N-doped GeTe as performance booster for embedded phase-change memories. In: International electron devices meeting
24. Shelby RM, Raoux S (2009) Crystallization dynamics of nitrogen-doped $Ge_2Sb_2Te_5$. J Appl Phys 105:104902
25. Chen YC, Che CT, Yu JY, Lee CY, Chen CF, Lung SL, Liu R (2003) 180 nm Sn-doped $Ge_2Sb_2Te_5$ chalcogenide phase-change memory device for low power, high speed embedded

memory for SoC applications. In: Proceedings of custom integrated circuits conference (CICC), pp 395–398
26. Matsuzaki N, Kurotsuchi K, Matsui Y, Tonomura O, Yamamoto N, Fujisaki Y, Kitai N, Takemura R, Osaka K, Hanzawa S, Moriya H, Iwasaki T, Kawahara T, Takaura N, Terao M, Matsuoka M, Moniwa M (2005) Oxygen-doped GeSbTe phase-change memory cells featuring 1.5 V/100 μA standard 0.13-μm CMOS operations. In: IEDM technical digest, pp 757–780
27. Czubatyj W, Lowrey T, Kostylev S, Asano I (2006) Current reduction in ovonic memory devices. In: Proceedings of the European symposium on phase change and ovonic science, Grenoble, France
28. Hwang YN, Lee SH, Ahn SJ, Lee SY, Ryoo KC, Hong HS, Koo HC, Yeung F, Oh JH, Kim HJ, Jeong WC, Park JH, Horii H, Ha YH, Yi JH, Koh GH, Jeong HS, Kim K (2003) Writing current reduction for high-density phase-change RAM In: IEDM technical digest, pp 893–896
29. Breitwisch M (2009) Phase change random access memory integration. In: Raoux S, Wuttig M (eds) Phase change materials: Science and applications. Springer, New York
30. Shportko K, Kremers S, Woda M, Lencer D, Robertson J, Wuttig M (2008) Resonant bonding in crystalline phase-change materials. Nat Mater 7:653–658
31. Wuttig M, Lüsebrink D, Wamwangi D, Welnic W, Gillessen M, Dronskowski R (2007) The role of vacancies and local distortions in the design of new phase-change materials. Nat Mater 6:122–128
32. Lencer D, Salinga M, Grabowski B, Hickel T, Neugebauer J, Wuttig M (2008) A map for phase-change materials. Nat Mater 7:972–977
33. Lencer D, Salinga M, Wuttig M (2010) Design rules for phase-change materials in storage applications. Adv Mater 2011:23
34. Matsunaga T, Yamada N, Kubota Y (2004) Structures of stable and metastable $Ge_2Sb_2Te_5$, an intermetallic compound in GeTe-Sb_2Te_3 pseudobinary systems. Acta Cryst B60:685–691
35. Karpinsky O, Shelimova L, Kretova M, Fleurial J (1998) An X-ray study of the mixed-layered compounds of $(GeTe)_n(Sb_2Te_3)_m$ homologous series. J Alloys Compd 268:112–117
36. Chattopadhyay T, Boucherle J, Vonschnering H (1987) Neutron-diffraction study on the structural phase-transition in GeTe. J Phys C 20:1431–1440
37. Raoux S, Munoz B, Cheng H-Y, Jordan-Sweet JL (2009) Phase transitions in Ge–Te phase change materials studied by time-resolved X-ray diffraction. Appl Phys Lett 95:143118
38. Lee B-S, Abelson JR, Bishop SG, Kang D-H, Cheong B, Kim K-B (2005) Investigation of the optical and electronic properties of $Ge_2Sb_2Te_5$ phase change material in its amorphous, cubic, and hexagonal phases. J Appl Phys 97:093509
39. Siegrist T, Jost P, Volker H, Woda M, Merkelbach P, Schlockermann C, Wuttig M (2011) Disorder-induced localization in crystalline phase change materials. Nat Mater 10:202
40. Raoux S, Jordan-Sweet JL, Kellock AJ (2008) Crystallization properties of ultra-thin phase change films. J Appl Phys 103:114310
41. Simpson RE, Krbal M, Fons P, Kolobov AV, Tominaga J, Uruga T, Tanida H (2010) Toward the ultimate limit of phase change in $Ge_2Sb_2Te_5$. Nano Lett 10:414
42. Lee S-H, Jung Y, Agarwal R (2007) Highly scalable non-volatile and ultra-low power phase-change nanowires memory. Nat Nanotechnol 2:626
43. Caldwell MA, Raoux S, Wang RY, Wong H-SP, Milliron DJ (2010) Synthesis and size-dependent crystallization of colloidal germanium telluride nanoparticles. J Mater Chem 20:1285
44. Xiong F, Liao AD, Estrada D, Pop E (2011) Low-power switching of phase-change materials with carbon nanotube electrodes. Science 332:568
45. Raoux S, Cheng Y-Y, Jordan-Sweet J, Munoz B, Hitzbleck M (2009) Influence of interfaces and doping on the crystallization temperature of Ge–Sb. Appl Phys Lett 94:183114
46. Wong HSP, Raoux S, Kim S, Liang J, Reifenberg JP, Rajendran B, Asheghi M, Goodson KE (2010) Phase change memory. In: Proceedings of IEEE vol 98, p 2201

47. Pirovano A, Lacaita AL, Benvenuti A, Pellizzer F, Hudgens S, Bez R (2003) Scaling analysis of phase-change memory technology. In: Electron devices meeting 2003, IEDM '03 technical digest. IEEE international, pp 29.6.1–29.6.4
48. Sasago Y, Kinoshita M, Morikawa T, Kurotsuchi K, Hanzawa S, Mine T, Shima A, Fujisaki Y, Kume H, Moriya H, Takaura N, Torii K (2006) Cross-point phase change memory with $4F^2$ cell size driven by low-contact-resistivity poly-Si diode. In: Proceedings of very large scale integration (VLSI) technology Symposium, pp 24–25
49. Pellizzer F, Benvenuti A, Gleixner B, Kim Y, Johnson B, Magistretti M, Marangon T, Pirovano A, Bez R, Atwood G (2006) A 90 nm phase change memory technology for stand-alone non-volatile memory applications. In: Proceedings of very large scale integration (VLSI) technology symposium, pp 122–123
50. Raoux S, Burr GW, Breitwisch MJ, Rettner CT, Chen YC, Shelby RM, Salinga M, Krebs D, Chen SH, Lung H-L, Lam C (2008) Phase-change random access memory: a scalable technology. IBM J Res Dev 52:465–479
51. Pellizzer F, Pirovano A, Ottogalli F, Magistretti M, Scaravaggi M, Zuliani P, Tosi M, Benvenuti A, Besana P, Cadeo S, Marangon T, Morandi R, Piva R, Spandre A, Zonca R, Modelli A, Varesi E, Lowrey T, Lacaita A, Casagrande G, Cappelletti P, Bez R, Novel μ-trench phase-change memory cell for embedded and stand-alone non-volatile memory applications. In: Proceedings of very large scale integration (VLSI) technology symposium, pp 18–19
52. Lacaita AL (2006) Phase change memories: state-of-the-art, challenges and perspectives. Solid State Electron 50:24–31
53. Oh GH, Park YL, Lee JI, Im DH, Bae JS, Kim DH, Ahn DH, Horii H, Park SO, Yoon HS, Park IS, Ko YS, Chung UI, Moon JT (2009) Parallel multi-confined (PMC) cell technology for high density MLC PRAM In: Proceedings of very large scale integration (VLSI) technology symposium, pp 220–221
54. Giraud V, Cluzel J, Sousa V, Jacquot A, Dauscher A, Lenoir B, Scherrer H, Romer S (2005) Thermal characterization and analysis of phase change random access memory. J Appl Phys 98:013520
55. Servalli G (2009) 45 nm generation phase change memory technology. In: Proceedings of IEEE international electron devices meeting
56. Breitwisch M, Nirschl T, Chen CF, Zhu Y, Lee MH, Lamorey M, Burr GW, Joseph E, Schrott A, Philipp JB, Cheek R, Happ TD, Chen SH, Zaidr S, Flaitz P, Bruley J, Dasaka R, Rajendran B, Rossnagel S, Yang M, Chen YC, Bergmann R, Lung HL, Lam C (2007) Novel lithography-independent pore phase change memory. In: Proceedings of very large scale integration (VLSI) technology symposium, pp 100–101
57. Chen WS, Lee C, Chao DS, Chen YC, Chen F, Chen CW, Yen R, Chen MJ, Wang WH, Hsiao TC, Yeh JT, Chiou SH, Liu MY, Wang TC, Chein LL, Huang C, Shih NT, Tu LS, Huang D, Yu TH, Kao MJ, Tsai MJ (2007) A Novel cross-spacer phase change memory with ultra-small lithography independent contact area. In: Proceedings of IEEE international electron devices meeting, pp 319–322
58. Ha YH, Yi JH, Horii H, Park JH, Joo SH, Park SO, Chung U In, Moon JT (2003) An edge contact type cell for phase change RAM featuring very low power consumption. In: Proceedings of Very Large Scale Integration (VLSI) Technology Symposium, pp 175–176
59. Oh JH, Park JH, Lim YS, Lim HS, Oh YT, Kim JS, Shin J, Park JH, Song YJ, Ryoo KC, Lim DW, Park SS, Kim JI, Kim JH, Yu J, Yeung F, Jeong CW, Kong JH, Kang DH, Koh GH, Jeong GT, Jeong HS, Kim K (2006) Full integration of highly manufacturable 512 Mb PRAM based on 90 nm technology. In: Proceedings of IEEE international electron devices meeting
60. Ahn SJ, Hwang YN, Song YJ, Lee SH, Lee SY, Park JH, Jeong CW, Ryoo KC, Shin JM, Fai Y, Oh JH, Koh GH, Jeong GT, Joo SH, Choi SH, Son YH, Shin JC, Kim YT, Jeong HS, Kim K (2005) Highly reliable 50 nm contact cell technology for 256 Mb PRAM. In: Proceedings of very large scale integration (VLSI) technology symposium, pp 98–99

61. Song YJ, Ryoo KC, Hwang YN, Jeong CW, Lim DW, Park SS, Kim JI, Kim JH, Lee SY, Kong JH, Ahn SJ, Lee SH, Park JH, Oh JH, Oh YT, Kim JS, Shin JM, Fai Y, Koh GH, Jeong GT, Kim RH, Lim HS, Park IS, Jeong HS, Kim K (2006) Highly reliable 256 Mb PRAM with advanced ring contact technology and novel encapsulating technology. In: Proceedings of very large scale integration (VLSI) technology symposium, pp 118–119
62. Hwang YN, Lee SH, Ahn SJ, Lee SY, Ryoo KC, Hong HS, Koo HC, Yeung F, Oh JH, Kim HJ, Jeong WC, Park JH, Horii H, Ha YH, Yi JH, Koh GH, Jeong GT, Jeong HS, Kim K. Writing current reduction for high-density phase-change RAM. In: Proceedings of very large scale integration (VLSI) technology symposium, pp 37.1.1–37.1.4
63. Lee JI, Park H, Cho SL, Park YL, Bae BJ, Park JH, Park JS, An HG, Bae JS, Ahn DH, Kim YT, Horii H, Song SA, Shin JC, Park SO, Kim HS, Chung U-I, Moon JT, Ryu BI (2007) Highly scalable phase change memory with CVD GeSbTe for sub 50 nm generation. In: Proceedings of very large scale integration (VLSI) technology symposium, pp 1002–1003
64. Im DH, Lee JI, Cho SL, An HG, Kim DH, Kim IS, Park H, Ahn DH, Horii H, Park OS, Chung U-I, Moon JT (2008) A unified 7.5 nm dash-type confined cell for high performance PRAM device. In: Proceedings of very large scale integration (VLSI) technology symposium
65. Chen Y-C (2009) Phase change random access memory advanced prototype devices and scaling. In: Raoux S, Wuttig M (eds) Phase change materials: science and applications. Springer, New York
66. Bez R, Gleixner RJ, Pellizzer F, Pirovano A, Atwood G (2009) Phase change memory cell concepts and designs. In: Raoux S, Wuttig M (eds) Phase change materials: science and applications. Springer, New York
67. Milliron DJ, Huang Q, Zhu Y (2009) Novel deposition methods. In: Raoux S, Wuttig M (eds) Phase change materials: science and applications. Springer, New York
68. Huang Q, Kellock AJ, Raoux S (2008) Electrodeposition of SbTe phase-change alloys. J Electrochem Soc 155:D104
69. Milliron DJ, Raoux S, Shelby RM, Jordan-Sweet J (2007) Solution-phase deposition and nanopatterning of GeSbSe phase change materials. Nat Mater 6:352
70. Ritala M, Pore V, Hatanpää T, Heikkilä M, Leskelä M, Mizohata K, Schrott A, Raoux S, Rossnagel SM (1946) Atomic layer deposition of $Ge_2Sb_2Te_5$ thin films. Microelectron Eng 2009:86
71. Ritala M, Leskelä M (2001) Atomic layer deposition. In: Nalwa HS (ed) Handbook of thin film materials ,vol 1. Academic Press, San Diego, pp 103–159
72. Ritala M, Niinistö J (2008) Atomic layer deposition. In: Jones AC, Hitchman MJ (eds) Chemical vapour deposition: precursors, processes and applications. Royal Society of Chemistry, Cambridge, pp 158–206
73. Leskelä M, Pore V, Hatanpää T, Heikkilä M, Ritala M, Schrott A, Raoux S, Rossnagel SM (2009) Atomic layer deposition of materials for phase-change memories. ECS Trans 25(4):399–407
74. Lee J, Choi S, Lee C, Kang Y, Kim D (2007) GeSbTe deposition for the PRAM applications. Appl Surf Sci 253:3969–3976
75. Choi BJ, Choi S, Shin YC, Hwang CS, Lee JW, Jeong J, Kim YJ, Hwang S-Y, Hong SK (2007) Cyclic PECVD of $Ge_2Sb_2Te_5$ thin films using metalorganic sources. J Electrochem Soc 154:H318–H324
76. Choi BJ, Choi S, Shin YC, Kim KM, Hwang CS, Kim YJ, Son YJ, Hong SK (2007) Combined atomic layer and chemical vapour deposition, and selective growth of $Ge_2Sb_2Te_5$ films on TiN/W contact plug. Chem Mater 19:4387–4389
77. Choi BJ, Choi S, Eom T, Ryu SW, Cho D-Y, Heo J, Kim HJ, Hwang CS, Kim YJ, Hong SK (2009) Influence of substrates on the nucleation and growth behaviours of $Ge_2Sb_2Te_5$ films by combined plasma-enhanced atomic layer and chemical vapour deposition. Chem Mater 21:2386–2396
78. Choi BJ, Oh SH, Choi S, Eom T, Shin YC, Kim KM, Yi K-W, Hwang CS, Kim YJ, Park HC, Baek TS, Hong SK (2009) Switching power reduction in phase change memory cell using CVD $Ge_2Sb_2Te_5$ and ultrathin TiO_2 films. J Electrochem Soc 156:H59–H63

79. Choi S, Choi BJ, Eom T, Jang JH, Lee W, Hwang CS (2010) Growth and phase separation behaviour in Ge-doped Sb-Te thin film films deposited by combined plasma-enhanced chemical vapour and atomic layer deposition. J Phys Chem C 114:17899–17904
80. Eom T, Choi BJ, Choi S, Park TJ, Kim TJ, Seo M, Rha SH, Hwang CS (2009) $Ge_2Sb_2Te_5$ charge trapping nanoislands with high-k blocking oxides for charge trap memory. Electrochem Solid-State Lett 12:H378–H380
81. Pore V, Hatanpää T, Ritala M, Leskelä M (2009) Atomic layer deposition of metal tellurides and selenides using alkylsilyl compounds of tellurium and selenium. J Am Chem Soc 131:3478–3480
82. Knapas K, Hatanpää T, Ritala M, Leskelä M (2010) In situ reaction mechanism studies on atomic layer deposition of Sb_2Te_3 and GeTe from $(Et_3Si)_2Te$ and chlorides. Chem Mater 22:1386–1391
83. Sarnet T, Pore V, Hatanpää T, Ritala M, Leskelä M, Schrott A, Zhu Y, Raoux S, Cheng H-Y (2011) Atomic layer deposition and characterization of GeTe thin films. J Electrochem Soc 158:D694–D697
84. Pore V, Knapas K, Hatanpää T, Sarnet T, Kemell M, Ritala M, Leskelä M, Mizohata K (2011) Atomic layer deposition of antimony and its compounds using dechlorosilylation reactions of tris(triethylsilyl)antimony. Chem Mater 23:247–254
85. Maitrejean S, Lhostis S, Haukka S, Jahan C, Gourvest E, Matero R, Blomberg T, Toffoli A, Persico A, Jayet C, Veillerot M, Barnes JP, Pierre F, Fillot F, Perniola L, Sousa V, Sprey H, Boulanger F, de Salvo B, Billon T (2011) Demonstration of phase change memories devices using $Ge_2Sb_2Te_5$ films deposited by atomic layer deposition. In: 2011 IEEE international interconnect technology conference and 2011 materials for advanced metallization (IITC/MAM), Dresden, doi:10.1109/IITC.2011.5940298

Chapter 6
FeRAM

Susanne Hoffmann-Eifert and Takayuki Watanabe

6.1 Principle of FeRAM operation

6.1.1 Information storage and FeRAM configuration

In ferroelectric random access memory (FeRAM), nonvolatility is achieved by the remanent polarization of the ferroelectric (FE) material. The polarization vector P can be reversed by an applied electric field E resulting in a FE hysteresis loop which is shown in Fig. 6.1a. The states of the remanent polarization, $(+P_r)$ or $(-P_r)$, define the logic states '1' or '0' of the memory cell, respectively. The information is written or read by the application of defined voltage pulses. If the applied electrical field is in the same direction as the remanent polarization, no switching occurs, whereas switching occurs if the applied field and remanent polarization are of opposite directions. The respective changes of polarization, either ΔP_{ns}, which is solely due to the dielectric response of the material, or in the case of switching ΔP_s, result in a different transient current behavior of the FE capacitor (see Fig. 6.1b). The integration of the current yields the amount of charge $\Delta Q = A\ \Delta P$ which is proportional to the change in polarization $\Delta P = 2\ P_r$, where A defines the area of the capacitor. The difference in charge, either ΔQ_{ns} or ΔQ_s, enables a distinction to be made between the logic states of the FE capacitor. One important criterion for FeRAM device application is the differentiation between the switching 's' and non-switching 'ns' state of the FE capacitor during read-out. In order to avoid read failures, the switched charge ΔQ_s should not fall below a critical value of $\sim$30 fC. Thus, when the cell size of the FeRAM capacitor

S. Hoffmann-Eifert (✉)
Peter Grünberg Institute, Forschungszentrum Jülich, Jülich, Germany
e-mail: su.hoffmann@fz-juelich.de

T. Watanabe
Corporate R&D Headquarters, CANON Inc, Tokyo, Japan

C. S. Hwang (ed.), *Atomic Layer Deposition for Semiconductors*,
DOI: 10.1007/978-1-4614-8054-9_6,

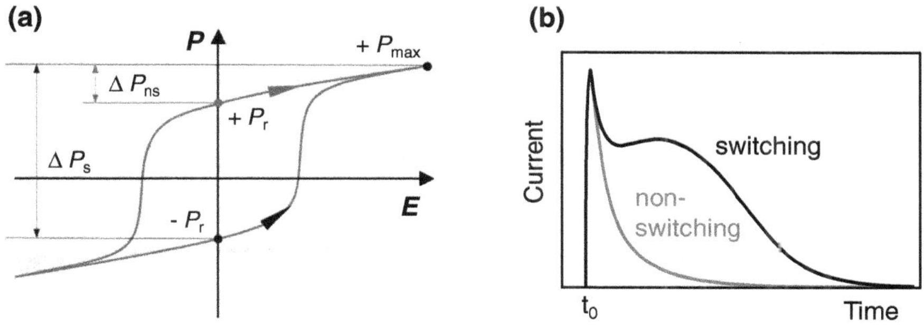

Fig. 6.1 **a** Change of the polarization of a ferroelectric capacitor for a non-switching (ns) and a switching (s) event. **b** Current response of non-switching and switching case of the FE polarization

is reduced in order to increase the integration density, the sensing margin must be ensured by increasing either A or ΔP. The capacitor area A can be increased by modification of the capacitor's design from a planar structure to a three-dimensional (3D) structure, such as deep trench. The switching charge ΔQ_s can be increased by controlling the crystal orientation or by substituting a small ΔP FE material with a large ΔP one.

FE memory cells are introduced in various configurations that differ in the number of transistors and capacitors used. Regarding the FeRAM designs, the one transistor-one capacitor (1T-1C) cell exhibits the smallest size and is therefore the most extensively investigated [3, 4]. The single 1T cell represents the ferroelectric field effect transistors (FeFET). Figure 6.2 shows a sketch of a simplified 1T-1C FeRAM cell, which is more complex than a corresponding 1T-1C dynamic random access memory (DRAM) cell as it is not only addressed by word lines and bit lines, but needs one additional drive line (or cell plate) for switching purposes. The function of each line is defined in the schemes for writing and reading which are explained for example in Refs. [5, 6].

For 'writing' it has to be ensured that the voltage across the FE capacitor is larger than the coercive voltage. The 'reading' of the logic state of the storage capacitor is done by means of a sense amplifier that compares the voltage on the bit line, which depends on the values of the capacitors, C_{BL} and C_{FE}, to a reference voltage. For logic state "1" the signal is amplified, for logic state "0" it is set to 0 V. This operation is *destructive* and the data must be restored after reading.

6.1.2 Ferroelectric Materials for FeRAMs

FE oxides which are intensively investigated for application in memory devices mostly exhibit either a *perovskite-type structure* such as lead zirconate titanate, $Pb(Zr_xTi_{1-x})O_3$ (PZT) [7], or *layered perovskite structure* as for example strontium bismuth tantalate, $SrBi_2Ta_2O_9$ (SBT), and bismuth lanthanum titanate,

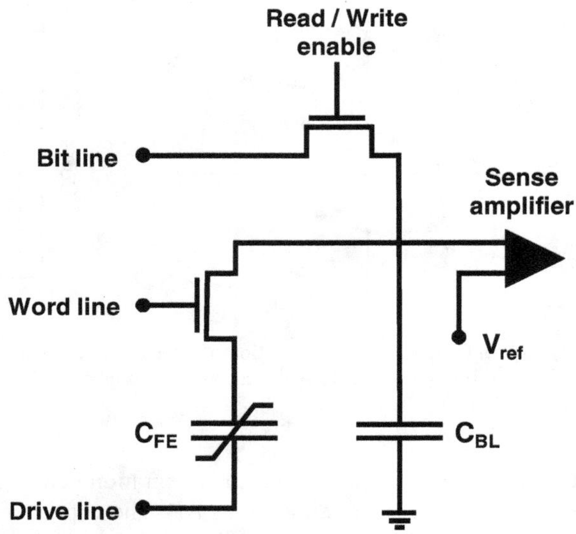

Fig. 6.2 Simplified 1T-1C FE memory circuit

$(Bi_{1-x}La_x)_4Ti_3O_{12}$ (BLT) (see e.g. [2, 8–10]). For more details see also topical books [11–13].

Oxide FEs must attain a crystalline perovskite-type structure in order to display hysteretic polarization behavior. Thus, crystallization temperature and atmosphere are critical parameters for the integration of the thin films. In addition, due to the non-centro-symmetric structures of the unit cells, which are inherent to FEs, a certain texture of the thin films is required in order to switch the polarization vector in the electric field applied by the electrodes. Figure 6.3 summarizes the differences in the cell structures of PZT and SBT, which to some extent determine the characteristics of the FE materials. PZT (see Fig. 6.3a) exhibits different FE properties depending on the type of structural distortion [14]. Below the Curie temperature, the material undergoes a morphotropic phase transition from the tetragonal to rhombohedral state when the composition is changed from Ti-rich to Zr-rich, respectively. In the *tetragonal phase*, e.g., PZT (30/70), the polarization vector lies in the [001] direction, while for the *rhombohedral* phase, e.g., PZT (60/40), the polarization vector is oriented along the [111] direction. For FE memory applications, the tetragonal phase of PZT is favored over the rhombohedral phase because of its higher P_r values and its squarer hysteresis loops [15]. In addition, the crystallization temperature of the Ti-rich phase is lower than that of the Zr-rich composition which makes the integration of Ti-rich PZT thin films easier. The sensitivity of the FE properties of PZT to the composition and texture makes a precise deposition process indispensable, especially if PZT is to be deposited on narrow 3D structures [16, 17].

$SrBi_2Ta_2O_9$ (SBT; see Fig. 6.3b) is a prominent representative of FE material with an Aurivillius-type phase. [2] This structure can be visualized as pseudo-perovskite layers separated by $(Bi_2O_2)^{2+}$ layers. The bismuth layer structure can be

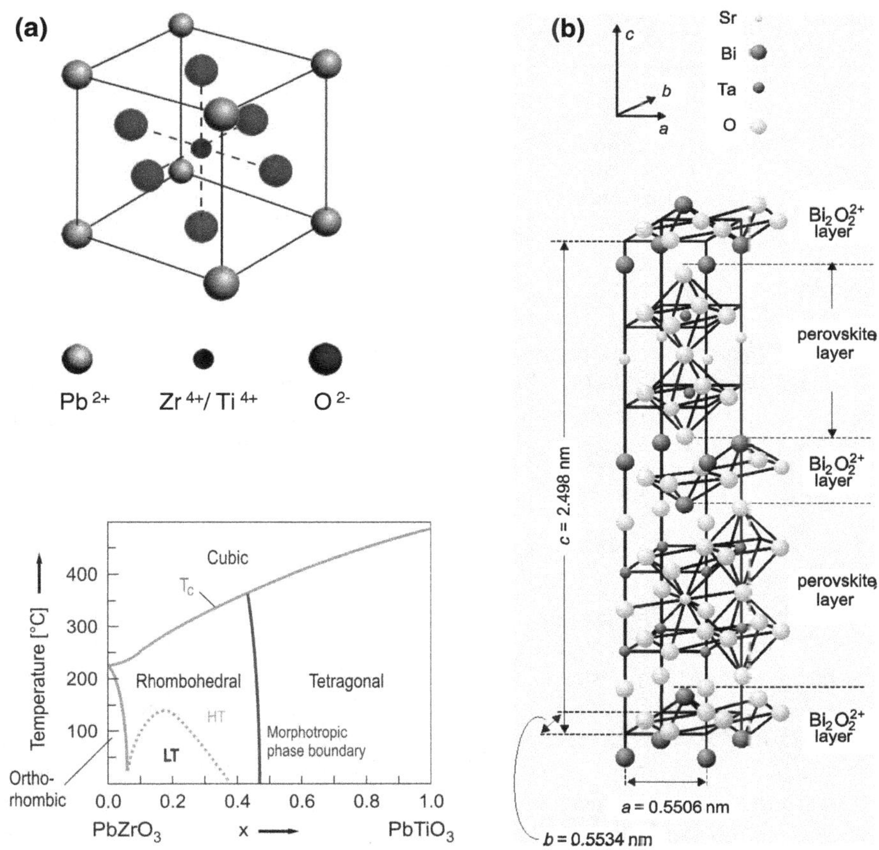

Fig. 6.3 Typical FE materials; **a** Perovskite structure and phase diagram of $Pb(Zr_xTi_{1-x})O_3$ and **b** layered structure of Aurivillius-type $SrBi_2Ta_2O_9$ (Reprinted with permission from [21]. Copyright 2005, WILEY-VCH Verlag GmbH & Co. KGaA, Berlin)

stabilized in various layer configurations of different stoichiometry [18]. Due to the comparatively few directions allowed for spontaneous polarization, the switchable polarization of polycrystalline bismuth layer structured films is rather small compared to lead-based perovskite films.

The operation voltage of the device has to be greater than the coercive voltage, which is inherent to the integrated FE material. In this respect, FEs of the Aurivillius family are superior to lead-based perovskites because of the lower coercive field [19, 20]. For further reduction of the operation voltage, the FE layer has to be thinned, while any inhomogeneities like variation of film thickness, small pores, or contamination by a non-FE impurity phase due to the off-stoichiometry will be exaggerated and can readily cause serious failure. Consequently, as the FE layer becomes thinner, a more precise deposition process which can produce

Table 6.1 Properties of typical ferroelectrics used for FeRAM taken from Ref. [6]

Ferroelectric material		$Pb(Zr_xTi_{1-x})O_3$	$SrBi_2Ta_2O_9$	$(Bi,La)_4Ti_3O_{12}$
Remanent polarization [μC/cm^2]		10–40	5–10	10–15
Coercive field [kV/cm]		50–70	30–50	30–50
Crystallization temperature [°C]		450–650	650–800	650–750
Endurance	On Pt electrode	Poor	Good	
	On oxide electrode	Good		

homogenous thin FE layers is necessary. Note that the FE film is grown on 3D structures in order to maintain sufficient switching charge for sensing.

Some representative properties of these materials utilized in FeRAM devices are summarized in Table 6.1. More data on commercially available FeRAM products of medium integration density can be found from e.g., Ramtron [22], Symetrix [23], Texas Instruments [24], and Fujitsu Ltd. Electronic Devices [25].

6.1.3 Electrode Materials

In general, the FE oxides are deposited in oxygen-rich atmosphere at temperatures between 550 and 800 °C. Therefore, the electrode as well as the plug materials to be used for FeRAM devices must be resistant to oxidation and inert in contact with the FE material, which means that no uncontrolled interdiffusion of metal ions may occur and that no intermediate layers may be formed deteriorating device performance. This requirement is fulfilled by noble metals such as Pt, by metals which form a conductive oxide, such as Ir–IrO_2 and Ru–RuO_2, and by conductive oxides such as $SrRuO_3$. Like the FE layer, these electrode layers must be deposited uniformly in order to avoid any electrical errors. For deposition of metals and conducting oxides by atomic layer deposition (ALD) see articles such as [26–28].

6.2 Trends in Ferroelectric Random Access Memories

6.2.1 Planar Ferroelectric Capacitor Stacks for Medium Integration Density

Most of the applications of commercialized FeRAMs are in the area of low-density memory for NV portable applications, as for example electronic passports and other types of e-cards [29]. Today's 'standard' method for the growth of PZT thin films in integrated FeRAM cells of up to 64 Mb density is metal organic chemical vapor deposition (MOCVD) [30–34]. Figures 6.4a and b show cross sections of planar FE capacitors fully integrated into a 1T-1C structure and a planar FE capacitor after optimization of the etching procedure.

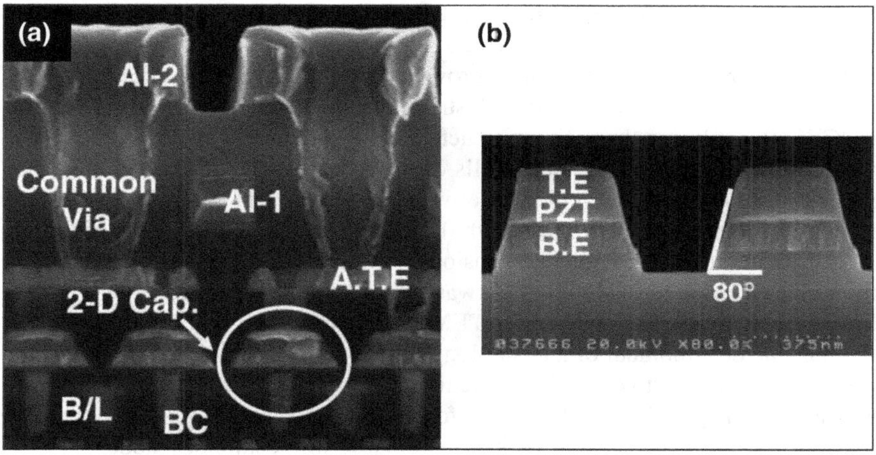

Fig. 6.4 Vertical SEM cross sections of planar FE capacitors (*2D-Cap.*) from metal/MOCVD PZT/metal stacks: **a** Fully integrated in a 1T-1C FeRAM structure and **b** Standalone cells with optimized side wall steepness (Reprinted with permission from [34]. Copyright 2006, The Japan Society of Applied Physics)

For higher integration densities, the FeRAM cells are expected to 'follow' the trend of the DRAM cells by utilizing multi-stacked or trench cells in order to increase the effective capacitor area at a fixed cell size. A graphical summary of a possible scaling scenario of future FeRAMs, suggested in Ref. [1] based on the ITRS road map [35], is shown in Fig. 6.5. The feature size for which a 3D structure of the FE capacitor will become necessary has been estimated and the result is expressed in the 3D folding factor in Fig. 6.5. From this it is expected that FeRAMs above the 256 Mb generations would be based on nonplanar FE capacitor technologies.

6.2.2 Transition from 2D to 3D FE Capacitor for High-Density Integration

Researchers at Samsung Advanced Institute of Technology (SAIT) demonstrated a prototype 3D FE capacitor cell in 2004 [36, 37]. The device structure was built in SiO_2 capacitor holes of 0.32 μm and 0.25 μm in diameter, respectively, with a depth of about 0.6 μm and consisted of an ALD Ir bottom electrode, a 60 nm thick MOCVD PZT film, and an ALD Ir top electrode (see Fig. 6.6a) [34]. The 0.32 μm trench FE capacitor showed FE properties similar to the corresponding planar stack with a $2P_r$ value of 24 μC/cm^2 at an external bias voltage of 2.1 V. However, in the 0.25 μm trench diameter capacitor $2P_r$ was decreased to 19 μC/cm^2 (see Fig. 6.6b) [6].

A careful and comprehensive microstructure analysis of the device cross sections [38] revealed a possible inherent limitation of the MOCVD processing, namely the inability of conformal growth of multicomponent thin films with homogeneous stoichiometry into 3D structures. In detail, it was found for the MOCVD PZT films in the 0.32 μm trench that the columnar perovskite PZT grains were well established on the side walls of the trench, while at the bottom a fine-grained non-FE pyrochlore-type phase was detected as depicted in Fig. 6.7a. The decrease of remanent polarization with decreasing trench diameter (see Fig. 6.6b) was attributed to an incomplete extension of the columnar perovskite-type grains on the 0.25 μm trench side walls as was proven from synchrotron X-ray micro-diffraction analysis [38] and careful TEM studies [34] (see Fig. 6.7b).

Thus, the development of a technology for conformal deposition of perovskite-type FE PZT films with controlled stoichiometry and microstructure was identified as one of the key technical challenges for the development of a 3D FE capacitor essential for 6–8 F^2 cell FeRAMs beyond the 100 nm technology node.

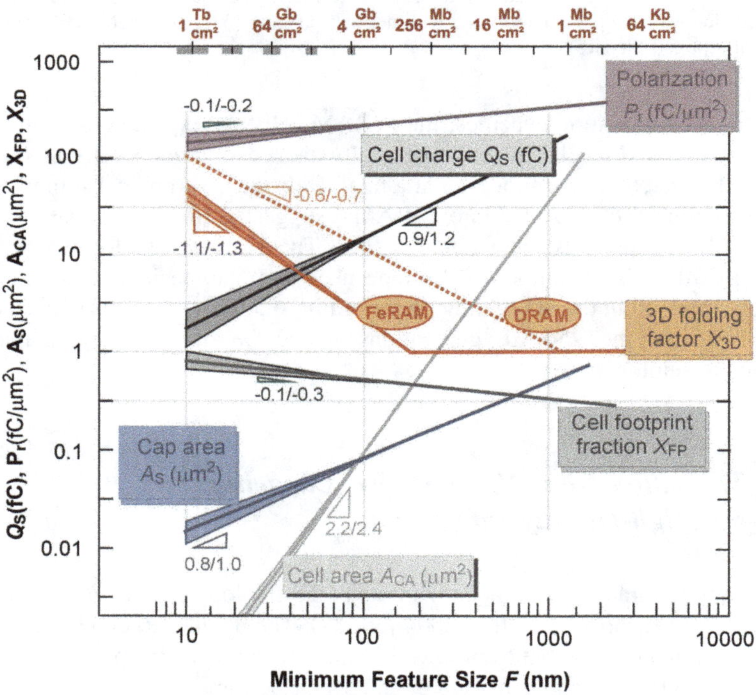

Fig. 6.5 Roadmap of future FeRAM devices [1]. Copyright Wiley-VCH Verlag GmbH & Co. KGaA. Reproduced with permission

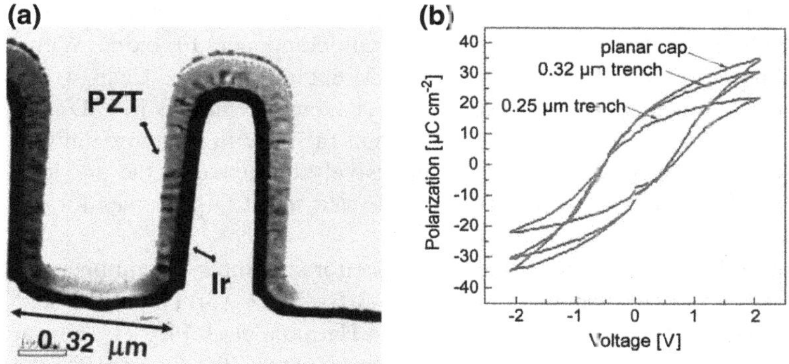

Fig. 6.6 **a** TEM image of a 3D MOCVD PZT/ALD Ir capacitor deposited in a trench hole of 0.32 μm diameter (Reprinted with permission from [34]. Copyright 2006, The Japan Society of Applied Physics). **b** Measured (P–V) loops for 3D FeRAM cell structures with different trench diameters [6]. Copyright Wiley-VCH Verlag GmbH & Co. KGaA. Reproduced with permission

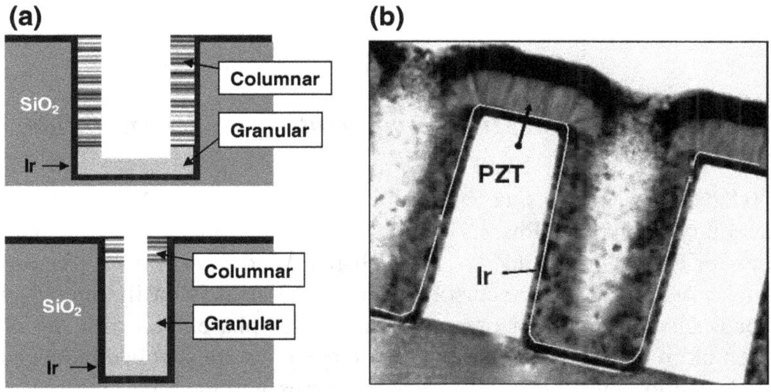

Fig. 6.7 Grain structure in 3D FE capacitors from MOCVD-grown PZT layer and ALD Ir electrodes. **a** Schematic drawing of the distribution of perovskite- and pyrochlore-type phase. **b** TEM image of cross section of ALD Ir/MOCVD PZT/ALD Ir capacitor on SiO_2 trench holes of 0.25 μm in diameter (Reprinted with permission from [34]. Copyright 2006, The Japan Society of Applied Physics)

6.3 Atomic Layer Deposition of Multicomponent Oxides on 3D Structures

6.3.1 State of the Art of Multicomponent Oxides ALD

The intensive research on the integration of FE capacitors into 3D structures for high-density NV memory devices clearly revealed that for the realization of Fe-RAMs beyond the 100 nm technology node a deposition process has to be developed, which offers the ability of conformal growth of stoichiometric films onto challenging 3D structures, even for a multicomponent FE oxide. With respect to the integration of high-*k* oxides for DRAM applications (see Chap. 4), the ALD technique has become established as the standard technique for 3D growth of functional oxides. The principle of ALD and the criteria for the definition of an ALD-type growth process have been intensively discussed in the previous chapters. Therefore, this paragraph will be restricted to ALD processes for FE oxide thin films for FeRAM applications.

Despite the strong interest in 3D FE capacitor structures, the number of studies on ALD of multicomponent FE oxides as $SrBi_2Ta_2O_9$ (SBT) [39] or lead-based perovskites [40–50] is rather limited. While Harjuoja et al. [40, 41], Hwang et al. [42], and Lee et al. [43] reported on ternary systems $PbTiO_3$ and $PbZrO_3$, Watanabe et al. combined all three binary oxide processes, namely PbO, ZrO_2, and TiO_2, to develop an ALD process for the quaternary $Pb(Zr,Ti)O_3$ compound [46–50].

In the following, we will recapitulate important steps in the development of the thermal ALD process for PZT thin films and their integration into 3D structures for future FeRAM devices.

6.3.2 ALD Process for Multicomponent Ferroelectric Oxide

Figure 6.8 depicts a special source gas supply sequence for ALD of thin films from the quaternary compound $Pb(Zr,Ti)O_3$, which is a combination of binary ALD processes for PbO, TiO_2, and ZrO_2. The self-regulated growth is very sensitive to the ligand structure of the precursor molecule, while it is easily altered once a precursor is mixed with other precursors in the gas phase. To avoid any type of interaction of precursors in the gas phase, the reactants are supplied independently into the deposition chamber. The binary processes will each be repeated several times to adjust the cation composition as necessary. Hence, the self-regulated growth, which is the most characteristic benefit to ALD, must be achieved on both another pre-deposited cation layer and the layer itself.

6.3.3 Precursors for ALD

FEs considered for FeRAM applications are multicomponent oxides that typically consist of three kinds of metals and oxygen. Consequently, an ideal precursor would have all component elements in one molecule. The metal composition in the molecule would be designed according to the target composition of the deposited film. This idea has been partially achieved, for example, for the Sr–Ta double alkoxide precursors used in MOCVD and even ALD processes [51]. However, such ideal precursors are not always available. Consequently, the ALD process for

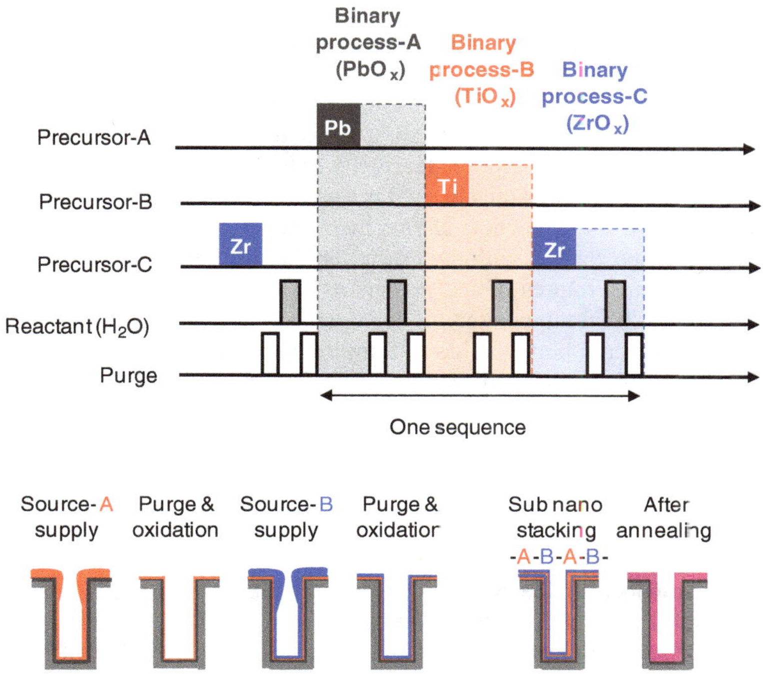

Fig. 6.8 Gas supply sequence for ALD of a quaternary oxide film (*top*) and schematic picture of the film growth process for a ternary film (*bottom*)

FEs will become a combination of ALD processes of binary or ternary oxides, and the selection of appropriate precursors will be of central importance. There are some key parameters to be considered in selecting precursors for ALD processes. Knowledge in MOCVD methods may be assumed.

(1) Vapor pressure

As in the case of chemical vapor deposition processes in general, one of the important parameters for selecting precursors is the vapor pressure. In ALD processes, a sufficiently high amount of precursor must be supplied to the reactor chamber to fully cover the total surface of the substrate. Hence, precursors with high vapor pressures may be preferred for ALD, especially when the precursor is carried due to its vapor pressure or by bubbling. If the precursor is maintained at a high temperature to increase the vapor pressure, a long-term stability of the precursor at the high temperature is simultaneously required. Typically, metal halides have a high vapor pressure. However, they are used with some reservation because a halide vapor generated after a reaction with an oxygen source may etch the growing thin film or damage the equipment. The priority of the vapor pressure in the selection of a precursor may decrease if liquid delivery (LD) or liquid injection (LI) systems are utilized which can vaporize the precursor as available.

(2) Thermal stability and compatibility with other precursors

On the one hand, the higher the substrate temperature is, the higher deposition rate is expected to be. On the other hand, a self-regulated process will be lost once the precursor molecule decomposes thermally at too high deposition temperature. The substrate temperature is thus limited below the lowest thermal decomposition temperature of any of the precursors. Therefore, precursors that decompose at high temperature may be chosen, with the restriction that for multicomponent ALD the decomposition temperatures of the precursors are within the same regime. The thermal decomposition temperature of each precursor should be determined via a pre-study ALD of their binary oxide or from thermal decomposition analysis. The required FE film thicknesses will be around 10–30 nm, which is thicker compared with gate dielectric, so that the process time will be longer. In order to shorten the process time, it is advantageous to use precursors with a high thermal decomposition temperature.

(3) Self-regulated growth

In multicomponent oxide ALD processes the precursors should be introduced into the chamber separately in order to avoid any cross-reactions in the gas phase. Therefore, the self-regulated growth of the component binary ALD will have to be established on at least two kinds of surfaces. One is the layer of the predeposited other metal oxide, and the second is the layer of the material itself.

(4) Reactivity with oxygen source

Before the pulse of oxygen source is introduced into the chamber, the chemisorbed surface metal elements still hold a part of their ligands. Therefore, the surface layer must react with the oxygen source to form the metal oxide layer and to remove the unnecessary ligand part. If the reaction with the oxygen source is insufficient, the precursor may leave impurities such as carbon in the film. The contamination may remain even after a crystallization annealing process and thus degrade the insulation property of the FEs. Consequently, the precursor must have a high reactivity with a chosen oxygen source such as H_2O, O_2, and O_3 gases.

6.4 Case Study

6.4.1 Example of ALD-PZT

An ALD system equipped with a pulsed LI vaporizer with three independent liquid precursors and a horizontal gas flow reactor was used for this study. The discrete precursor injection prevents the precursors from pre-reacting in the gas phase. Due to the narrow selectivity of Pb-, Ti-, and Zr-precursors and their limited thermal stability, LD source injection methods were favored, where precursor solutions are pulse sprayed into a vaporizer kept at 150–200 °C and thermally evaporated. This method is well established for MOCVD processes, but it is not very often applied in ALD-type processes [52]. The LI principle is free from long-term thermal degradation of precursors, which allows the operating lifetime of the precursors to be lengthened in comparison to bubbler-based techniques. All precursors were dissolved in ethylcyclohexane ($C_6H_{11}C_2H_5$) with a concentration of about 0.05–0.1 mol/l. In order to avoid lead loss during the PZT film growth by formation of the volatile and toxic higher oxidized species of lead oxide, water vapor was chosen as the oxygen source in preference to either ozone or oxygen plasma. The reactor pressure was set to 1.0 Torr and argon was used as carrier gas.

6.4.2 ALD-Type Processes for PZT Films Utilizing Different Precursor Combinations

ALD-type processes for the three binary oxides, PbO, TiO_2, and ZrO_2, were set-up and characterized. PbO films were grown by pulsed LD ALD processes from $Pb(C_{12}H_{21}O_2)_2$ [$Pb(TMOD)_2$; bis(2,2,6,6-tetramethyl-3,5-octanedionato)lead] [44] and from $Pb(C_{11}H_{19}O_2)_2$ [$Pb(TMHD)_2$; bis(2,2,6,6-tetramethyl-3,5-heptanedionato)lead] [45]. ALD-type processes were also set-up for TiO_2 films using a stabilized Ti-alkoxide compound $Ti(OC_3H_7)_2(C_{11}H_{19}O_2)_2$ [$Ti(Oi\text{-}Pr)_2(TMHD)_2$; diisopropoxide bis(2,2,6,6-tetramethyl-3,5-heptanedionato)titanium] [45] as well as the pure Ti-alkoxide precursor $Ti(OC_3H_7)_4$ [$Ti(Oi\text{-}Pr)_4$, TTIP: tetrakis(isopropoxy)titanium] [53]. For the ZrO_2 films, different β-diketonate-based precursors were studied, namely, $Zr(C_{11}H_{19}O_2)_4$ [$Zr(TMHD)_4$; tetrakis(2,2,6,6-tetramethyl-3,5-heptanedionato)zirconium] [46]; and a variant $Zr(C_9H_{15}O_2)_4$ [$Zr(DIBM)_4$; tetrakis(2,6-dimethyl-3,5-heptanedionato)zirconium] [50].

Table 6.2 summarizes the combinations of precursors which were utilized for LI-ALD of PZT films giving the formula, structure, decomposition temperature as determined from thermogravimetric analyses, and the ALD window as analyzed from deposition experiments.

Amorphous Pb–Zr–Ti–O-containing layers were deposited at 240 °C, which was evaluated as a stable ALD temperature for all the binary-oxide-type processes (see Table 6.2). Subsequent postdeposition annealing should yield the desired perovskite phase at moderate sintering temperature, which limits the possibility of intermixing the cations to short length scales. Therefore, the combination of the ALD sequences for the binary oxides should result in a reproducible sequence of nano-laminate layers. From the chemisorption-controlled character of ALD it becomes clear that in the ALD growth of a multicomponent material, the deposition rate of a certain metal oxide depends on the growth conditions and surface properties of the previously deposited metal oxide layer, and thus on the sequence

Table 6.2 Summary of the precursor combinations tested for the ALD PZT process comprising the structure and thermal properties of the single precursors and findings for the multicomponent ALD processes

	System 1	System 2	System 3
Lead		$Pb(TMHD)_2$	
Source			
Structure		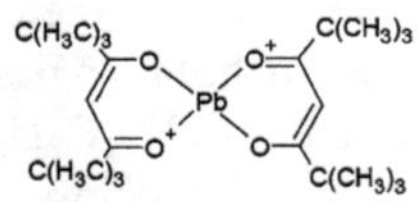	
Decomposition		$T_{decomp} \sim 300$ °C [54]	
ALD of PbO	$T_{ALD} < \sim 270$ °C [45]	$T_{ALD} < \sim 270$ °C [45]	$T_{ALD} < \sim 270$ °C [45]
Titanium	$Ti(Oi–Pr)_2(TMHD)_2$		$Ti(Oi–Pr)_4$
Source			
Structure	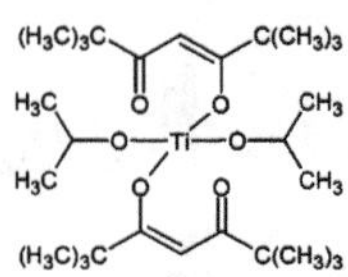		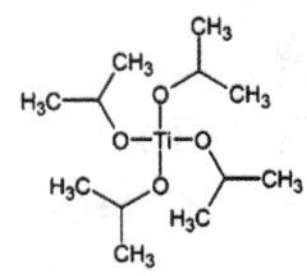
Decomposition	$T_{decomp} \sim 290$ °C [54]		$T_{decomp} \sim 210$ °C [55]
ALD of TiO_2	$T_{ALD} < \sim 380$ °C [56]	$T_{ALD} < \sim 250$ °C [57]	$T_{ALD} < \sim 250$ °C [57]
Zirconium		$Zr(TMHD)_4$	$Zr(DIBM)_4$
Source			
Structure		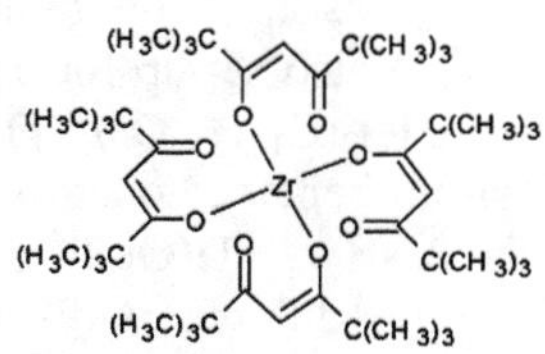	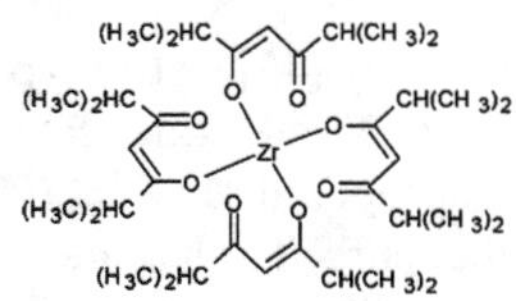
Decomposition		$T_{decomp} > \sim 380$ °C [58] [59]	$T_{decomp} \sim 350$ °C [60]
ALD of ZrO_2	$T_{ALD} \sim 360$ °C [50]	$T_{ALD} \sim 360$ °C [50]	$T_{ALD} \sim 320$ °C [50]
ALD of PZT	$T_{ALD} \sim 240$ °C [46]	$T_{ALD} \sim 240$ °C [46]	$T_{ALD} \sim 240$ °C [50]
Zr/(Zr + Ti) ratio	<0.1	<0.1	$0.2 < x < 0.5$
Pb/(Ti + Zr) ratio	not stable [46]	~0.8 [46]	~1.2 [50]

of the ALD steps as well as on the precursor. Therefore, the Pb-, Ti-, and Zr-precursors listed above were tested in three different combinations. Special attention was therefore paid to the saturation growth behavior of each constituent cation in the multicomponent ALD process. The results for the three systems can be summarized as follows:

System 1. $Pb(TMHD)_2$ was chosen from the standard precursors for lead(II). $Zr(TMHD)_4$ is a standard oxygen-containing precursor used in LI-MOCVD-type processes [59]. In order to have good compatibility of the transition metal sources, $Ti(Oi–Pr)_2(TMHD)_2$ was chosen for the Ti source. Although ALD processes were established for all three binary oxides, the combination of the ALD processes into a nano-laminate process for PZT at about 240 °C suffered from two drawbacks. First, the metal sources displayed a strong interdependence of the deposition rates of the cations in the deposited amorphous layer. This further resulted in a variation of the cation composition when a Pb–Zr–Ti–O-containing film was grown into a 3D pinhole structure (see Fig. 6.9a). In addition, the films were lead-deficient and the Zr content could not be raised much above 10 at % [46].

System 2. The interdependence of the deposition rates among the three sources was primarily attributed to a possible cross reaction between the lead and the titanium source, because the Zr content of the grown layers was below 10 %. Thus, a titanium source was sought which is more compatible with the standard lead source $Pb(TMHD)_2$ with respect to its thermal properties. The replacement of the stabilized titanium source by the pure alkoxide $Ti(Oi–Pr)_4$, while maintaining $Pb(TMHD)_2$ and $Zr(TMHD)_4$ as the respective cation precursors, significantly improved the stability of the ALD process by reducing correlation effects in the growth of the binary oxide layers. The lead content in the films was slightly below the stoichiometric composition. Amorphous Pb–Zr–Ti–O-containing films deposited into 3D hole structures revealed good conformity and constant stoichiometry even for hole diameters of 0.18 µm (see Fig. 6.9b). The drawback of the process was the low deposition rate of Zr–O under the compatible process conditions, keeping the Zr content of the films below 10 at %.

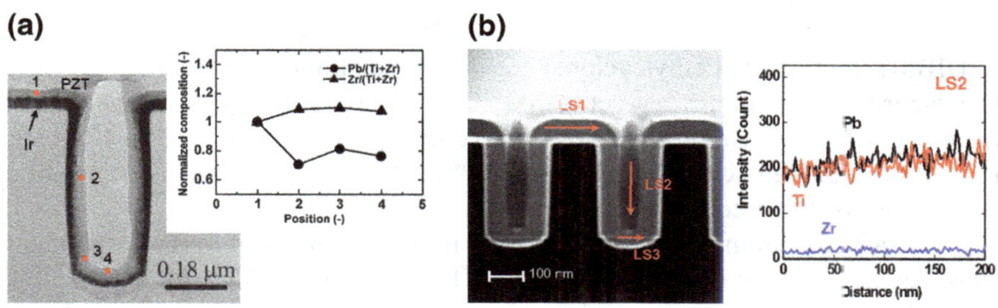

Fig. 6.9 **a** TEM cross-sectional image of an ALD PZT film grown from $Pb(TMHD)_2$, $Ti(Oi–Pr)_2(TMHD)_2$, $Zr(TMHD)_4$, and H_2O on an ALD Ir-coated SiO_2 pinhole structure (*top*) and the local cation composition of the PZT film normalized to the composition at the top flat surface. **b** STEM cross-sectional image of an ALD PZT film grown from $Pb(TMHD)_2$, $Ti(Oi–Pr)_4$, $Zr(TMHD)_4$, and H_2O on an ALD Ir film-coated SiO_2 pinhole structure of 0.18 µm diameter (*top*) and EDS line scan profiles taken along the sidewall (Reprinted with permission from [46]. Copyright 2007, The Electrochemical Society.)

Crystallized Zr-doped $PbTiO_3$ films showed piezoelectric properties but the leakage current of the films was too high to measure the FE properties [46].

System 3. The third change addressed the zirconium precursor in order to achieve a higher Zr content in the resulting PZT films. For β-diketonate compounds, the thermal decomposition temperature and the vapor pressure change with a variation of the ligand structure. The β-diketonate Zr compounds with symmetric ligands, namely $Zr(TMHD)_4$ and $Zr(DIBM)_4$, show good thermal stability, while the vaporization rate of the latter was found to be higher by approximately a factor of eight compared to the standard TMHD compound [60]. The higher vaporization rate of the $Zr(DIBM)_4$ precursor indeed resulted in an increase of the growth rate of LI-ALD ZrO_2 thin films by a factor of three [50]. The ALD multicomponent oxide process from $Pb(TMHD)_2$, $Ti(Oi–Pr)_4$, $Zr(DIBM)_4$, and water at 240 °C substrate temperature showed stable saturation behavior for all binary processes irrespective of whether the other respective compounds were varied. The deposition process allowed the [Zr]/([Ti] + [Zr]) ratio to be adjusted up to 0.5, while the [Pb]/([Ti] + [Zr]) composition was stable at a slightly Pb-rich value of about 1.2. Details of the deposition process are described in Ref. [50]. Scanning transmission electron microscopy (STEM) analysis of amorphous ALD PZT thin films grown into 3D pinhole structures of 0.18 μm in diameter showed an excellent step coverage and conformity and a constant [Pb]/([Ti] + [Zr]) stoichiometry along the top, side wall, and bottom of the pinhole structure (see Fig. 6.10) [50].

Hence, an ALD-type process for the conformal deposition of Pb–Zr–Ti–O-containing films was successfully developed. This process involves pulsed spray evaporation of the liquid precursors, which are $Pb(TMHD)_2$, $Ti(Oi–Pr)_4$, and $Zr(DIBM)_4$ dissolved in ethylcyclohexane at a concentration of 0.05 mol per liter. The vaporizer temperature was set to 200 °C. Water vapor was used as the oxygen source, and Ar was used for purging. The reactor pressure was fixed at 1.0 Torr, and the substrate temperature was set to 240 °C which is below the decomposition temperature of the precursors.

The exact stoichiometry of the ALD PZT films was analyzed for corresponding films deposited on planar $Pt(111)/AlO_x/SiO_x/Si$ substrates. The formation of a $PbPt_x$ alloy [61] was not observed for the pulsed LD ALD PZT process. This is attributed to the low deposition temperature of 240 °C, which is lower than the stability regime of the $PbPt_x$ phase. The composition of the ALD PZT films grown on the planar substrates was analyzed by means of X-ray fluorescence spectroscopy (XRF). The [Zr]/([Zr] + [Ti]) composition in the Pb–Zr–Ti–O films was

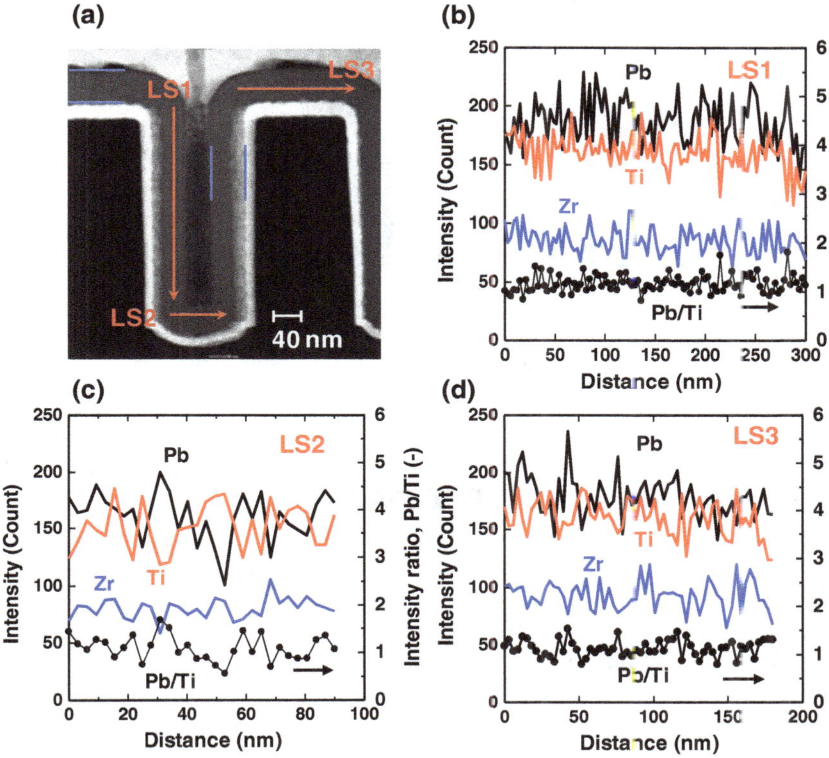

Fig. 6.10 **a** STEM cross-sectional image of an ALD PZT film grown from $Pb(TMHD)_2$, $Ti(Oi–Pr)_4$, $Zr(DIBM)_4$, and H_2O on an ALD Ir-coated SiO_2 pinhole structure of 0.18 μm diameter and (**b–d**) EDS line scan profiles taken at three specific positions shown in (**a**) (Reprinted with permission from [50]. Copyright 2008, The Electrochemical Society.)

adjusted by the number of Zr–O cycles during one total Pb–O/Zr–O/Ti–O cycling sequence, while the [Pb]/([Zr] + [Ti]) ratio stabilized at a value of about 1.2 irrespective of the Zr content (see Figs. 6.11 a, b). This process window, which was attributed to the limited chemisorption sites on the growing Pb–O surface for the $Zr(DIBM)_4$ and $Ti(Oi–Pr)_4$ molecules, enables an adjustment of the Zr/Ti composition of the ALD PZT films as required by the application. The excess of PbO of about 20 % in the as-deposited amorphous films is advantageous because it can compensate the PbO loss during high-temperature annealing for crystallization of the film into the FE perovskite phase. A PbO-rich amorphous phase may help to suppress crystallization into the Pb-deficient non-FE pyrochlore-type phase.

Another report on ALD of PbO and $PbTiO_3$ films using a nitrogen containing lead precursor, $Pb(DMAMP)_2$ (DMAMP = bis(3-*N*,*N*-dimethylamino-2-methyl-2-propoxide)), and $Ti(Oi–Pr)_4$ was given by Lee et al., where an unusual growth behavior depending on the cation composition ratio was found [43].

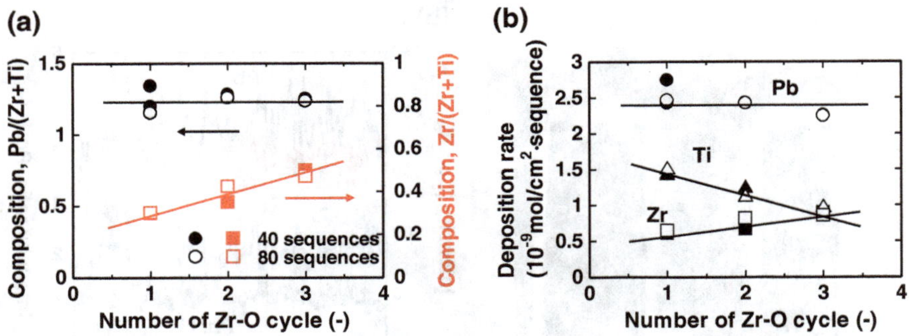

Fig. 6.11 Film composition (**a**) and deposition rates (**b**) as a function of the repetition numbers of the Zr–O cycles in one Pb–O/n Zr–O/Ti–O sequence. The ALD PZT film was grown from $Pb(TMHD)_2$, $Ti(Oi–Pr)_4$, $Zr(DIBM)_4$, and H_2O. (Reprinted with permission from [50]. Copyright 2008, The Electrochemical Society.)

6.4.3 Crystallization of ALD PZT Films

Amorphous ALD PZT films with 70 nm thickness were grown on planar Pt-coated Si substrates. The compositional ratios measured by XRF were [Pb]/([Zr] + [Ti]) = 1.2 and [Zr]/([Zr] + [Ti]) = 0.47. The ALD PZT films were crystallized in a two-step annealing process, i.e., 400 °C for 30 min and subsequently 700 °C for 10 nm in oxygen [50]. X-ray diffraction analysis revealed the formation of a single-phase perovskite structure with almost random crystal orientation (see Fig. 6.12a). No peaks were detected from the pyrochlore phase. Platinum circular electrodes with a diameter of 250 μm were deposited by sputtering through a shadow mask. After postmetallization annealing at 400 °C for 10 min the FE hysteresis of the planar Pt/70 nm ALD-PZT/Pt capacitor was recorded using a 100 Hz triangular wave (see Fig. 6.12b). For the remanent polarization ($2P_r$) and the coercive field ($2E_c$), values of about 24 μC/cm^2 and 200 kV/cm were determined, respectively [50].

6.4.4 Nanostructure Analysis of ALD PZT Films

Although the ALD PZT thin films in the as-deposited state were amorphous under X-ray diffraction analysis, they might contain nano-size crystallites which are too small to be detected by standard XRD techniques. However, the information on the nanostructure of the multicomponent ALD films in the as-deposited state is of great importance for understanding the ALD process and for optimizing the crystallization process into the FE perovskite structure. Therefore, a high-resolution transmission electron microscopy (HRTEM) study and STEM analysis were performed on an ALD PZT film grown from $Pb(TMHD)_2$, $Ti(Oi–Pr)_4$, $Zr(DIBM)_4$, and H_2O at 240 °C on ALD Ir-coated SiO_2. These studies on the nanoscale

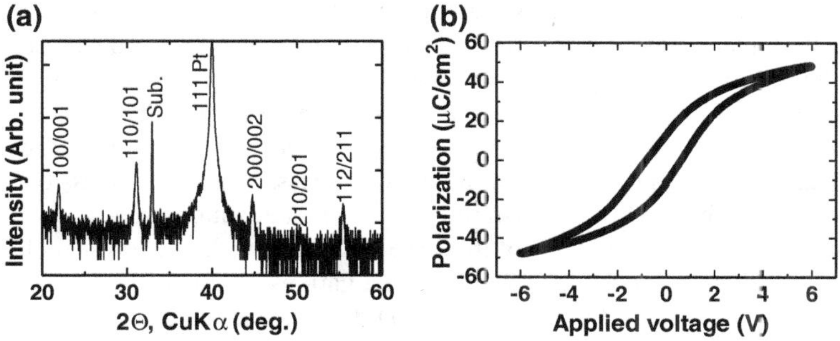

Fig. 6.12 **a** XRD pattern of an ALD PZT film (t = 70 nm, [Pb]/([Zr] + [Ti]) = 1.2, [Zr]/([Ti] + [Zr]) = 0.47 deposited at 240 °C) after annealing at 400 °C for 30 min and at 700 °C for 10 min. **b** P–V hysteresis loop of the PZT film in (**a**) with Pt circular top electrodes of 0.25 mm diameter. (Reprinted with permission from [50]. Copyright 2008, The Electrochemical Society.)

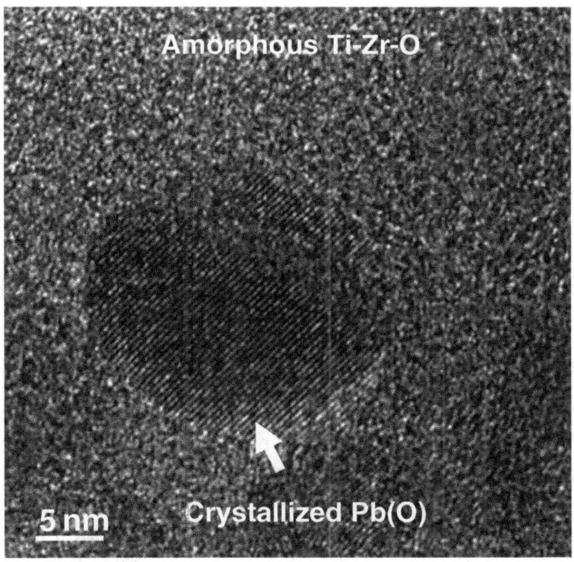

Fig. 6.13 HRTEM image of an as-deposited ALD PZT film grown at 240 °C from $Pb(TMHD)_2$, $Ti(Oi–Pr)_4$, $Zr(DIBM)_4$, and H_2O on ALD Ir-coated SiO_2. The image reveals a nano composite-type structure of PbO_x nano-crystals embedded in an amorphous Ti–Zr–O matrix. (Reprinted with permission from [50]. Copyright 2008, The Electrochemical Society.)

revealed that the as-deposited ALD PZT film contained an amorphous Ti–Zr–O matrix with a homogeneous distribution of titanium and zirconium with embedded nanoclusters of PbO_x [50]. Figure 6.13 shows the HRTEM micrograph of a single PbO_x nanocrystallite with a diameter of about 10 nm embedded in the amorphous Ti–Zr–O matrix.Unfortunately, the size of the crystallites was too small to identify the PbO_x phase, i.e., to determine of x. It has to be mentioned that HRTEM studies on an as-deposited ALD PZT film grown from $Pb(TMHD)_2$, $Ti(Oi–Pr)_4$, $Zr(TMHD)_4$, and H_2O at 240 °C on ALD Ir-coated SiO_2 did not reveal a comparable nano composite structure. Thus, the comparison of the film nanostructures with the film compositions shows that the film from ALD system 2 with excess Ti content and under stoichiometric Pb content exhibits an amorphous nanostructure, while the film from ALD system 3 with nearly the same Ti and Zr content and over stoichiometric Pb content shows a nano composite structure. These observations may point toward key features for a further understanding of the growth and crystallization of ALD PZT thin films. Nevertheless, a deeper discussion on the possible role of the PbO_x nanocrystallites in the crystallization process of the ALD PZT films was beyond the scope of this case study and would be too speculative. It must therefore be postponed for future investigations together with a crystallization study of the ALD PZT thin films in 3D device-type structures.

Within the scope of a case study, ALD-type processes for the multi-component FE oxide $Pb(Zr_xTi_{1-x})O_3$ based on a combination of binary oxide processes utilizing different alkoxide and β-diketonate-type precursors were successfully demonstrated. Additionally, conformality and compositional homogeneity were demonstrated for films grown into capacitor pinhole structures of 0.18 μm in diameter.

6.5 Trends in Ferroelectric Field Effect Transistors

FeFET utilizes a thin FE oxide layer to replace the gate oxide. This enables the source-drain current to be controlled by varying the polarization state in the FE oxide. Hence, the FeFET represents a nonvolatile FE memory device with highest integration density, which enables a nondestructive read-out of the stored information, i.e., the orientation of the polarization [62]. Recently, research on FeFET devices experienced renewed interest after the successful integration of high-*k* oxides, such as HfO_2, as gate dielectrics. One of the key factors for this evolution is the availability of ALD processes enabling the deposition of an oxide layer without a severe degradation of the transistor channel properties. A modified ALD PZT process was recently reported for the fabrication of integrated metal-FE-insulator-semiconductor diodes [63]. Another exciting trend was the discovery of ferroelectricity in silicon-doped HfO_2 and ZrO_2 thin films. A closer investigation revealed that the doping stabilizes an orthorhombic state of the binary oxide which is FE [64]. Especially, the proven compatibility of these oxides makes them attractive for integration into FeFETs, which is reflected by increased activities in this field [65].

Acknowledgments The authors thank Prof. C.S. Hwang (Seoul National University, Korea) and Prof. R. Waser and Dr. U. Böttger (RWTH Aachen University, Germany) for fruitful discussions. Thanks is also to S. Carella, L. Cataneo (SAES Getters SEA, Italy), and Dr. Y. Tasaki (Toshima MfG Co., Ltd., Japan) for precursor supply and discussion, Samsung Advanced Institute of Technology (SAIT, Korea) for supply of the 3D substrates, Dr. A. Besmehn, Dr. S. Willbold, and W. Krumpen (ZCH, FZJ, Germany) for chemical analysis, M. Gebauer and M. Gerst (PGI-7, FZJ) for technical support, Dr. H.-J. Penkalla and D. Esser (IEK-2, FZJ) for FIB lamellae preparation, and Dr. T. Weirich and F. Dorn (GFE, RWTH Aachen University) for STEM and HRTEM analysis. T.W. also extends thanks to the Alexander von Humboldt Stiftung (AvH) for awarding him a research fellowship.

References

1. Böttger U, Summerfelt SR (2005) Ferroelectric random access memories. In: Waser R (ed) Nanoelectronics and Information Technology, WILEY-VCH Verlag GmbH & Co. KGaA, Berlin, pp 565–589
2. Scott JF, Paz-de-Araujo CA (1989) Ferroelectric memories. Science 246:1400–1405
3. Koike H, Otsuki T, Kimura T, Fukuma M, Hayashi Y, Maejima Y, Amanuma K, Tanabe N, Matsuki T, Saito S, Takeuchi T, Kobayashi S, Kunio T, Hase T, Miyasaka Y, Shohata N, Takada M (1996) A 60 ns 1 Mb nonvolatile ferroelectric memory with non-driven cell plate line write/read scheme. In: 1996 IEEE international solid-state circuits conference, Digest of Technical Papers 39, pp 368–369
4. Takashima D, Kunishima I (1998) High-density chain ferroelectric random access memory (chain FRAM). IEEE J Solid State Circ 33:787–792
5. Sheikholeslami A, Gulak PG (2000) A survey of circuit innovations in ferroelectric random-access memories. Proc IEEE 88:667–689
6. Park SO, Bae BJ, Yoo DC, Chung U-I (2008) Ferroelectric random access memory. In: Waser R (ed) Nanotechnology, vol 3: Information Technology I. Wiley-VCH Verlag GmbH & Co. KGaA, Weinheim, pp 397–418
7. Kim HH, Song YJ, Lee SY, Joo HJ, Jang NW, Jung DJ, Park YS, Park SO, Lee KM, Joo SH, Lee SW, Nam SD, Kim K (2002) Novel integration technologies for highly manufacturable 32 Mb FRAM. In: 2002 symposium on VLSI technology, Digest of Technical Papers, IEEE, pp 210–211
8. Lee H, Hesse D, Zakharov N, Goesele U (2002) Ferroelectric $Bi_{3.25}La_{0.75}Ti_3O_{12}$ films of uniform a-axis orientation on silicon substrates. Science 296:2006–2009
9. Watanabe T, Funakubo H (2006) Controlled crystal growth of layered-perovskite thin films as an approach to study their basic properties, J. Appl. Phys. 100:51602/1–11
10. DeAraujo CAP, Cuchiaro JD, McMillan LD, Scott MC, Scott JF (1995) Fatigue-free ferroelectric capacitors with platinum electrodes. Nature 374:627–629
11. Ramesh R (1997) Thin film ferroelectric materials and devices. Kluwer Academic Publishers, Boston
12. Scott JF (2000) Ferroelectric memories. Springer, Berlin
13. Okuyama M, Ishibashi Y (2005) Ferroelectric thin films, Basic properties and device physics for memory applications. Springer, Berlin
14. Waser R, Boettger U, Tiedke S (2005) Polar oxides: properties, characterization, and imaging. WILEY-VCH Verlag GmbH & Co. KGaA, Weinheim
15. Yokoyama S, Honda Y, Morioka H, Okamoto S, Funakubo H, Iijima T, Matsuda H, Saito K, Yamamoto T, Okino H, Sakata O, Kimura S (2005) Dependence of electrical properties of epitaxial $Pb(Zr,Ti)O_3$ thick films on crystal orientation and Zr/(Zr + Ti) ratio. J Appl Phys 98:94106-1-8
16. Pertsev NA, Kukhar VG, Kohlstedt H, Waser R (2003) Phase diagrams and physical properties of single-domain epitaxial $Pb(Zr_{1-x}Ti_x)O_3$ thin films. Phys Rev B (Condens Matter Mater Phys) 67:54107-1-10
17. Qiu QY, Mahjoub R, Alpay SP, Nagarajan V (2010) Misfit strain-film thickness phase diagrams and related electromechanical properties of epitaxial ultra-thin lead zirconate titanate films. Acta Mater 58:823–835
18. Newnham R, Wolfe R, Dorrian J (1971) Structural basis of ferroelectricity in bismuth titanate family. Mater Res Bull 6:1029

19. Wouters DJ, Maes D, Goux L, Lisoni JG, Paraschiv V, Johnson JA, Schwitters M, Everaert J, Boullart W, Schaekers M, Willegems M, Vander Meeren H, Haspeslagh L, Artoni C, Caputa C, Casella P, Corallo G, Russo G, Zambrano R, Monchoix H, Vecchio G, Van Autryve L (2006) Integration of $SrBi_2Ta_2O_9$ thin films for high density ferroelectric random access memory. J Appl Phys 100:51603/1–16
20. Goux L, Maes D, Lisoni JG, Vander Meeren H, Paraschiv V, Haspeslagh L, Artoni C, Russo G, Zambrano R, Wouters DJ (2006) Scaling potential of pin-type 3-D SBT ferroelectric capacitors integrated in 0.18 μm CMOS technology. Microelectron Eng 83:2027–2031
21. Richter D, Trolier-McKinstry S (2005) Ferroelectrics. In: Waser R (ed) Nanoelectronics and information technology. WILEY-VCH Verlag GmbH & Co. KGaA, Berlin, pp 59–78
22. Ramtron and http://www.ramtron.com. Nonvolatile F-RAM Memory
23. Symetrix and http://www.symetrixcorp.com. Trinion non-volatile memory chip in smart cards or smart tags
24. Texas Instruments and http://www.ti.com. FRAM
25. Fujitsu Ltd. (2005) Electronic devices and http://www.fujitsu.com. FRAM Guide Book
26. Kim H (2003) Atomic layer deposition of metal and nitride thin films: current research efforts and applications for semiconductor device processing. J Vac Sci Technol B 21:2231–2261
27. Aaltonen T, Ritala M, Tung Y, Chi Y, Arstila K, Meinander K, Leskela M (2004) Atomic layer deposition of noble metals: exploration of the low limit of the deposition temperature. J Mater Res 19:3353–3358
28. Hamalainen J, Hatanpaa T, Puukilainen E, Sajavaara T, Ritala M, Leskela M (2011) Iridium metal and iridium oxide thin films grown by atomic layer deposition at low temperatures. J Mater Chem 21:16488–16493
29. Udayakumar KR, Moise TS, Summerfelt SR, Boku K, Remack K, Rodriguez J, Arendt M, Shinn G, Eliason J, Bailey R, Staubs P (2008) Manufacturable high-density 8 Mbit one transistor-one capacitor embedded ferroelectric random access memory. Jpn J Appl Phys 47:2710–2713
30. Gilbert S, Hunter S, Ritchey D, Chi C, Taylor D, Amano J, Aggarwal S, Moise T, Sakoda T, Summerfelt S, Singh K, Kazemi C, Carl D, Bierman B (2003) Preparation of $Pb(Zr, Ti)O_3$ thin films by metalorganic chemical vapor deposition for low voltage ferroelectric memory. J Appl Phys 93:1713–1717
31. Aratani M, Oikawa T, Ozeki T, Funakubo H (2001) Epitaxial-grade polycrystalline $Pb(Zr, Ti)O_3$ film deposited at low temperature by pulsed-metalorganic chemical vapor deposition. Appl Phys Lett 79:1000–1002
32. Tokita K, Aratani M, Funakubo H (2002) Effect of atmosphere during heating of substrate on the low temperature deposition of metalorganic chemical vapor deposited $Pb(Zr_xTi_{1-x})O_3$ thin films. Appl Phys Lett 81:898–900
33. Asano G, Oikawa T, Funakubo H, Saito K (2003) Good ferroelectricity of $Pb(Zr, Ti)O_3$ thin films fabricated by highly reproducible deposition on bottom Ir electrode at 395 °C. Jpn J Appl Phys 42:L1083–L1086
34. Lee S, Kim K (2006) Current development status and future challenges of ferroelectric random access memory technologies. Jpn J Appl Phys 45:3189–3193
35. ITRS (2007) The International Technology Roadmap for Semiconductors. 2007 edn. http://www.itrs.net/links/2007ITRS/
36. Park Y, Lee JH, Koo JM, Kim SP, Shin S, Cho CR, Lee JK (2004) Preparation of $Pb(Zr_xTi_{1-x})O_3$ films on trench structure for high-density ferroelectric random access memory. Integr Ferroelectr 66:85–95
37. Koo J, Seo B, Kim S, Shin S, Lee J, Baik H, Lee J, Lee J, Bae B, Lim J, Yoo D, Park S, Kim H, Han H, Baik S, Choi J, Park Y, Park Y (2005) Fabrication of 3D trench PZT capacitors for 256Mbit FRAM device application. In: IEEE international electron devices meeting 2005, Technical Digest, pp 351–354
38. Shin S, Han H, Park YJ, Choi JY, Park Y, Baik S (2007) Characterization of 3D trench PZT capacitors for high density FRAM devices by synchrotron X-ray microdiffraction, Synchrotron Radiat Instr, Pts 1 and 2 879:1554–1556

39. Shin W, Ryu S, You I, Yoon S, Cho S, Lee N, Kim K, Yu B, Lee W, Choi K, Yoon S (2004) Low voltage switching characteristics of 60 nm thick $SrBi_2Ta_2O_9$ thin films deposited by plasma-enhanced ALD. Electrochem Solid State Lett 7:F31–F34
40. Harjuoja J, Kosola A, Putkonen M, Niinisto L (2006) Atomic layer deposition and post-deposition annealing of $PbTiO_3$ thin films. Thin Solid Films 496:346–352
41. Harjuoja J, Vayrynen S, Putkonen M, Niinisto L, Rauhala E (2007) Atomic layer deposition of $PbZrO_3$ thin films. Appl Surf Sci 253:5228–5232
42. Hwang GW, Lee HJ, Lee K, Hwang CS (2007) Atomic layer deposition and electrical properties of $PbTiO_3$ thin films using metallorganic precursors and H_2O. J Electrochem Soc 154:G69–G76
43. Lee HJ, Park MH, Min Y–S, Clavel G, Pinna N, Hwang CS (2010) Unusual growth behavior of atomic layer deposited $PbTiO_3$ thin films using water and ozone as oxygen sources and their combination. J Phys Chem C114:12736–12741
44. Watanabe T, Hoffmann-Eifert S, Hwang CS, Waser R (2006) Liquid-injection atomic Layer deposition of TiO_x and Pb-Ti-O films. J Electrochem Soc 153:199–204
45. Watanabe T, Hoffmann-Eifert S, Mi S, Jia C, Waser R, Hwang CS (2007) Growth of ternary $PbTiO_x$ films in a combination of binary oxide atomic layer depositions. J Appl Phys 101:14114/1-8
46. Watanabe T, Hoffmann-Eifert S, Peter F, Mi S, Jia C, Hwang CS, Waser R (2007) Liquid injection ALD of $Pb(Zr,Ti)O_x$ thin films by a combination of self-regulating component oxide processes. J Electrochem Soc 154:G262–G269
47. Watanabe T, Hoffmann-Eifert S, Waser R, Hwang CS (2007) Atomic layer deposition of $Pb(Zr,Ti)O_x$ thin films by a combination of binary atomic layer deposition. In: Proceedings of the 2006 15th IEEE international symposium on applications of ferroelectrics. IEEE, Piscataway, pp 89–92
48. Watanabe T, Hoffmann-Eifert S, Waser R, Hwang CS (2007) Liquid injection atomic layer deposition of $Pb(Zr,Ti)O_3$ thin films on three dimensional structures, In: Proceedings of the 2007 16th ieee international symposium on applications of ferroelectrics. IEEE, Piscataway, pp 156–158
49. Hoffmann-Eifert S, Watanabe T, Hwang CS, Waser R (2008) Liquid injection atomic layer deposition of perovskite-type multi-component oxide thin films for ferroelectric and higher-k three dimensional capacitor structures. 2008 17th IEEE international symposium on the applications of ferroelectrics, IEEE, Piscataway, pp 446–447
50. Watanabe T, Hoffmann-Eifert S, Hwang CS, Waser R (2008) Growth behavior of atomic-layer-deposited Pb(Zr, Ti)Ox thin films on planar substrate and three-dimensional hole structures. J Electrochem Soc 155:D715–D722
51. Vehkamaki M, Ritala M, Leskela M, Jones AC, Davies HO, Sajavaara T, Rauhala E (2004) Atomic layer deposition of strontium tantalate thin films from bimetallic precursors and water. J Electrochem Soc 151:F69–F72
52. Schumacher M, Baumann P, Seidel T (2006) AVD and ALD as two complementary technology solutions for next generation dielectric and conductive thin-film processing. Chem Vapor Depos 12:99–108
53. Watanabe T, Hoffmann-Eifert S, Yang L, Ruediger A, Kuegeler C, Hwang CS, Waser R (2007) Liquid injection atomic layer deposition of TiO_x films using $Ti[OCH(CH_3)_2]_4$. J Electrochem Soc 154:G134–G140
54. Shin J, Hong S, Lee J, Cho H, Kim K, Hwang C, Kim H (1998) Preparation and characterization of $(Pb, La)TiO_3$ thin films by solid source mixture metal organic chemical vapor deposition. J Kor Phys Soc 32:S1529–S1531
55. Kim SK, Hoffmann-Eifert S, Reiners M, Waser R (2011) Relation between enhancement in growth and thickness-dependent crystallization in ALD TiO_2 thin films. J Electrochem Soc 158:D6–D9
56. Kim SK, Hoffmann-Eifert S, Mi S, Waser R (2009) Liquid injection atomic layer deposition of crystalline TiO_2 thin films with a smooth morphology from $Ti(O\text{-}iPr)_2(DPM)_2$. J Electrochem Soc 156:D296–D300

57. Aarik J, Aidla A, Uustare T, Ritala M, Leskela M (2000) Titanium isopropoxide as a precursor for atomic layer deposition: characterization of titanium dioxide growth process. Appl Surf Sci 161:385–395
58. Jones AC, Hitchman ML (2009) Chemical vapor deposition: precursors, processes and applications. Royal Social of Chemistry, RSC Publishing, Cambridge; ISBN: 978-0-85404-465-8
59. Si J, Desu S, Tsai C (1994) Metal-organic chemical-vapor-deposition of ZrO_2 films using $Zr(THD)_4$ as precursors. J Mater Res 9:1721–1727
60. Cattaneo L, Carella S (2007) SAES Getters Inc. and Italy, (unpublished data)
61. Zhao JS, Lee HJ, Lee K, Sim JS, Hwang CS (2006) Influence of the Pt top electrode annealing procedure on the ferroelectric property of MOCVD $Pb(Zr_{0.2}Ti_{0.8})O_3$ thin films. Electrochem Solid-State Lett 9:F69–F72
62. Kohlstedt H, Ishiwara H (2005) Ferroelectric field effect transistors. In: Waser R (ed) Nanoelectronics and information technology. WILEY-VCH Verlag GmbH & Co. KGaA, Berlin, pp 387–405
63. Zhang F, Perng YC, Choi JH, Wu T, Chung TK, Carman GP, Locke C, Thomas S, Saddow SE, Chang JP (2011) Atomic layer deposition of $Pb(Zr,Ti)O_x$ on 4H-SiC for metal-ferroelectric-insulator-semiconductor diodes. J Appl Phys 109:124109/1-4
64. Boescke TS, Mueller J, Braeuhaus D, Schroeder U, Boettger U (2011) Ferroelectricity in hafnium oxide. Appl Phys Lett 99:102903
65. Mueller J, Boescke TS, Schroeder U, Hoffmann R, Mikolajick T, Frey L (2012) Nanosecond polarization switching and long retention in a novel MFIS-FET based on ferroelectric HfO_2. IEEE Electron Device Lett 33:185–187

Part IV
ALD for Logic Devices

Chapter 7
Front End of the Line Process

Jeong Hwan Han, Moonju Cho, Annelies Delabie, Tae Joo Park and Cheol Seong Hwang

7.1 Introduction

The gate dielectric oxide plays a key role in the performance and reliability of Metal–Oxide–Semiconductor Field Effect Transistors (MOSFETs), a typical logic device in modern ultra large-scale integrated chips. The interface state density (N_{it}) and the charge trapping behavior at the interface between the gate dielectric film and a Si substrate determine the channel mobility degradation, threshold voltage (V_{TH}) instability, and life time of the device. SiO_2/Si gate stack fabricated by thermal oxidation of the Si substrate has provided the outstanding interface properties for several decades. However, the continuous scaling of the devices has driven the development of the alternative gate dielectric materials and processes. In recent years, atomic layer deposition (ALD) HfO_2-based gate dielectrics with a metal gate have been implemented in a mass production [1].

ALD of metal oxides is based on sequential self-limiting chemisorption reactions of a metal-containing precursor and an oxygen source [1, 2]. Both precursors affect the quality of high-k films as well as the interfacial layer (IL) at the interface with a substrate. Therefore, appropriate choice of metal precursor and oxygen source, and understanding of their reaction mechanisms on the various surfaces are important to obtain the promising performance and reliability of MOSFETs.

J. H. Han (✉) · M. Cho · A. Delabie
Imec, Kapeldreef 75, 3001 Leuven, Belgium
e-mail: hanj@imec.be

A. Delabie
Department of Chemistry, University of Leuven, Celestijnenlaan 200F, 3001 Leuven, Belgium

T. J. Park
Department of Materials Engineering, Hanyang University, Ansan 426-791, Korea

C. S. Hwang
Department of Materials Science and Engineering and Inter-university Semiconductor Research Center, Seoul National University, Seoul 151-744, Korea

C. S. Hwang (ed.), *Atomic Layer Deposition for Semiconductors*,
DOI: 10.1007/978-1-4614-8054-9_7,

Furthermore, evolutions of new channel materials (SiGe, Ge, and III–V compound semiconductors such as GaAs, InGaAs, GaSb, InP, etc.) [3, 4] and multi-dimensional devices including FinFETs or nanowire devices [5–7] further increased the importance of appropriate metal precursor and the oxygen source for the ALD process.

In this chapter, the ALD characteristics and properties of Hf-based high-k films from different types of metal precursors, such as $HfCl_4$, $Hf(N(C_2H_5)(CH_3))_4$, $(HfN(CH_3)_2)_4$, and $HfO^tBu(NEtMe)_3$ are compared, and the influences of oxygen sources type, H_2O and O_3, are discussed. The discussion covers the ALD high-k films on not only Si substrate but also the assorted alternative substrates such as Ge and III–Vs. Finally, the performance and reliability of the sub 1 nm equivalent-oxide-thickness (EOT) planar Si channel devices, SiGe- or III–Vs channel devices, and FinFET devices with HfO_2-based gate oxides grown by ALD are discussed in detail. In fact, there have been many studies of other high-k dielectrics, such as lanthanide-based oxides and ternary oxides, but dealing with all the other high-k layers than HfO_2 is not feasible for the given space of this chapter. However, this does not undermine the importance of other high-k and higher-k dielectrics for futuristic MOSFET devices.

7.2 Metal Precursor

7.2.1 Hafnium Chlorides ($HfCl_4$)

$HfCl_4$ has been the most commonly employed inorganic Hf precursor to fabricate HfO_2-based dielectrics because it has the advantages of smaller molecule size and high thermal stability (>600 °C) providing less steric hindrance and a wider temperature window for self-limited growth compared with the metal-organic (MO) Hf precursors. In addition, HfO_2 film from $HfCl_4$ is free from carbon contamination which might deteriorate the physical/chemical and electrical properties of HfO_2. HfO_2 ALD using $HfCl_4$ is typically performed in combination with either H_2O or O_3. In combination with H_2O as the oxygen source, the process demonstrates fluent ALD saturation behavior at the reactor temperatures between 200 and 600 °C.

For $HfCl_4/H_2O$ process, reaction pathway of ALD HfO_2 is explained by the ligand exchange reaction between $HfCl_4$ and surface hydroxyl (OH) groups, releasing HCl [8, 9]. Thus, the amount of OH groups existing on the surface can strongly influence the initial growth behavior as well as the steady-state growth. The steady-state growth per cycle of ALD HfO_2 films at the deposition temperatures of 200 and 400 °C are ~0.13 and ~0.044 nm/cycle, respectively, as shown in Fig. 7.1a [10]. Here, decrease in growth rate as increasing growth temperature is attributed to lower OH density on the film surface at higher reactor temperature [9–11]. Initial HfO_2 growth behavior is also very sensitive to the status of starting

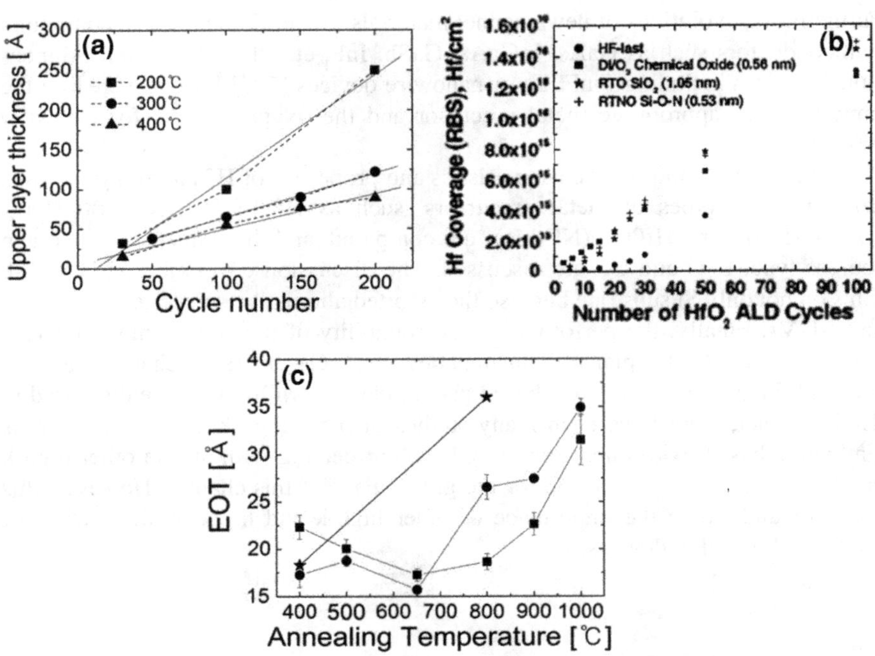

Fig. 7.1 (**a**) Change in the HfO_2 (*upper layer*) thicknesses as a function of the number of ALD cycles at the reactor temperatures from 200 to 400 °C [10]. (**b**) The Hf coverage as a function of the number of ALD $HfCl_4/H_2O$ cycles on the various substrates [12]. (**c**) The EOT of MIS stacks after PDA at the various temperatures. The HfO_2 films were deposited at the temperatures of 300 (*square symbol*) and 400 °C (*circle symbol*), respectively. *Star symbol* corresponds to HfO_2/Si stack without a RBL [15]

surfaces. Figure 7.1b showed the variation of Hf-coverage at the early growth stage with the number of ALD cycles on the variously prepared Si surfaces, such as chemical SiO_2, thermal SiO_2, and diluted HF-cleaned Si [12]. To achieve excellent EOT scalability of gate stack, thick interfacial low-k SiO_2 layer is unfavorable in spite of the improved initiation behavior of ALD on it. However, on HF-cleaned Si surface (H-terminated surface) an incubation period of ~20 cycles was observed, because of the lack of OH functional group, while linear growth behavior without the incubation step was exhibited on the chemical SiO_2 surface. H-terminated Si hinders nucleation of ALD HfO_2 resulting in rough, three-dimensional (3D), and nonlinear growth during first few ALD cycles due to the retarded chemisorptions of $HfCl_4$. Beyond the influence on the nucleation and growth rates of HfO_2 films, wafer temperature of ALD process affected the crystalline structure of resulting HfO_2 and thickness of the interfacial SiO_x layer which is formed between Si and HfO_2 layer during HfO_2 ALD process. The microstructure of HfO_2 films deposited on Si was investigated with varying the growth temperatures from 200 to 370 °C [13]. As-deposited HfO_2 film grown at

200 °C showed amorphous phase, whereas HfO_2 films grown at 300 and 370 °C appeared polycrystalline with monoclinic and tetragonal phases. With increasing post deposition annealing (PDA) temperature, the portion of tetragonal phase was decreased, resulting in the monoclinic-rich phase [13]. The formation of the interfacial layer (IL) is influenced by the process temperature of HfO_2 ALD, but PDA and pre-metal degas conditions play a critical role. Cho et al. [10] reported that an IL is spontaneously growing at all investigated growth temperatures ranging from 200 to 400 °C when ALD HfO_2 films using $HfCl_4/H_2O$ were deposited on the H-terminated Si. At 200 °C, thickness of the interfacial SiO_x layer increased with the increasing number of ALD HfO_2 cycles from 1.4 nm at 30 cycles to 5.5 nm at 200 cycles. However, IL thickness decreased as increasing growth temperature over 300 °C due to the dissolution of the IL into the growing HfO_2 layer. It was reported that the use of a pre-metallization degas improves EOT scaling [14]. The origin of the improvement is in the removal of H_2O from the high-*k* dielectric. Since HfO_2 is a poor diffusion barrier for H_2O, and H_2O is a strong oxidizer of silicon, H_2O adsorbed to the dielectric during air exposure can diffuse down to the silicon surface and contribute to interfacial layer regrowth during subsequent high temperature steps in the process. As a result, the interfacial oxide thickness and hence the EOT increases. For aggressive EOT scaling, it is therefore important to reduce the amount of H_2O released at temperatures where oxidation of the silicon is favorable.

In spite of thin IL layer of $\sim$1–2 nm at the high growth temperatures of 300 or 400 °C, the low-k $HfSiO_x$ layer formed after PDA over 800 °C by intermixing of HfO_2 and SiO_x which increased the EOT.

To prevent the increase in EOT resulting from IL formation, a thin ALD Al_2O_3 layer was introduced as a reaction barrier layer (RBL) [15]. By interposing RBL, EOT increase of $HfO_2/Al_2O_3/SiO_2$ stack after PDA can be remarkably reduced indicating Si diffusion into HfO_2 layer is suppressed as shown in Fig. 7.1c. Although ALD Al_2O_3 films grown with either O_2 plasma or H_2O as oxygen source were adopted, Al_2O_3 layer from O_3 showed the best electrical properties regarding the charge injection, stability against flat band voltage (V_{FB}) shift, and increase in leakage current density due to stoichiometric O/Al composition originating from strong oxidation power of O_3. Al_2O_3 RBL, however, brings about high density of negative interfacial fixed charge between Al_2O_3 and SiO_2 which induces positive V_{FB} shift. Nitridation of RBL layer greatly improved the thermal stability of the capacitance–voltage (C–V) characteristics, providing ideal V_{FB} and very small hysteresis for HfO_2/Al_2O_3 stack [16]. As the N incorporation into the Al_2O_3 layer produced positive fixed charges, the negative fixed charges at Al_2O_3/SiO_2 could be compensated. Another way to optimize V_{FB} is the combination of the thicknesses of the Al_2O_3 and HfO_2 with an appropriate ratio which results in the ideal V_{FB} value because positive fixed charge at the HfO_2/Al_2O_3 interface can be compensated by the negatively fixed charge at Al_2O_3/SiO_2.

Unlike HfO_2 film deposited from MO precursors, effect of Cl residue in the HfO_2 on the electrical properties of HfO_2-based MOSFETs should be considered in case of $HfCl_4/H_2O$ process. Therefore, time-dependent dielectric breakdown

(TDDB) characteristics of HfO_2 films grown with different H_2O pulse time were examined [17]. Cl concentration in the deposited HfO_2 films is decreasing as increasing H_2O pulse time from 0.3 to 90 s. It was found that one order higher magnitude of Cl concentration does not exacerbate the TDDB characteristic of the film. Furthermore, first-principles calculation proved that additional trap energy level is not formed inside the HfO_2 band gap when Cl content of HfO_2 film increases.

As MOSFET on high-mobility channels such as Ge and III–V compound semiconductors would offer significant improvements in the electrical performances over Si-based MOSFET, ALD HfO_2 on Ge, GaAs, and InGaAs substrates has obtained intensive interests [18–20]. ALD HfO_2 on HF-cleaned Ge surface was investigated using $HfCl_4$ and H_2O. The available reaction sites existing on HF-cleaned Ge for $HfCl_4$ chemisorptions are OH groups and possibly also oxygen bridges (Ge–O–Ge). These remain present on the Ge surface after HF-cleaning, in contrast to Si surfaces, which provide Si–H termination which is a poor reaction site for chemisorptions of $HfCl_4$. HF-cleaned Ge is, therefore, a more favorable surface for initiation of HfO_2 film than HF-cleaned Si. For this reason, substrate-enhanced ALD HfO_2 growth was obtained at 300 °C for first a few ALD cycles on Ge surface. The steady-state growth rate is $\sim$0.04 nm/cycle which is comparable with that on Si. The optimized growth condition allowed promising scalability of HfO_2/Ge stack with thin interfacial GeO_2 layer ($< \sim$0.4 nm) and uniform/smooth HfO_2 film which was as thin as 1.6 nm [18]. In contrast, additional oxidation of Ge occurred when O_3 was employed as the oxygen source in $HfCl_4$-based ALD process [21]. HfO_2 ALD process with $HfCl_4/O_3$ induces thicker interfacial GeO_2 compared to HfO_2 ALD from $HfCl_4/H_2O$. Interestingly, an even thinner IL was formed during Al_2O_3 ALD process from trimethylaluminum (TMA)/O_3 on HF-cleaned Ge than $HfCl_4/O_3$ process. This is related to the difference in required O_3 pulse time for ALD saturation according to the types of metal precursor. ALD process using $HfCl_4$ precursor requires higher O_3 dose than other processes using MO precursor due to strong Hf–Cl bond. Change in the thickness of interfacial GeO_2 layer before and after ALD of HfO_2 and Al_2O_3 films on 1 nm-thick GeO_2 and HF-cleaned Ge is summarized in Fig. 7.2a.

Ge channel layer, however, has still suffered from inadequate interface properties with high-k oxide due to the high electrically active N_{it} originating from the lack of stable passivating native oxide in contrast to Si channel. In the case of Ge substrate, the volatilization and desorption of GeO from the interfacial GeO_2 layer [as described by $GeO_2 + Ge \rightarrow 2GeO(g)$] occurs at rather low temperatures ($\sim$400 °C) and leaves a large amount of interface states and charge trapping sites [22, 23]. This can be suppressed using the high-k materials to form a stable germanate such as La_2O_3, ZrO_2, Y_2O_3, etc., [22, 24–29] or the interfacial reaction barriers by various surface treatments [30–39]. Prior to high-k deposition, therefore, surface nitridation, oxidation, or sulfur (S) treatments have been performed on Ge to reduce N_{it} by passivating the defective sites on surface [37–39] Among them, effect of S-passivation of the Ge channel using $(NH_4)_2S$ on the interface quality of HfO_2/S/Ge stack deserves detailed explanation due to its high

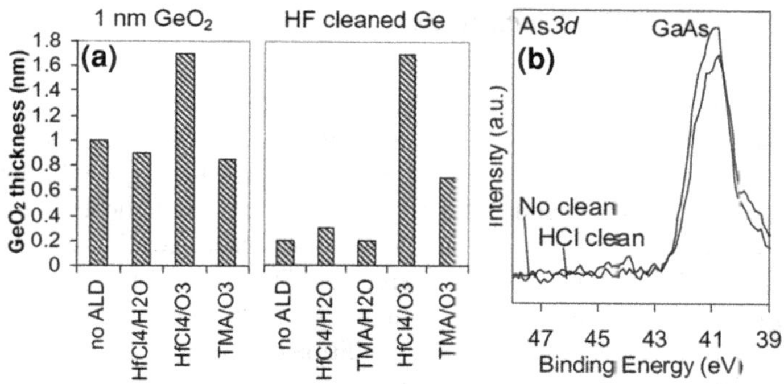

Fig. 7.2 (**a**) Thickness of GeO_2 layer before and after ALD of 2 nm thick HfO_2 or Al_2O_3 on 1 nm thick GeO_2 and HF-cleaned Ge [21]. (**b**) XP spectra of As 3d after 40 cycles $HfCl_4/H_2O$ process on GaAs with native oxide and HCl-cleaned GaAs [19]

effectiveness [39]. When the Ge surface is treated with $(NH_4)_2S$ prior to ALD HfO_2 or ALD Al_2O_3, interfaces between high-k oxides and Ge channel show improved IL properties that is free from defective GeO_x, suggesting S-treatment is very promising for EOT scaling as well as interfacial quality. Nevertheless, the midgap N_{it} of $10^{13}/cm^2$-eV for HfO_2/S-treated Ge is not sufficiently low compared to N_{it} for Al_2O_3/S-treated Ge. Therefore, bi-layer of HfO_2/Al_2O_3 was fabricated to reduce N_{it} and EOT. Consequently, high mobility of >200 cm^2/Vs at an EOT of 1.5 nm was obtained using 2 nm-HfO_2/2 nm Al_2O_3/S-treated Ge stack.

Another important channel material for future CMOS is III–V compound semiconductors, such as GaAs, InGaAs, etc., because of its higher carrier mobility even though their density of states is lower compared with Si. Initial reaction of $HfCl_4$ on GaAs substrate is quite different from that of Ge. ALD HfO_2 growth using $HfCl_4$ and H_2O was performed on native GaAs oxide and HCl-cleaned GaAs [19]. Here, "self-cleaning" of native oxide of GaAs was observed during the early growth stage. In X-ray photoelectron spectroscopy (XPS) of Fig. 7.2b, as 3d spectra corresponding to AsO_x native oxide disappeared after ALD $HfCl_4/H_2O$ process indicating that interfacial "self-cleaning" occurred. When an ALD HfO_2 process using tetrakis[diethylamino]hafnium (TDEAH, $Hf(N(C_2H_5)_2)_4$) precursor was employed to fabricate HfO_2 film on GaAs surface, the self-cleaning of IL was not found. On the other hand, Chang et al. [40] reported the removal of native AsO_x on InGaAs during HfO_2 deposition using tetrakis[ethylmethylamino]hafnium (TEMAH, $(HfN(C_2H_5)(CH_3))_4$). The "self-cleaning" of GaAs native oxide was also achieved from ALD Al_2O_3 growth where TMA and H_2O precursors are used [41–43]. These results suggest that the reduction of IL on III–V channel is strongly related to not only process parameters but also the types of metal precursors, emphasizing the importance of appropriate selection of metal precursor for native oxide free interface.

7.2.2 Metal–Organic Precursors

In spite of many advantages of $HfCl_4$ precursor mentioned in the previous section, the low reactivity and low vapor pressure of $HfCl_4$ lead to undesired growth behavior and film properties such as high substrate-dependent growth, low growth rate, and presence of Cl residue. Furthermore, formation of the HCl by-product might result in film etching as well as corrosion of reactor wall and increases the risk of failure of abatement system. Therefore, ALD growth of HfO_2 film from MO precursors has been investigated for achieving better growth characteristics and more ALD-hardware-friendly growth conditions. Hf–amide and Hf–alkoxide compounds are the most widely used among a wide variety of MO Hf precursors. In this section, growth behaviors and properties of ALD HfO_2 film from MO precursors are described.

7.2.2.1 Tetrakis[ethylmethylamino]hafnium ($Hf(N(C_2H_5)(CH_3))_4$, TEMAH) and Tetrakis[dimethylamino]hafnium ($Hf(N(CH_3)_2)_4$, TDMAH)

The nucleation behavior of ALD HfO_2 using TEMAH precursor is less dependent on the types of starting surfaces due to its higher reactivity as compared to the $HfCl_4$ precursor. The metal-nitrogen bond of the amide precursors such as Hf–N has relatively weak bond strength compared to metal-halide bond. Figure 7.3a showed the number of Hf atom on chemical SiO_2 and HF-cleaned Si as a function of the number of ALD TEMAH/H_2O cycles [44]. Although there are small incubation cycles on HF-cleaned Si, relatively less substrate dependency of initial behavior were obtained compared to $HfCl_4/H_2O$ ALD process shown in Fig. 7.1b. This allowed smoother surface morphology even on low-OH containing surface.

Influences of the growth temperature on the resulting HfO_2 properties were investigated on Si and Ge substrates using TEMAH and O_3 at the growth temperatures ranging from 160 to 360 °C [45]. Because of limited thermal stability of TEMAH precursor, high impurity concentration, too high growth rate, and non-uniformity of HfO_2 film were resulted in when the deposition temperature is over 300 °C due to thermal decomposition of Hf precursor (Fig. 7.3b). Growth rate of ALD HfO_2 at 160 °C is ~0.12 nm/cycle and decreases with increasing growth temperature to 280 °C indicating that the density of reaction sites decreases with increasing growth temperature, whereas film density increases with temperature, and saturated at 280 °C. Generally, lower growth temperature induces higher C impurity in the film which adversely affects the physical/electrical properties of the film. Jung et al. [46] reported the effects of the C concentration on the dielectric property and leakage current density of HfO_2. Crystalline structure of HfO_2 film grown at 200 °C showed a tetragonal phase after PDA at 600 °C which has higher dielectric constant than amorphous and monoclinic phase. The residual C can stabilize the tetragonal phase because defects induced by C impurity lower the

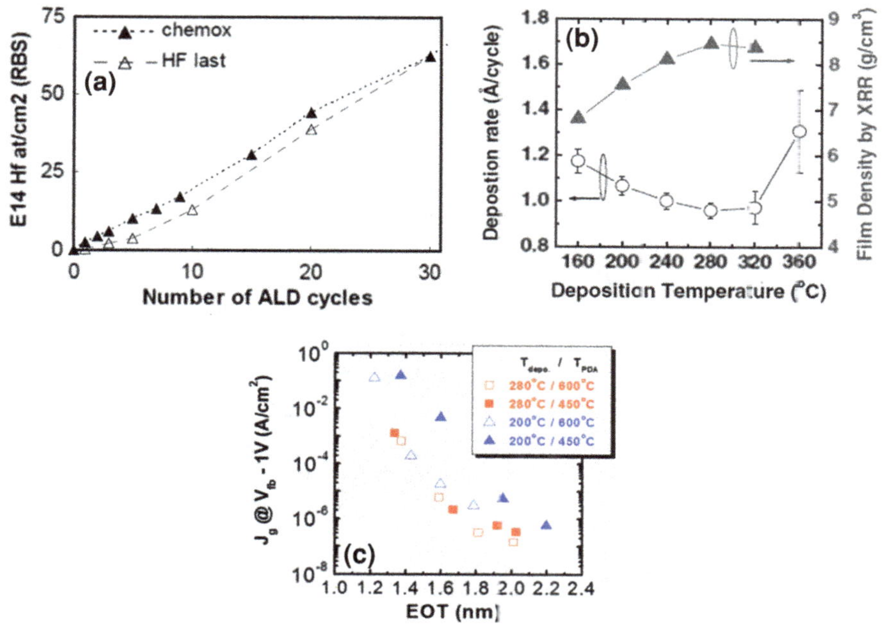

Fig. 7.3 (**a**) Hf coverage as a function of the number of ALD TEMAH/H_2O cycles on the chemical SiO_2 and HF-cleaned Si [44]. (**b**) Variations in the growth and film density of ALD HfO_2 films on Si as a function of deposition temperature [45]. (**c**) Variations in the leakage current density (at a voltage of V_{FB}—1 V) as a function of EOT of the HfO_2 films grown at 200 and 280 °C after PDA at 450 and 600 °C [46]

phase transition energy from monoclinic to tetragonal. However, gain on the dielectric performance with low process temperature is nullified by deteriorated leakage current property of HfO_2. Figure 7.3c showed the changes in the leakage current density as a function of EOT of HfO_2 film grown at 200 and 280 °C. Although HfO_2 film deposited at low temperature (200 °C) showed higher dielectric constant (and thus lower EOT), no improvement in gate current density-EOT (J_g-EOT) curve was found.

To obtain EOT of sub 1 nm on Ge channel, TiO_2/HfO_2 gate stacks was implemented on ultra-thin GeO_2 which is formed by O_2 plasma treatment prior to ALD high-k process [47]. HfO_2 film is deposited with TEMAH and O_2 plasma. Presence of stable and uniform GeO_2 IL prevents intermixing of HfO_2 and Ge during ALD HfO_2 process and resulted in significantly reduced hysteresis of <30 mV, whereas C–V hysteresis of 900 mV was achieved without GeO_2 passivation layer. Furthermore, very low interfacial trap density of $N_{it} \sim 5 \times 10^{11}$ eV^{-1} cm^{-2} indicates HfO_2/GeO_2 provides high interface quality for high-k TiO_2 on Ge channel. Finally, an EOT of 0.9 nm with low leakage current of 2×10^{-7} A/cm^2 at $V_{FB} \pm 1$ V was achieved for TiO_2(3 nm)/HfO_2(1.2 nm)/GeO_2(0.7 nm)/Ge capacitor.

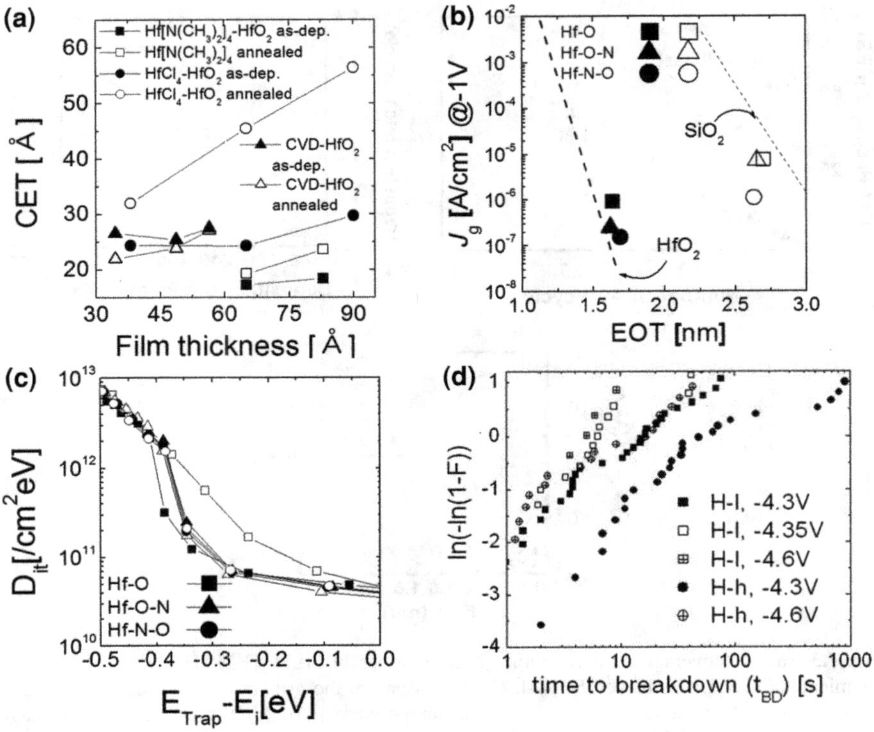

Fig. 7.4 (**a**) Variations in the capacitance equivalent thickness (CET) of the HfO_2 films grown from the $Hf[N(CH_3)_2]_4$ and $HfCl_4$ precursors, respectively, after PDA as a function of the physical thickness. The CVD HfO_2 film data were also included for comparison [48]. (**b**) The leakage current density as a function of EOT for Hf–O, Hf–O–N, and Hf–N–O films. The *filled* and *open symbols* correspond to before and after PDA, respectively [49]. (**c**) The variations in N_{it} as a function of E_{trap}-E_i before (*closed*) and after (*open*) CVS (5 MV/cm, 300 s) [49]. (**d**) Weibull distribution from TDDB analysis deposited as H–h and H–l samples [50]

HfO_2 films were also grown by ALD using Tetrakis[dimethylamino]hafnium ($(HfN(CH_3)_2)_4$, TDMAH) and H_2O at 300 °C. TDMAH precursor belongs to amide precursor group like as TEMAH. The growth rate of HfO_2 film from TDMAH was 0.078 nm/cycle and less substrate-dependent initial growth was observed compared to $HfCl_4/H_2O$ process [48]. As TDMAH contains nitrogen atoms in the precursor, SiN_x layer was simultaneously formed on Si during HfO_2 deposition. It was found that this spontaneously formed SiN_x layer that plays a role as a RBL which prevents diffusion of Si into the HfO_2. Figure 7.4a showed the variations in the EOTs of the TDMAH-HfO_2 film and $HfCl_4$-HfO_2 film after PDA as a function of the physical HfO_2 thickness. Interestingly, the increase in the EOT of HfO_2 film after PDA is much smaller for the TDMAH-HfO_2 than $HfCl_4$-HfO_2 film due to suppression of Si diffusion into HfO_2 by the SiN_x RBL.

To effectively incorporate N into the HfO_2 and IL layers for obtaining better electrical properties of metal–oxide–semiconductor capacitor (MOSCAP), modified ALD HfO_2 processes with in situ NH_3 injection, where the sequence of TDMAH/purge-NH_3/purge-H_2O/purge (Hf–N–O) or TDMAH/purge-H_2O/purge-NH_3/purge (Hf–O–N) was adopted, were demonstrated [49]. Interfacial SiN_x layer is formed at the HfO_2/Si interface for both HfO_2 ALD processes. The in situ NH_3 pulse leads to reduced C and increased N contents in the HfO_2 and IL layers compared to conventional ALD HfO_2 (Hf–O). Decreased C concentration might be ascribed to enhanced desorption of C by NH_3 injection. Figure 7.4b showed the J_g-EOT curve of Hf–O, Hf–O–N, and Hf–N–O samples before and after PDA. The leakage current density of HfO_2 film was improved by NH_3 injection. This might be attributed to the formation of SiN_x IL and the lower density of electrical defects induced by C residue. The variation in the N_{it} was measured by the conductance method before and after constant-voltage stress (CVS) at 5 MV/cm. Figure 7.4c showed that larger degradation in the N_{it} was observed after CVS in case of the HfO_2 film without NH_3 injection compared to in case of the HfO_2 film with NH_3 injection. It is believed that SiN_x IL suppresses the degradations of N_{it}. The effect of C impurity concentration on the reliability of HfO_2 was also examined by varying concentration of O_3 to make HfO_2 films with different *C* concentration [50]. Figure 7.4d shows Weibull distribution from TDDB analysis for HfO_2 films from high O_3 concentration (H–h) and low O_3 concentration (H–l). The films grown with higher O_3 concentration has a lower amount of C residue. It was confirmed that C impurity in the HfO_2 film produces deep acceptor-like trap states in the band gap, and results in inferior leakage current and poor TDDB properties.

7.2.2.2 Tert-butoxytris[ethylmethylamido]hafnium ($HfO^tBu(NEtMe)_3$, BTEMAH)

In this section, the growth behavior and electrical properties of ALD HfO_2 film deposited using heteroleptic tert-butoxytris(ethylmethylamido)hafnium (BTEMAH) precursor and O_3 at a deposition temperature of 300 °C are described [51]. The structure of BTEMAH is slightly modified from that of TEMAH precursor by replacing one of four amido ligands in TEMAH with a tert-buthoxy ligand. This buthoxy ligand largely increases the volatility and reactivity of Hf precursor which results in not only improved growth rate (0.16 nm/cycle) but also 20 % higher Hf density of the HfO_2 film compared with the HfO_2 film grown with TDMAH precursor. Higher Hf density induces more amorphous-like nature of HfO_2 film and the amorphous phase at the as-deposited state is maintained up to $\sim$15 nm while it changes to crystalline (monoclinic) phase at $< \sim 10$ nm for TDMAH (or TEMAH) case. Changes in the microstructure of HfO_2 and thickness of IL between Si and HfO_2 are observed by high-resolution transmission electron microscopy (HRTEM) in comparison with the TDMAH HfO_2 film (Fig. 7.5). After PDA at 700 °C, HfO_2 from BTEMAH remains amorphous phase while HfO_2 film from TDMAH is fully crystallized. Both HfO_2 films are crystallized after

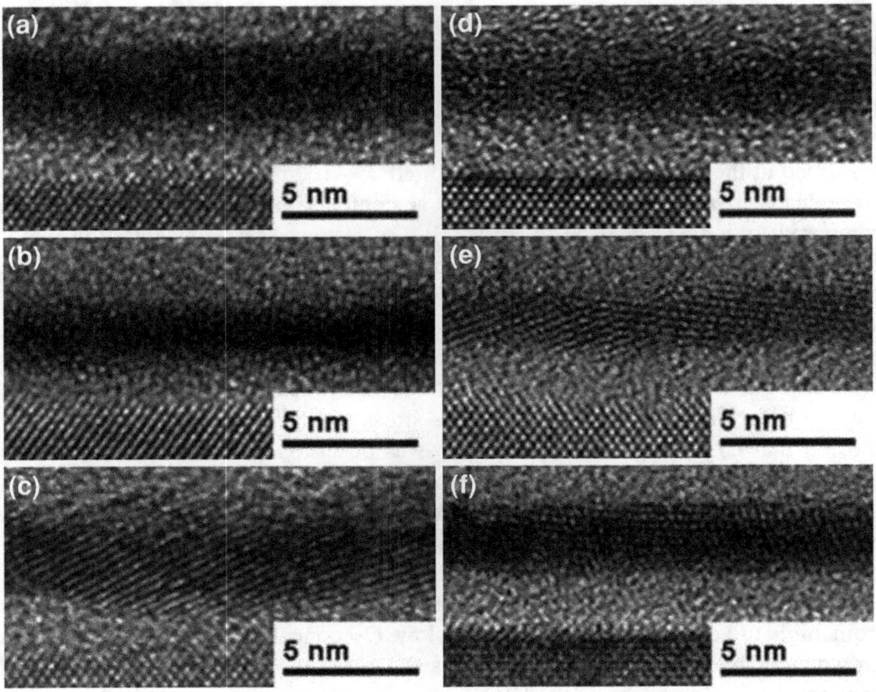

Fig. 7.5 HRTEM images of HfO_2 films grown on Si from the BTEMAH precursor (**a**) in the as-deposited state, and after annealing at (**b**) 700 °C and (**c**) 1,000 °C. (**d–f**) show the corresponding films from TDMAH [51]

PDA at 1,000 °C. It should be noted that thicknesses of interfacial SiO_2 layers are different according to the types of Hf precursor. The thickness of IL layer for BTEMAH-HfO_2 is thinner than that for TDMAH-HfO_2, and there is significantly smaller increase in the thickness of IL after annealing up to 1,000 °C in the case of BTEMAH which might be attributed to higher density of HfO_2 film grown from BTEMAH.

7.3 Oxygen Sources

7.3.1 Effect of Oxygen Source on Properties of ALD High-k Films on Si

The type of oxygen source crucially affects the various properties of ALD high-k metal-oxide films for the give n metal precursor. From the early stages of the ALD process development, H_2O has been extensively used as an oxygen source which provides near-perfect ALD reaction (ligand exchange). This can be most typically observed from the ALD of Al_2O_3 films using TMA as the Al-precursor. As various metal precursors have been developed for more stable and efficient ALD process, another oxygen source than H_2O, such as O_3, NO_2 and H_2O_2 has been required for the better reactivity because they also functions as a reaction agent which removes or exchanges the ligand molecules in the metal precursor during ALD reaction [52, 53]. However, the understanding of the detailed chemical reaction routes for ALD processes is limited. The well-known chemical reactions between TMA and H_2O, and diethylzinc and H_2O are two examples for the well-understood ALD mechanisms. Nevertheless, it seems that the detailed chemical reaction mechanism of oxygen source during ALD is directly related with the physical density and the impurities level of the film, which determine the electrical properties of the ALD high-k film such as the permittivity and gate leakage current density. In addition, interface properties of the ALD high-k films with a substrate are also greatly influenced by the oxygen source, because the surface reaction of the oxygen source with a substrate in the initial stage of the film growth determines the IL growth, nucleation behavior, incubation time (cycle), etc. H_2O and the representative alternative, O_3 as an oxygen source for ALD high-k film process on Si are mainly discussed in this section.

The growth rate is one of the most important physical factors for the ALD process, which is influenced by the oxygen source as well as the metal precursor. In general, O_3 provides the higher growth rate of the ALD high-k film due to the stronger oxidation power and higher reactivity compared to H_2O, which diminishes the steric hindrance effect originating from the incompletely reacted ligand molecules during the surface reaction. Liu et al. reported the higher growth rate of ALD HfO_2 film using TEMAH with O_3 than H_2O as shown in Fig. 7.6a. Similar results have been reported for the cases with other precursors [54–59]. However, it also depends on the process conditions such as the chemical structure of the metal precursor, deposition temperature, etc. [60, 61].

The residual impurities in ALD high-k films are unavoidable, because it is either difficult to remove perfectly the ligand molecules in the metal precursor or

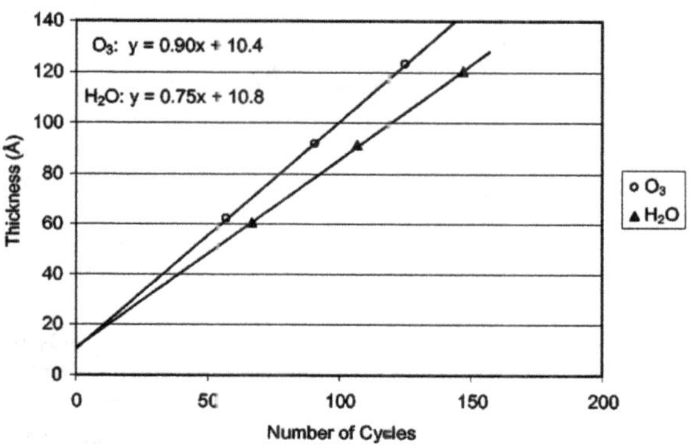

Fig. 7.6 Thickness of ALD Al_2O_3 film grown using TMA with H_2O and O_3 as an oxygen source as a function of the number of cycles. The slope in the fitting equation is the deposition rate, and the intercept is the thickness of the native oxide of the Si wafer [56]

to avoid the incorporation of by-product gas molecules into the film during ALD process. The device performance and reliability could be significantly deteriorated by the residual impurities in the gate dielectric high-k film, even in minute quantities, because the impurities can act as active electrical defects during device operation. In addition, too high impurity concentrations result in a lower physical density of the film, which can enhance the interfacial reactions with a Si substrate such as inter-mixing and Si out-diffusion [62]. In general, the concentration of residual impurities from the precursor molecules, such as Cl, C, H, N, etc., can be decreased by adopting O_3 due to the higher reactivity and stronger oxidation power compared with H_2O [62–64]. Park et al. recently reported the O_3 in ALD process largely decreased the C- and N-related residual impurities in La_2O_3 high-k film compared to H_2O, which was observed by high-resolution in situ XPS at the each half-ALD cycle as shown in C 1s core level spectra of Fig. 7.7a, b. While the residual impurities such as C–N, C–O, C–H, and C–O–H were accumulated with

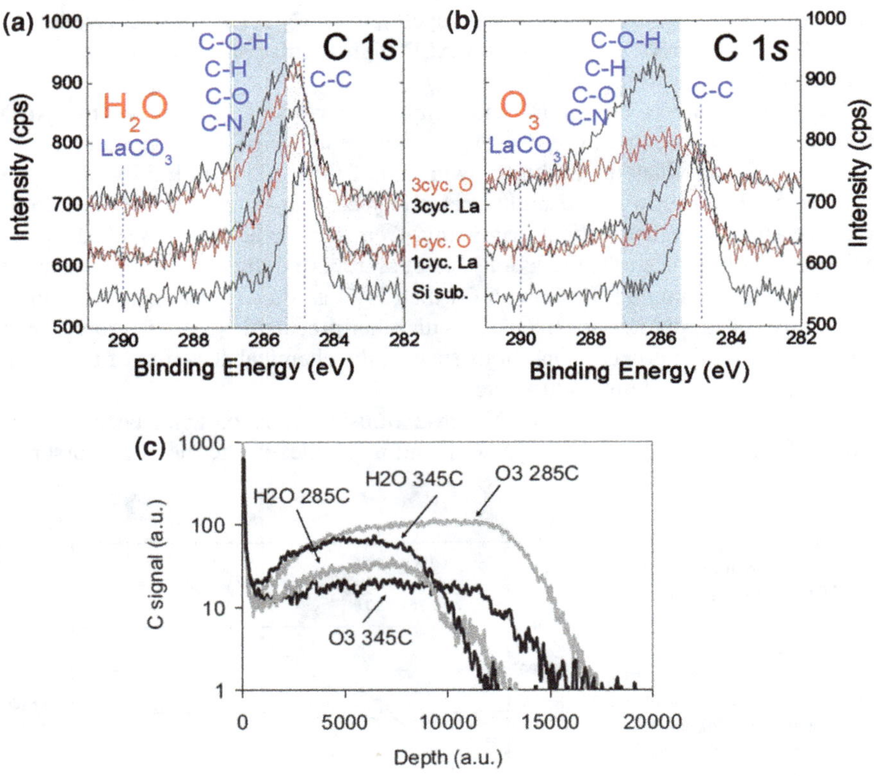

Fig. 7.7 C 1s core level XPS spectra of the ALD La_2O_3 films grown on Si using Tris(N,N′-diisopropylformamidinato)lanthanum with (**a**) H_2O and (**b**) O_3 for first and third ALD half-cycles [62]. (**c**) TOF–SIMS depth profiles of the C impurities in 10 nm thick ALD HfO_2 films grown using TEMAH with H_2O and O_3 at low and higher temperatures [55]

increasing ALD cycle number from 1 to 3 cycle in H_2O process (Fig. 7.7a), the accumulated impurities were removed considerably at each O_3 pulse in O_3 process as indicated in red (Fig. 7.7b). Therefore, O_3 with a higher concentration was even more effective to decrease the impurity level, which improved the reliability of the high-k films, such as breakdown voltage (time-zero dielectric breakdown, TZDB) and TDDB [63, 65–68]. However, it also depends on the process conditions such as the feeding time and purging time of oxygen source, process temperature, etc. The different temperature dependence of C impurity between O_3 and H_2O was reported for HfO_2 ALD using TEMAH, whereas the O_3 process effectively decreased the C concentration in the ALD high-k film grown at 345 °C compared to H_2O process, but it result in a higher C concentration in the film grown at 285 °C as shown in Fig. 7.7c [55, 56]. The lower impurity concentration and higher physical density of the ALD high-k film grown using O_3 results in the higher permittivity and higher breakdown voltage (field) compared to H_2O process [63, 64, 69–73]. Nevertheless, there are several other reports revealing the reversed trend making this issue still somewhat controversial [54, 55, 65, 66].

Interfacial reactions such as initial oxidation of a Si substrate and the inter-mixing of the related elements by diffusion-out of Si from a substrate are inevitable during the ALD film growth and various post-deposition processes. Especially, the initial oxidation of a Si substrate affects even the crystallinity of the film bulk as well as IL formation. Figure 7.8a showed the comparisons in the initial growth behavior of ALD HfO_2 on various substrates using either H_2O or O_3 as the oxygen source. While H_2O process on the HF-last Si substrate induces the long incubation time (or cycle) and the island-type film growth due to a lack of the reactive sites, O_3 process forms a thin surface SiO_2 layer during the initial stages of ALD film growth to induce a more 2D growth without incubation time. As a result, the larger grain size and flat surface with thicker IL were observed in the ALD high-k film grown with O_3 compared to H_2O as shown in the HRTEM image of Fig. 7.8c [54, 64]. However, the thicker IL of the film grown using O_3 is the critical drawback with respect to the scaling of EOT of the film.

The initial oxidation status of a Si surface during ALD process, which is mainly determined by the oxygen source, influences significantly the N_{it} of the high-k film affecting the carrier mobility in the MOSFETs. This results in a very delicate balance, because the improvement in N_{it} can be achieved not only by the thicker IL in O_3 process (more SiO_2/Si-like interface), but also by H-passivation at the interface with Si in H_2O process [74, 75]. However, it seems obvious that the severe interfacial oxidation by O_3 with a high concentration deteriorates the N_{it}, because the excess concentration of oxygen in the film grown in a higher O_3 concentration induced excess bonding with interfacial Si so as to decrease the thickness of interfacial sub-oxide (SiO_{2-x}) releasing the interfacial stress [65, 66, 74]. Moreover, the influence of oxygen source on the fixed charge and charge trap density of the ALD high-k film causing the V_{FB} shift and hysteresis, respectively, heavily depends on the process conditions.

The diffusion-out of Si from a substrate degrades the permittivity, device reliability, and interface properties of the ALD high-k film. While the thicker IL

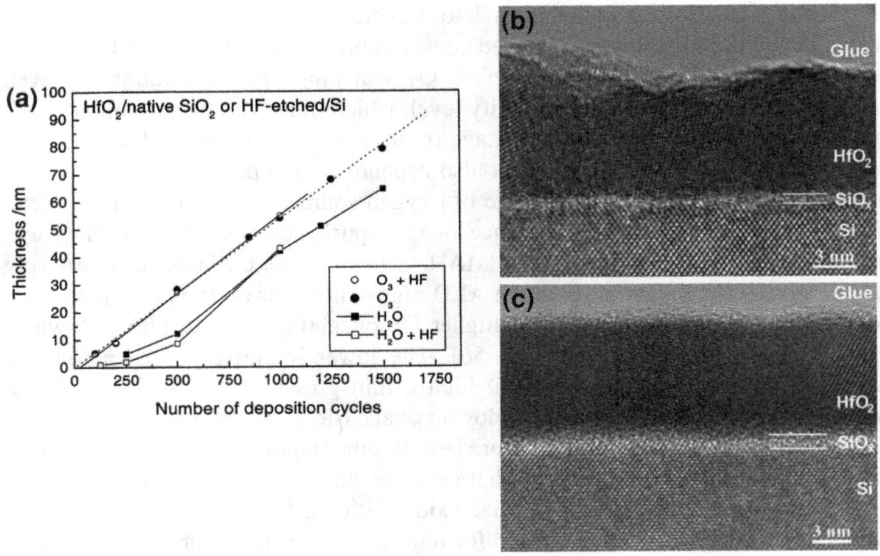

Fig. 7.8 (**a**) The thickness of ALD HfO_2 film using $Cp_2Hf(CH_3)_2$ with H_2O and O_3 as a function of the number of cycles. HF indicates a surface pretreatment with HF etching. Otherwise, the films were deposited on Si covered by native oxide. Linear fitting curves are provided when ozone was used. The deposition temperature was 350 °C. Cross-sectional HRTEM images of HfO_2 films grown on HF-etched Si(100) using (**b**) H_2O and (**c**) O_3 as oxygen source [54]

grown during ALD using O_3 decreases the permittivity of the film compared to H_2O, it plays as a good diffusion barrier for Si diffusion-out from a substrate during PDA [76]. Hence, the very thin SiO_2 can be intentionally grown with various methods prior to the ALD high-k film growth for achieving the thermal stability of the film.

Consequently, O_3-process provides the better physical properties as well as the related electrical properties of the ALD high-k film on Si compared to H_2O. However, identifying the various aspects of the O_3-process for high-k deposition including the structural and chemical changes occurring during the whole CMOSFET fabrication processes requires further study.

7.3.2 Effect of Oxygen Source on Interface Properties of ALD High-k Films on Ge and III–Vs

While the alternative substrates such as Ge and III–V compound semiconductors provide the enhanced device performances due to their high intrinsic carrier mobilities, the reliabilities such as V_{TH} instability, TDDB, etc., were deteriorated compared to the Si-based devices. This is because the interface properties of ALD high-k film/substrate are degraded, partly by complicated interfacial reactions during ALD process of high-k films resulting in the high N_{it} and high density of the charge trapping sites near the interface. Therefore, the interfacial reactions between the ALD high-k film and Ge or III–V compound semiconductors occurred during ALD and the post-deposition processes should be carefully controlled. On the other hand, the bulk properties of the high-k films were reported similar with those of the film grown on a Si substrate [77].

In the case of Ge substrates, the oxygen-deficient GeO_x IL deteriorates the interface properties as mentioned in the preceding sections [22, 23]. Therefore, regarding the oxygen source, the understanding of the properties of IL grown simultaneously on a Ge substrate during ALD process of high-k film is most important [77]. Figure 7.9 shows the cross-sectional HRTEM image of ALD HfO_2 film grown with (Fig. 7.9a, b) H_2O and (Fig 7.9c, d) O_3 on Ge substrate [78]. While HfO_2 film grown with O_3 has flat HfO_2 film and uniform IL, that is grown with H_2O is locally in direct contact with the Ge substrate and has very thin non-uniform IL resulting in the rough surface. This is similar with the ALD film grown with H_2O on a Si substrate but more serious, because the suppressed (negligible) GeO_2 IL growth compared to SiO_2 [21, 79]. Therefore, the high-k film grown with O_3 shows the higher EOT than that grown with H_2O due to the thicker IL, but the N_{it} is obviously lower for O_3 process than H_2O process [78, 80], which is quite different from the case of Si substrate. It was also reported that the band gap energy of the interfacial GeO_2 layer grown during H_2O-based ALD process ($\sim$4.3 eV) is lower than that during O_3-based ALD process ($\sim$5.7 eV) due to the incorporated hydroxyl group, which would affect the gate leakage current density through the film and charge trapping characteristics of the device [78, 81–83].

The interfacial reactions at the interface between ALD high-k film and III–V substrates, such as GaAs, InGaAs, InAs, etc., are even more complicated as compared to Ge substrate because the selective oxidation and reduction of the substrate elements and the volatile group V oxides induces the high N_{it} in the band gap energy of semiconductor. Although the "self-cleaning" of IL on III–Vs by ALD process was reported [84–87], the interfacial reaction is still a key restriction for the adoption of III–Vs because even the tiny quantity of elemental As, specific Ga oxidation state, etc., at the interface deteriorates significantly the interface

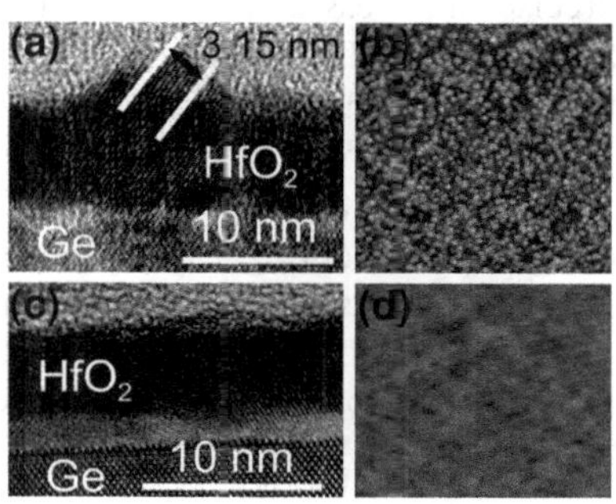

Fig. 7.9 Cross-sectional HRTEM images and corresponding AFM surface morphology of HfO_2 films grown using $HfCl_4$ with (**a**) and (**b**) H_2O and (**c**) and (**d**) O_3 as oxygen sources [78]

properties [88–91]. Many surface passivation (or cleaning) methods using various chemicals, such as H_2SO_4, HCl, HF, NH_4OH, H_2O_2, H, $(NH_4)_2Sx$, Na_2S, H_2S, etc., along with ALD process [92–105], hardly provide the sufficient improvement in the interface properties. The interfacial RBL such as Si and Ge grown by various methods suppressed interfacial reaction efficiently, but this buried channel structure has an EOT scaling limit of the gate dielectrics [106–111].

Although "self-cleaning" [84–87] during ALD of high-k film suggests the reactivity of the metal precursor affects the interfacial reaction at high-k/III–V substrate, the oxygen source mainly determines the interfacial reaction. H_2O has been popular oxygen source for ALD of high-k on III–Vs because it minimizes the interfacial reaction (thinner IL) during ALD compared to O_3 [112]. In situ XPS study by Brenan et al. revealed that ALD Al_2O_3 grown using O_3 induced the thick IL with a high oxidation state (As^{5+}) on $In_{0.53}Ga_{0.47}As$ substrate, which was hardly eliminated by self-cleaning effect of TMA pulse [113]. The thick IL formed by O_3 generated the larger amount of elemental As and Ga-oxide at the interface than the case of H_2O, which degraded the electrical properties including interface property [113]. Madan et al. also reported that ALD Al_2O_3 process using O_3 resulted the 1.5 time higher N_{it} and an order of magnitude higher capture cross section of the midgap trap at the interface with $In_{0.53}Ga_{0.47}As$ substrate than that using H_2O. The typical C–V curves of Al_2O_3 grown using H_2O and O_3 on $In_{0.53}Ga_{0.47}As$ substrate in Fig. 7.10 show that the H_2O process improved the interface property with the less frequency dispersion compared to O_3 process [114]. For the same reason, the lower O_3 concentration for the ALD HfO_2 process on GaAs substrate allowed the better electrical properties than the case with a higher O_3 concentration [115].

7.4 Transistor Characteristics with ALD Gate Oxides

7.4.1 Sub 1 nm EOT Devices

Since continuous MOSFET downscaling is inevitable, investigating the performance and reliability of sub 1-nm EOT regime devices became an important topic. The sub 1 nm EOT has been mainly obtained by ALD high-k oxides, thanks to the precise thickness control across the wafer. However, reduction of the interfacial SiO_2 layer (or IL with slightly different chemical composition) while at the same time maintaining a good quality of the Si/oxide interface is not straightforward.

Several approaches have been followed to obtain the sub 1 nm EOT devices, such as depositing $HfSiO_x$ IL before ALD-HfO_2 deposition [116], controlling oxygen source during ALD-HfO_2 deposition [117], or mixing exotic materials to HfO_2 to increase the dielectric constant [118]. One popular method introduced recently is remote scavenging of oxygen by a very thin TiN metal layer on top of the ALD gate oxide [119, 120]. The oxygen transfer from the SiO_2 IL to the TiN

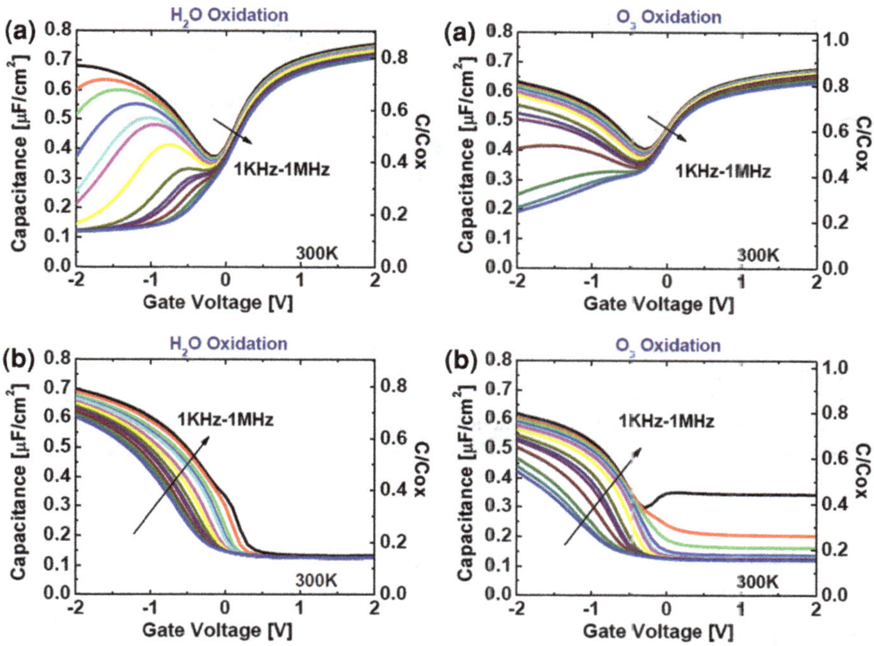

Fig. 7.10 C–V characteristics as a function of frequency of MOSCAP with ALD Al_2O_3 grown using H_2O and O_3 (**a**) on n-type and (**b**) p-type $In_{0.53}Ga_{0.47}As$ [114]

metal through the high-k layer decreases the physical SiO_2 IL thickness and as such the EOT. The IL thickness can be controlled by the TiN layer thickness.

HRTEM images showed the thickness of the IL decreases in SiO_2/HfO_2 gate stacks when a TiN metal gate is present instead of TaN [5], while the HfO_2 layer thickness does not change. Since the initial SiO_2/HfO_2 stacks are deposited by exactly the same process, this scavenging effect is coming from the metal gate difference only. Finally thinner SiO_2 layer generates smaller EOT as shown in Fig. 7.11a [5]. Note that a Lanthanum (La) oxide capping layer is used for the V_{TH} tuning in this case, but this does not disturb the scavenging effect. Figure 7.11b shows variation in J_g as a function of EOT. Thinner EOT devices, which experienced severe scavenging process, still show a reasonable J_g trend in both 1.2- and 1.8 nm-thick ALD HfO_2 devices. However, the mobility significantly decreases because the thinner SiO_2 layer enhances remote phonon and charge scattering from the HfO_2 layer to the channel region (Fig. 7.11c) [5]. Ando et al. [119] also showed that scavenged devices, with HfO_2 dielectric layers with or without La-capping layer show decreased mobility in the sub 1-nm EOT range.

Next, the device reliability in sub 1nm EOT devices is focused. The negative bias temperature instability (NBTI) degradation mechanism was investigated for metal-gated ALD high-k devices as shown in Fig. 7.12a [121]. The NBTI degradation was studied from 2 down to 0.5 nm EOT with a severe criterion of 30 mV

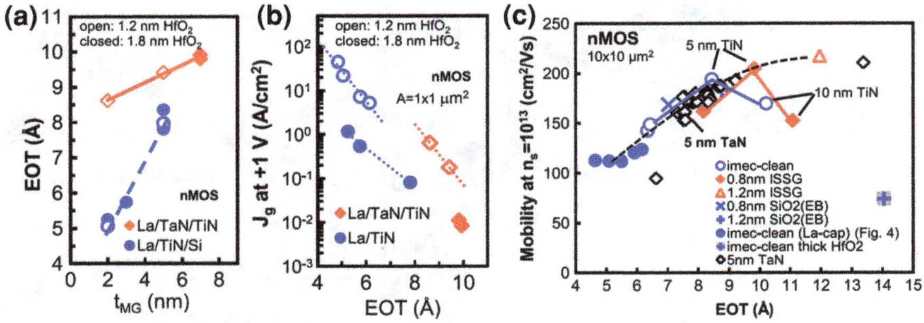

Fig. 7.11 (**a**) EOT extracted from C-V measurements on n-type MOSCAPs with ALD-HfO_2 and La capping layer as a function of the metal gate thickness for TiN and TaN. (**b**) The variation in J_g as a function of EOT for TaN and TiN gates and for two different ALD-HfO_2 thicknesses. Thinner EOTs are found with TiN. (**c**) The long channel electron mobility as a function of EOT for stacks with different interface layers, metal electrodes (TiN, TaN) and TiN thicknesses. IL has no impact on the mobility. TiN thickness is 2 nm when not indicated [5]

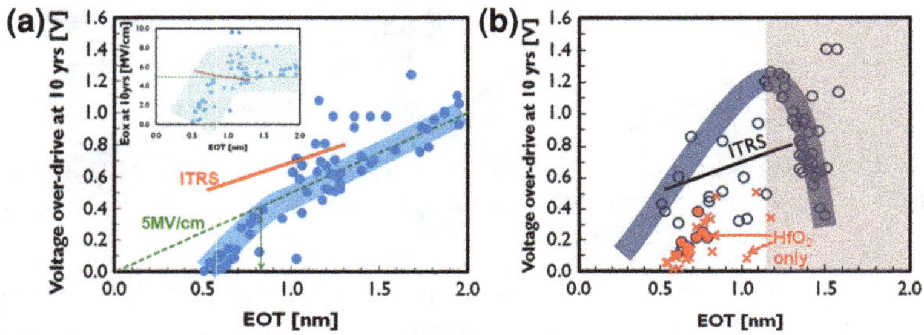

Fig. 7.12 (**a**) Over-drive voltage at 10 years extracted from NBTI on 70 different p-type MOSFET devices. Below 1 nm EOT, the over-drive voltage decreases more than the iso-electric field at 5 MV/cm. ITRS data is also added. The *inset* shows electric field at 10 years. In sub 1 nm EOT, ITRS suggests increased electric field, but real devices shows even more decreased NBTI reliability. (**b**) Over-drive voltage at 10 years extracted from PBTI on 63 different n-type MOSFET devices is shown (*circles*). The straight line is from ITRS. Over-drive voltage at 10 years from N/PBTI is shown together, with only HfO_2 devices. N/PBTI does not show clear difference (*closed circles* for PBTI, "x" symbols for NBTI) [121]

in V_{TH} difference (ΔV_{TH}). The international technology roadmap for semiconductors (ITRS) [122] suggests that satisfying the demand for a rapidly increasing electric field in the sub 1 nm EOT regime will become difficult to meet even with a higher quality oxide because the hole trapping into the bulk defects has been increased. Therefore, an *iso*-electric field target at 5 MV/cm has also been considered in Fig. 7.12a together with the ITRS standard. In the EOT regime higher than 1 nm, the NBTI degradation followed an *iso*-electric field model because the Si/SiO_2 interface degradation mechanism is dominant [123–126]. However, in the sub 1 nm EOT regime, NBTI was in addition affected by hole trapping into bulk defects in the high-k dielectric. The kinetics for V_{TH} shift according to the stress time showed much lower time exponent ($\sim$0.13) in the 0.5 nm EOT device than in the 2 nm EOT device ($\sim$0.24) [121], indicating a reduced interface degradation contribution to the total NBTI degradation. Charge pumping analysis combined with NBTI showed that bulk trap degradation increases to $\sim$two orders of magnitude higher than the interface trap generation during NBTI stress in a 0.58 nm EOT device [121]. The activation energy was lower in the 0.58 nm EOT device (0.49 eV) than in the 2 nm EOT (0.68 eV) due to the increased bulk trapping component including direct tunneling, which is independent of temperature. The bulk defects affecting NBTI are mostly preexisting defects in the ALD high-k layer. Therefore, decreasing high-k bulk defects can improve NBTI in sub 1 nm EOT regime, at least up to the *iso*-electric field limitation from interface degradation.

In case of positive BTI (PBTI) reliability, the degradation in the EOT regime higher than 1 nm is known to be driven by electron trapping into defects in the high-k layer [127–129]. Figure 7.12b shows lower voltage over-drive because thicker high-k dielectric layers contain more defects to be filled (the IL thickness is fixed at 1 nm). However, the PBTI lifetime decreases when the EOT decreases below 1 nm, because the thinner SiO_2 thickness increases the electron tunneling probability from the Si substrate into the high-k defects. Indeed, slight transconductance (G_m) degradation was also observed after stress in a 0.61 nm device with ALD HfO_2 gate oxide [130], which indicates trap generation near the Si/oxide interface. However, Fig. 7.12b clearly shows that PBTI degradation largely depends on the process and the high-k material quality at a given EOT. Depositing a La-oxide capping layer on top of the ALD HfO_2 dielectric or applying Ar/As implantation can improve PBTI by shifting up the trap levels away from the Si conduction band. As such, these levels cannot be reached by injected electrons during PBTI stress [128–131]. In addition, adding Gadolinium in the HfO_2 dielectric by ALD can also improve PBTI because a trap level aligned to the Si conduction band is absent in contrast to $HfLaO_x$ [132].

TDDB is also a large concern in sub 1 nm EOT devices, because the very thin oxide layer can be broken down by a percolation path containing just a few generated defects. Figure 7.13 shows an All-in-one TDDB map on a 0.63 nm EOT pMOSFET device with ALD-HfO_2 layer [133]. The All-in-one TDDB map is obtained by extracting soft breakdown (SBD) and post-SBD wear-out parameters from measuring the time to hard breakdown (HBD) [134]. Although the EOT is extremely small, the device shows a 10 years lifetime at $V_G = 0.5$ V from the SBD, and $V_G = 1.0$ V from the HBD. Since the operation voltage is lower than 1.0 V in this EOT regime, the device is not limited by the HBD until 10 years. This shows TDDB is a lower concern than NBTI in sub 1 nm EOT devices with ALD-HfO_2 dielectric layers.

On the other hand, the impact of impurities resulting from the ALD process needs to be studied carefully, because they can generate interface or bulk defects and affect device performance and reliability. Fortunately, Cl residue from $HfCl_4$ precursor does not degrade mobility and TDDB reliability when combined with

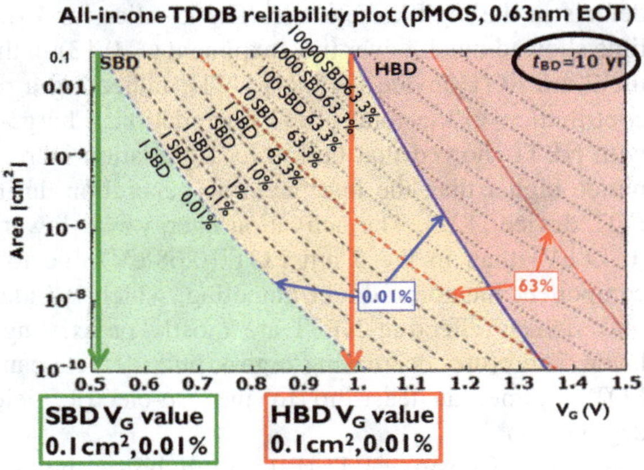

Fig. 7.13 'All-in-one' TDDB reliability for a 0.63 nm EOT p-type MOSFET device. For a fixed lifetime of 10 years three different regions of area/V_G combinations can be distinguished: SBD free, SBD & wear-out, and H_{BD}. For this very thin ALD-HfO_2 devices, a $V_{G,SBD} = 0.5$ V and $V_{G,HBD} = 1$ V is extracted [133]

H_2O oxidant in the 0.9 nm EOT devices as mentioned earlier in this chapter [17]. This observation was further confirmed by first-principles calculations, which proved that residual chlorine does not form additional trap energy levels inside the HfO_2 band gap [17]. In case of O_3 with $HfCl_4$, optimizing the O_3 concentration during ALD process is important to improve the interface trap density and mobility [135].

Study on carbon residue related to the ALD-HfO_2 dielectric was also reported regarding the TDMAH and O_3 precursors. The 10 years TDDB lifetime of high density O_3 condition was guaranteed at -1.0 V, whereas that of low density O_3 was only obtained at voltages lower than -0.8 V [50], which may imply higher O_3 successfully removed carbon residue-related defects. The first-principles calculations showed that the interstitial carbon atoms in the HfO_2 films produced deep acceptor-like trap states in the band gap, which may enhance the electrical conduction by a trap-mediated conduction mechanism. TEMAH precursor has been studied also, which showed using D_2O oxidant instead of H_2O oxidant decreases interface and bulk trap generation after CVS, and improves TDDB reliability [136].

7.4.2 SiGe and III–V Channel Devices

pMOSFETs with Ge or SiGe channels have been demonstrated to be a viable option for future logic device applications thanks to the high hole mobility and the possible integration onto silicon substrates [137–139]. However, a high N_{it} between Ge substrate and high-k materials is a big concern [140]. Adopting a thin Si layer on top of the Ge substrate can reduce the N_{it} significantly as mentioned in previous section [141]. When the Si capping layer thickness increases, the carrier scattering by the Si/oxide interface reduces, however, the carrier density in the Ge channel also decreases. Therefore, a careful optimization of the Si capping layer thickness is important for the mobility enhancement [142, 143].

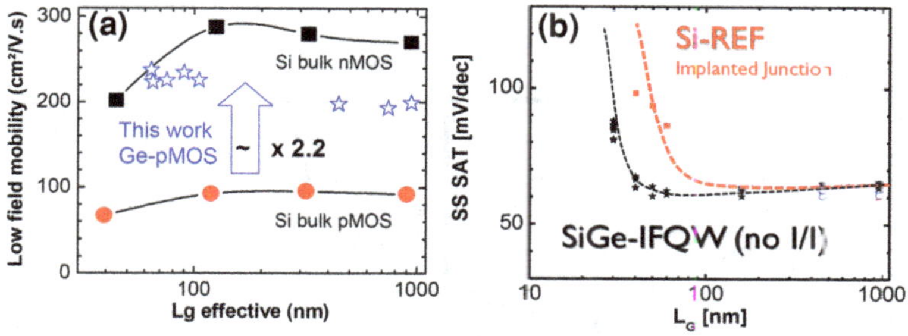

Fig. 7.14 (**a**) Low field hole mobility as a function of effective gate length for Ge p-type MOSFETs. About twice higher and stable mobility are found than in Si p-type MOSFETs [137]. (**b**) Enhanced short channel effect control with SiGe channel p-type FETs is shown. SiGe channel p-type FETs are compared to control Si planar devices (with implants) [145]

Figure 7.14a shows a good example of enhanced hole mobility in ALD-HfO_2 devices on Ge substrate, regardless of channel length (L_G) [137]. The low field hole mobility in pMOSFET devices increases about twice higher for the Ge substrate as compared to the Si substrate. ALD-HfO_2/Al_2O_3 structures also showed high hole mobility of 546 cm^2/Vs at 0.76 nm of EOT [144]. The mobility enhancement continues as Si_xGe_y substitutes Ge channel [3]. Furthermore, there is a report that the short channel effect has been improved with Si_xGe_y channel as shown in Fig. 7.14b [145].

NBTI reliability on the Ge channel device with a 0.5 nm of Si layer and an ALD-HfO_2 high-k dielectric showed similar 10 years lifetime as the Si substrate device [146], even though the initial N_{it} was about two orders of magnitude higher in the Ge device. Further NBTI study on the Si_xGe_y channel devices suggested a possible solution to obtain reliable pMOS devices in sub 1 nm EOT regime [147]. Figure 7.15a shows improved NBTI with the $Si_{0.45}Ge_{0.55}$ channel, and the ALD-HfO_2 device is more robust when the Si capping layer thickness on the $Si_{0.45}Ge_{0.55}$ channel decreases. This improved NBTI is due to the charge-injection level adjustment [147]. By adopting a Si_xGe_y channel and a thin Si capping layer, the Fermi level of the Si_xGe_y channel moves further from the oxide valence band edge, and less high-k bulk defects can be charged during the NBTI stress. Therefore, the reduction of bulk defects in the ALD high-k layer can improve further the NBTI reliability in the Si_xGe_y channel devices.

Hot carrier reliability studies have been also performed on the Si_xGe_y channel devices with ALD high-k layers. Loh et al. [148] showed that a higher Ge content and thinner Si capping layer can help to improve the channel hot carrier reliability. However, the Si_xGe_y channel devices showed more degradation than Si substrate devices after stressing at maximum impact ionization or ($V_G - V_{TH} = V_D$) conditions as shown in Fig. 7.15b, c. Franco et al. [149] showed less degradation after hot carrier stress than NBT stress in ALD-HfO_2 devices, due to the reduced charge trapping into the pre-existing HfO_2 bulk defects. However, again the Si_xGe_y

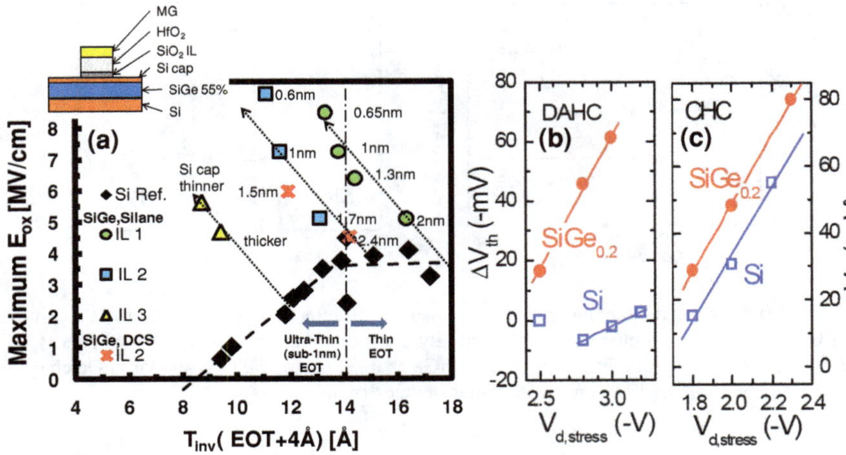

Fig. 7.15 (**a**) The operating electric field (E_{ox}) for 10 year NBTI reliability improves when reducing the Si cap thickness, which is observed consistently for several SiO_2 IL and for different Si precursors and epi-growths ('Prec. 1' and 'Prec. 2'). The *inset* shows a schematic diagram of a SiGe pFET [147]. In (**b**) and (**c**), voltage dependency of HC degradation at maximum impact ionization (DAHC) and $V_G = V_D$ (CHC) for Si and SiGe (Ge 20 %) devices are shown. HC stress performed for 100 s [148]

channel devices were more degraded than Si substrate devices after the hot carrier stress at $V_G = V_D$ (V_D is the drain voltage).

Although the III–V substrate for nMOSFETs is receiving intense research interests, the N_{it} between III–V substrate and gate oxide is a serious problem as in the Ge substrate case. The oxidation of the surface introduces stress at the surface, and reduction of the oxides does not decrease N_{it} generated by the oxidation [150]. Further, N_{it} in mid-gap is independent of the oxide, or different surface cleans and post-anneals, it is likely that the defects are originally due to the $In_{0.53}Ga_{0.47}As$ itself [151]. This mid-gap N_{it} generates the Fermi level pinning [152]. Figure 7.16a shows mobility from InGaAs or Ge substrate devices with different surface treatments, and sulfur treatment shows about 3 times higher peak mobility than HCl treatment for the InGaAs devices [153].

Another important factor in III–V devices is the border traps located in the high-k layer [153]. The spatial and energetic distribution of the traps inside the ALD-Al_2O_3 layer on InGaAs has been extracted by the TSCIS technique [154, 155] as shown in Fig. 7.16b [4]. 9 nm-thick ALD-Al_2O_3 were grown with TMA and H_2O on the $In_{0.53}Ga_{0.47}As$ channel after surface preparation for those devices. The border trap density inside the ALD Al_2O_3 layer was significantly reduced by $(NH_4)_2S$ treatment, and time-of-flight secondary ion mass spectroscopy showed lower In concentration inside the Al_2O_3 layer. Sulfur acted as an In-diffusion barrier during the ALD of Al_2O_3, lowered the border trap density in the oxide.

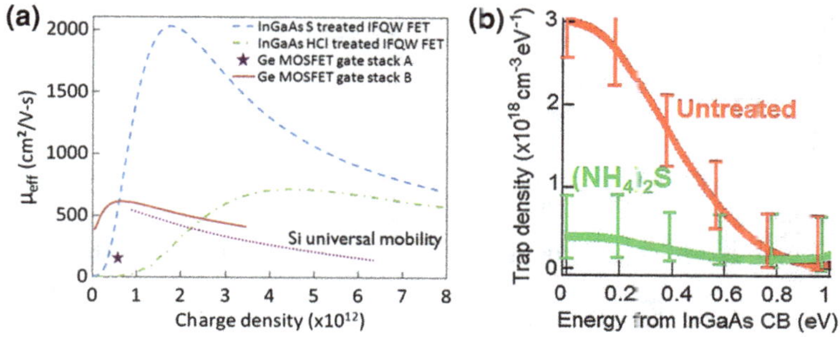

Fig. 7.16 (**a**) Mobility of InGaAs or Ge substrate devices with ALD-Al_2O_3 layer [153], (**b**) Al_2O_3 trap density as a function of energy at a depth of 1.5 nm inside the oxide measured from the interface of InGaAs/ALD-Al_2O_3 [4]

It is also reported that direct ALD of $TaSiO_x$ on (100) InP and (100) $In_{0.53}Ga_{0.47}As$ substrates results in low electron barriers that cannot prevent electron injection into the oxide [156]. This increases electron trapping in the $TaSiO_x$ and can degrade the device reliability.

7.4.3 Three-Dimensional Devices

The 3D device or FinFET is recently highlighted due to its superior electrostatic control to the devices with planar-geometry, which improves the short channel effect [157]. The conformality of ALD can be used to deposit a gate dielectric oxide and metal gate on the side- and top-walls of the fin structure.

The recent production of the 3D device in the 22 nm node [158] has accelerated related research. Indeed the 22 nm p-type FinFET device shows 27 % improved saturated drain current (I_{dsat}) characteristics from I_{on}–I_{off} analysis, and 13 % of I_{dsat} improvement was reported in n-type FinFET as compared to the 32 nm planar transistors. Another performance boost has been demonstrated with the conformal doping process using plasma (self-regulatory plasma doping, SRPD) [159]. The SRPD device with ALD HfO_2 gate stack shows 15 % improved I_{on}–I_{off} characteristic in n-type FinFETs by increasing the side-wall doping concentration relative to the ion implanted devices.

Replacement metal gate (RMG) process has been strongly considered recently, in order to lower thermal budget for the high-k/metal gate stacks [160, 161]. Figure 7.17a shows the 25 % higher I_{on} at 10^{-7} A/μm of I_{off} in the RMG p-type FinFET with ALD HfO_2 layer compared to the traditional gate first (GF) devices [162]. The high-k first (HKF) device where the deposited high-k layer is protected during gate removal or the high-k last (HKL) device where the high-k layer is deposited at the end of the process does not show a large difference in the I_{on}–I_{off} characteristics. Lower V_{TH} is also observed in the RMG devices compared to the

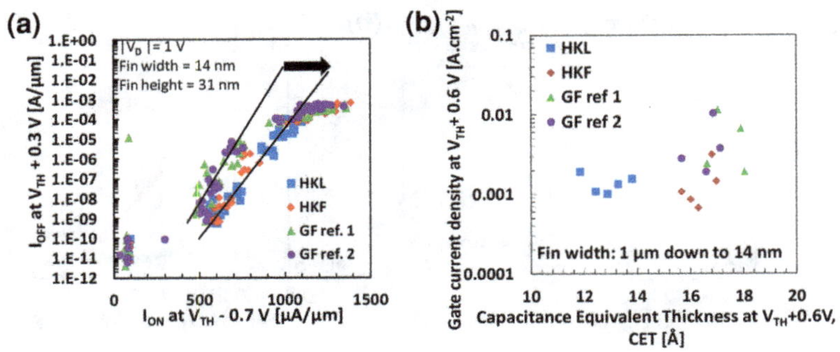

Fig. 7.17 (**a**) I_{on}–I_{off} for gate first and high-k first/high-k last bulk p-type FinFET devices at |VD| = 1 V, $(V_G - V_{TH}) = -0.7$ V (ON-state) and $(V_G - V_{TH}) = 0.3$ V (OFF-state). (**b**) Gate leakage current density versus CET at $V_{TH} + 0.6$ V for TiN/HfO_2 gate stack in gate first and high-k first/high-k bulk p-type FinFETs [162]

GF. Additionally, the HKL p-type FinFET devices revealed a lower EOT as compared to the HKF and GF devices at similar gate leakage (Fig. 7.17b). The lower thermal budget in HKL process can suppress the regrowth of low-k interfacial SiO_2.

A conventional FinFET device includes a top-wall with (100) direction, and side-walls with (110) direction. Especially, in this structure, the N_{it} from side-walls is an important factor to understand device performance and reliability, because the main channel is formed on the fin side-walls rather than the top-wall. In order to separate the top- and side-wall interface defect density N_{it}, charge pumping can be applied. By measuring charge pumping current in various fin width devices with a 2 nm ALD-HfO_2 gate oxide (Fig. 7.18a), N_{it} from top-wall and side-walls is calculated from the slope and y-intercept respectively [163].

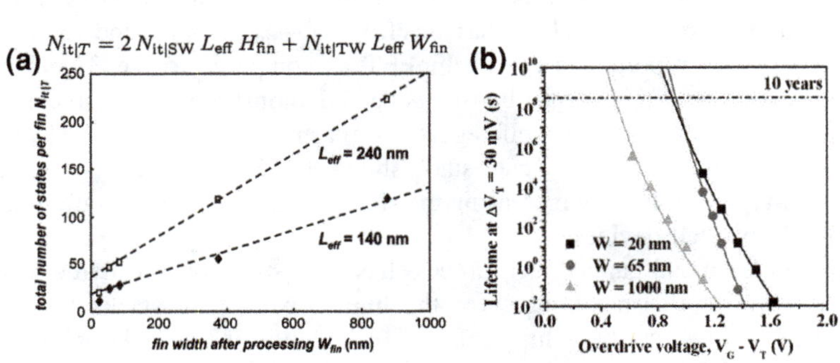

Fig. 7.18 (**a**) Total interface trap density per fin is plotted against the fin width for devices fabricated on a rotated-notch wafer. The slope of the *line* is a measure of N_{it} on top surface, while the intercept is a measure of the N_{it} on the sidewall [163]. (**b**) PBTI lifetime versus gate voltage over-drive for different W_{fin} (TiN gate thickness of 3 nm) is shown. PBTI improves in narrower FinFET devices with ALD-HfO_2 gate stack [167]

The main degradation mechanism for N/PBTI reliability on FinFETs remains the same as in planar devices, as described in the previous section. NBTI degradation is a combination of interface defect generation and hole trapping into the high-k bulk defects. Therefore, substrate rotation to change the side-wall orientation from (110) to (100) improves NBTI by reducing the number of interface states which can be broken during the NBTI stress [164–166]. PBTI degradation originates mainly from the electron trapping into the high-k bulk defects, and the parallel and close side-walls in a FinFET structure reduce the degradation effectively. Indeed PBTI improves in narrower FinFET devices with ALD-HfO_2 gate stack (Fig. 7.18b) [167], possibly due to the reduced injection charge density from the Si substrate by decreased electric field near the channel region [168]. This implies that NBTI in sub 1 nm EOT regime can be improved by reducing hole trapping, and ALD technique giving a high quality high-k layer can help to improve further both N and PBTI reliability. Additionally, it is reported that applying ALD-TiN metal gate instead of sputtered-TiN shows better NBTI [169], due to lower defect generation in the ALD HfO_2 high-k layer during the gate metal deposition.

TDDB reliability is strongly affected by the fin corners. When the fin corner is sharp, the Weibull distribution is much wider in the FinFET than planar devices [170], probably due to the non-uniform electric field at the fin corner in different devices. The corner rounding improves Weibull characteristic in FinFETs to a similar level as the planar devices [171]. As a result, TDDB is not a serious problem in FinFET devices compared to planar, when the corner rounding process is successfully introduced and conformal ALD oxide is deposited around the fin.

7.5 Summary

The influences of metal precursors and oxygen sources on the ALD high-k film growth, resulting film properties, and device performance were discussed focusing on HfO_2-based materials. They crucially affect the interface properties as well as the bulk properties of the film.

$HfCl_4$ was introduced as a Hf precursor for HfO_2 ALD due to its advantages of simple molecular structure, high thermal stability, and good HfO_2 dielectric quality. However, some drawbacks of $HfCl_4$ are low reactivity with H-terminated Si surface, low vapor pressure, strong Hf–Cl bond strength, low growth -per-cycle, Cl residue, and corrosive HCl by-product. Several metal–organic Hf precursors such as TEMAH and TDMAH have been employed to obtain comparable film properties as $HfCl_4$ precursor. They were characterized by their relatively weak Hf–N bond providing less substrate-dependent growth and higher growth rate compared to HfO_2 film grown using $HfCl_4$. In addition, these N-containing MO precursors simultaneously formed $SiON_x$ layer during ALD of HfO_2 on Si, which prevents out-diffusion of Si during ALD and PDA to result in low interfacial trap density. However, HfO_2 film grown from MO precursors is hardly free from C impurity which deteriorates interface and bulk properties. Therefore, process parameters such as precursor pulse time and reactor temperature should be carefully optimized according to the types of the metal precursors. Recently, BTEMAH was synthesized by replacing one of four amido ligands in TEMAH with a tert-butoxy ligand to obtain higher reactivity and volatility over amide precursors. Consequently, it showed improved growth-per-cycle and higher density of HfO_2 compared with HfO_2 grown from TDMAH.

O_3 provides the higher growth rate and lower impurity level of the ALD high-k film due to the stronger oxidation power and higher reactivity than those of H_2O. The thicker IL grown during ALD with O_3 improved the thermal stability against the Si out-diffusion from the substrate, but degrades the EOT of the film. The stable GeO_2 IL grown on Ge substrate during ALD with O_3 improves the interface property compared to ALD with H_2O. However, in case of III–V substrates, O_3 induced the severe interfacial reactions of the ALD high-k film with the substrates such as the selective oxidation and reduction of the substrate elements resulting in the degraded interface property.

The performance and reliability of various recently highlighted devices including ALD high-k film are also discussed. Adopting thin TiN metal layer decreased SiO_2 interface layer thickness and EOT of the total gate stack by scavenging effect, which enables to achieve sub 1 nm EOT but the mobility decreased due to the higher remote phonon and charge scattering from the ALD-HfO_2 layer to the channel region. NBTI degradation in sub 1 nm EOT devices was enhanced by increased hole trapping into the ALD high-k bulk defects. TDDB on a pMOSFET device having 0.63 nm EOT with ALD-HfO_2 layer showed the device was not limited by the HBD until 10 years of lifetime. NBTI reliability of the Si_xGe_y channel devices with ALD-HfO_2 showed improved lifetime as compared to Si channel devices due to the adjustment of the charge-injection level. For InGaAs substrates, both interface and bulk traps are crucial problems in device performance and reliability. The border trap density inside the ALD-Al_2O_3 layer was significantly reduced by $(NH_4)_2S$ treatment. ALD technique is essential for 3D FinFET devices due to the structural characteristic. RMG process with lower thermal budget for the ALD high-k gate stacks improved I_{on}–I_{off} characteristics and showed lower EOT at equal J_g.

It is evident that the usefulness of ALD for futuristic MOSFET devices will be ever-increasing as the device structure becomes more 3D and smaller thanks to its perfect conformality and atomic-level controllability.

References

1. http://www.intel.com/technology/index.htm
2. Suntola T (1984) Proceeding of the 16th international conference on solid state devices and materials, p 647
3. Mitard J, Witters L, Garcia Bardon M, Christie P, Franco J, Mercha A, Magnone P, Alioto M, Crupi F, Ragnarsson L-Å, Hikavyy A, Vincent B, Chiarella T, Loo R, Tseng J,

Yamaguchi S, Takeoka S, Wang W-E, Absil P, Hoffmann T (2010) Technical digest—International electron devices meeting, p 249
4. Alian A, Brammertz G, Degraeve R, Cho M, Merckling C, Lin D, Wang W-E, Caymax M, Meuris M, De Meyer K, Heyns M (2012) Elecron Device Lett 33:1544
5. Ragnarsson L-Å, Chiarella T, Togo M, Schram T, Absil P, Hoffmann T (2011) Microelectron Eng 88:1317
6. Kang CY, Choi R, Song SC, Choi K, Ju BS, Hussain MM, Lee BH, Bersuker G, Young C, Heh D, Kirsch P, Barnet J, Yang J-W, Xiong W, Tseng H–H, Jammy R (2006) Technical digest—International electron devices meeting, p 1
7. Thelander C, Fröberg LE, Rehnstedt C, Samuelson L, Wernersson L-E (2008) Elecron Device Lett 29:206
8. Delabie A, Caymax M, Brijs B, Brunco D, Conard T, Sleeckx E, Van Elshocht S, Ragnarsson L-Å, De Gendt S, Heyns M (2006) ECS Trans 1(5):433–446
9. Nyns L, Delabie A, Caymax M, Heyns MM, Van Elshocht S, Vinckier C, De Gendt S (2008) J Electrochem Soc 155:G269
10. Cho M, Park J, Park HB, Hwang CS, Jeong J, Hyun KS (2002) Appl Phys Lett 81:334
11. Ritala M, Leskelä M, Nykänen E, Soininen P, Niinistö L (1993) Thin Solid Films 225:288
12. Green ML, Ho M-Y, Busch B, Wilk GD, Sorsch T, Conard T, Brijs B, Vandervorst W, Räisänen PI, Muller D, Bude M, Grazul J (2002) J Appl Phys 92:7168
13. Triyoso D, Liu R, Roan D, Ramon M, Edwards NV, Gregory R, Werho D, Kulik J, Tam G, Irwin E, Wang X-D, La LB, Hobbs C, Garcia R, Baker J, White BE Jr, Tobin P (2004) J Electrochem Soc 151:F220
14. Ragnarsson L-Å, Brunco DP, Yamamoto K, Tökei Z, Pourtois G, Delabie A, Parmentier B, Conard T, Roussel P, Gendt SD, Heyns MM (2009) J Electrochem Soc 156:H416
15. Cho M, Park HB, Park J, Hwang CS, Lee J-C, Oh S-J, Jeong J, Hyun KS, Kang H-S, Kim Y-W, Lee J-H (2003) J Appl Phys 94:2563
16. Park HB, Cho M, Park J, Hwang CS, Lee J-C, Oh S-J (2003) J Appl Phys 94:1898
17. Cho M, Degraeve R, Pourtois G, Delabie A, Ragnarsson L-Å, Kauerauf T, Groeseneken G, De Gendt S, Heyns M, Hwang CS (2007) IEEE Trans Electron Dev 54:752
18. Delabie A, Puurunen RL, Brijs B, Caymax M, Conard T, Onsia B, Richard O, Vandervorst W, Zhao C, Heyns MM, Meuris M, Viitanen MM, Brongersma HH, de Ridder M, Goncharova LV, Garfunkel E, Gustafsson T, Tsai W (2005) J Appl Phys 97:0641004
19. Delabie A, Brunco DP, Conard T, Favia P, Bender H, Franquet A, Sioncke S, Vandervorst W, Elshocht SV, Heyns M, Meuris M, Kim E, McIntyre PC, Saraswat KC, LeBeau JM, Cagnon J, Stemmer S, Tsaie W (2008) J Electrochem Soc 155:H937
20. Goel N, Majhi P, Chui CO, Tsai W, Choi D, Harris JS (2006) Appl Phys Lett 89:163517
21. Delabie A, Alian A, Bellenger F, Caymax M, Conard T, Franquet A, Sioncke S, Van Elshocht S, Heyns MM, Meurisa M (2009) J Electrochem Soc 156:G163
22. Kita K, Takahashi T, Nomura H, Suzuki S, Nishimura T, Toriumi A (2008) Appl Surf Sci 254:6100
23. Suzuki S, Kita K, Nomura H, Takahashi T, Nishimura T, Toriumi A (2007) Ext. Abstract International conf on solid state device and materials, p 20
24. Chui CO, Ramanathan S, Triplett BB, McIntyre PC, Saraswat KC (2002) IEEE Electron Device Lett 23:473
25. Kamata Y, Kamimuta Y, Ino T, Nishiyama A (2005) Jpn J Appl Phys 44:2323
26. Chi D, Chui CO, Saraswat KC, Triplett BB, McIntyre PC (2004) J Appl Phys 96:813
27. Kamata Y et al (2005) Jpn J Appl Phys 44:2323
28. Nomura H et al (2006) Solid State Device Mater:406
29. Mavrou G, Galata S, Tsipas P, Sotiropoulos A, Panayiotatos Y, Dimoulas A, Evangelou EK, Seo JW, Dieker Ch (2008) J Appl Phys 103:014506
30. Chui CO, Kim H, McIntyre PC, Saraswat KC (2004) IEEE Electron Device Lett 25:274
31. Kim H, McIntyre PC, Chui CO, Saraswat KC, Cho MH (2004) Appl Phys Lett 85:2902
32. Oh J, Majhi P, Tseng H–H, Jammy R, Kelly DQ, Banerjee SK, Campbell JC (2008) Thin Solid Films 516:4107

33. Fukuda Y, Ueno T, Hirono S, Hashimoto S (2005) Jpn J Appl Phys 44:6981
34. Robertson J (2005) Solid State Electron 49:283
35. Maeda T, Nishizawa M, Morita Y, Takagi S (2007) Appl Phys Lett 90:072911
36. Delabie A, Bellenger F, Houssa M, Conard T, Van Elshocht S, Caymax M, Heyns M, Meuris M (2007) Appl Phys Lett 91:082904
37. Fukuda Y, Yazaki Y, Otani Y, Sato T, Toyota H, Ono T (2010) IEEE Trans Electron Devices 57:282
38. Xie Q, Musschoot J, Schaekers M, Caymax M, Delabie A, Qu XP, Jiang YL, Berghe SVD, Liu J, Detavernier C (2010) Appl Phys Lett 97:222902
39. Sioncke S, Lin HC, Brammertz G, Delabie A, Conard T, Franquet A, Meuris M, Struyf H, De Gendt S, Heyns M, Fleischmann C, Temst K, Vantomme A, Muller M, Kolbe M, Beckhoff B, Caymax M (2011) J Electrochem Soc 158:H687
40. Chang CH, Chiou YK, Chang YC, Lee KY, Lin TD, Wu TB, Hong M, Kwo J (2006) Appl Phys Lett 89:242911
41. Tallarida M, Adelmann C, Delabie A, Van Elshocht S, Caymax M, Schmeisser D (2011) Appl Phys Lett 99:042906
42. Ye PD, Wilk GD, Yang B, Kwo J, Chu SNG, Nakahara S, Gossmann H-JL, Mannaerts JP, Hong M, Ng KK, Bude J (2003) Appl Phys Lett 83:180
43. Frank MM, Wilk GD, Starodub D, Gustafsson T, Garfunkel E, Chabal YJ, Grazul J, Muller DA (2005) Appl Phys Lett 86:152904
44. Maes JW, Fedorenko Y, Delabie A, Ragnarsson L-Å, Swerts J, Nyns L, Van Elshocht S, Wang C-G, Wilk G (2007) ECS Trans 11(4):59–72
45. Jung H-S, Kim HK, Yu I-H, Lee SY, Lee J, Park J, Jang JH, Jeon S-H, Chung YJ, Cho D-Y, Lee N-I, Park TJ, Choi J-H, Hwang CS (2012) J Electrochem Soc 159:G33
46. Jung H-S, Jeon SH, Kim HK, Yu I-H, Lee SY, Lee J, Chung YJ, Cho D-Y, Lee N-I, Park TJ, Choi J-H, Han S, Hwang CS (2012) ECS J Solid State Sci Technol 1:N33
47. Xie Q, Deduytsche D, Schaekers M, Caymax M, Delabie A, Qu X-P, Detavernier C (2010) Appl Phys Lett 97:112905
48. Cho M, Park HB, Park J, Lee SW, Hwang CS, Jang GH, Jeong J (2003) Appl Phys Lett 83:5503
49. Kim JH, Park TJ, Cho M, Jang JH, Seo M, Na KD, Hwang CS, Won JY (2009) J Electrochem Soc 156:G48
50. Cho M, Kim JH, Hwang CS, Ahn H-S, Han S, Won JY (2007) Appl Phys Lett 90:182907
51. Seo M, Min Y-S, Kim SK, Park TJ, Kim JH, Na KD, Hwang CS (2008) J Mater Chem 18:4324
52. Shevjakov AM, Kuznetsova GN, Aleskovskii VB (1965) Proceedings of the second USSR conference on high-temperature chemistry of oxides, Leningrad, USSR, p 26
53. Suntola T, Antson J (1977) U.S. Patent No. 4,058,430
54. Niinistö J, Putkonen M, Niinistö L, Arstila K, Sajavaara T, Lu J, Kukli K, Ritala M, Leskelä M (2006) J Electrochem Soc 153:F39
55. Swerts J, Peys N, Nyns L, Delabie A, Franquet A, Maes JW, Elshocht SV, De Gendt S (2010) J Electrochem Soc 157:G26
56. Liu X, Ramanathan S, Longdergan A, Srivastava A, Lee E, Seidel TE, Barton JT, Pang D, Gordon RG (2005) J Electrochem Soc 152:G213
57. Won S-J, Suh S, Lee SW, Choi G-J, Hwang CS, Kim HJ (2010) Electrochem Solid-State Lett 13:G13
58. Choi G-J, Kim SK, Won S-J, Kim HJ, Hwang CS (2009) J Electrochem Soc 156:G138
59. Won S-J, Kim J-Y, Choi G-J, Heo J, Hwang CS, Kim HJ (2009) Chem Mater 21:4374
60. Cheng Y-L, Chang Y-L, Hsieh C-Y, Lin J-R (2013) JVST A 31(1)
61. Campabadal F, Beldarrain O, Zabala M, Acero MC, Rafi JM (2011) Proceeding og the 8th Spanish conference on electron devices, CDE
62. Park TJ, Sivasubramani P, Coss BE, Kim H-C, Lee B, Wallace RM, Kim J, Rousseau M, Liu X, Li H, Lehn J-S, Hong D, Shenai D (2010) Appl Phys Lett 97:092904

63. Cho M, Jeong DS, Park HB, Lee SW, Park TJ, Hwang CS, Jang GH, Jeong J (2004) Appl Phys Lett 85:5953
64. Park HB, Cho M, Park J, Lee SW, Hwang CS, Kim J-P, Lee J-H, Lee N-I, Lee J-C, Oh S-J (2003) J Appl Phys 94:3641
65. Kamiyama S, Miura T, Nara Y (2006) Electrochem Solid-State Lett 9:G285
66. Chung KJ, Park TJ, Sivasubramani P, Kim J, Ahn J (2010) ECS Trans 28:221
67. Degraeve R, Kerber A, Roussel Ph, Cartier E, Kauerauf T, Pantisano L, Groeseneken G (2003) IEEE technical digest of international electron devices meeting, p 935
68. Kyuno K, Kita K, Toriumi A (2005) Appl Phys Lett 86:063510
69. Lee C-K, Cho E, Lee H-S, Hwang CS, Han S (2008) Phys Rev B 78:012102
70. Fujimori H, Yashima M, Sasaki S, Kakihana M. Mori T, Tanaka M, Yoshimura M (2001) Chem Phys Lett 346:217
71. Bernay C, Ringuedé A, Colomban P, Lincot D, Cassir MJ (2003) J Phys Chem Solids 64:1761
72. Park TJ, Kim JH, Jang JH, Lee C-K, Na KD, Lee SY, Jung HS, Kim M, Han S, Hwang CS (2010) Chem Mater 22:4175
73. Kukli K, Ritala M, Sajavaara T, Keinonen J, Leskelä M (2002) Chem Vapor Depos 8:199
74. Park J, Cho M, Kim SK, Park TJ, Lee SW, Hong SH, Hwang CS (2005) Appl Phys Lett 86:112907
75. Baldovino S, Spiga S, Scarel G, Fanciulli M (2007) Appl Phys Lett 91:172905
76. Park TJ, Kim JH, Seo MH, Jang JH, Hwang CS (2007) Appl Phys Lett 90:152906
77. Delabie A, Puurenen RL, Brijs B, Caymax M, Conrad T, Onsia B, Richard O, Vandervorst W, Zhao C, Heyns MM, Meuris M, Viitanen MM, Brongersma HH, de Ridder M, Goncharova LV, Garfunkel E, Gustafsson T, Tsai W (2005) J Appl Phys 97:064104
78. Spiga S, Wiemer C, Tallarida G, Scarel G, Ferrari S, Seguini G, Fanciulli M (2005) Appl Phys Lett 87:112904
79. Kim H, Chui CO, Saraswat KC, McIntyre PC (2003) Appl Phys Lett 83:2647
80. Houssa M, De Jaeger B, Delabie A, Van Elshocht S, Afanasiev V, Autran J, Stesmans A, Meuris M, Heyns M, Non-Cryst J (2005) Solids 351:1902
81. Perego M, Scarel G, Fanciulli M, Fedushkin IL, Skatova AA (2007) Appl Phys Lett 90:162115
82. Broqvist J, Binder JF, Pasquarello A (2009) Appl Phys Lett 94:141911
83. Afanas'ev VV, Stesmans A, Delabie A, Bellenger F, Houssa M, Meuris M (2008) Appl Phys Lett 92:022109
84. Dalapati GK, Tong Y, Loh WY, Mun HK, Cho BJ (2007) IEEE Trans Electron Devices 54:1831
85. Shahrjerdi D, Tutuc E, Banerjee SK (2007) Appl Phys Lett 91:063501
86. Kim CY, Cho SW, Cho M-H, Chung KB, An C-H, Kim H, Lee HJ, Ko D-H (2008) Appl Phys Lett 93:192902
87. Hackley JC, Demaree JD, Gougousi T (2008) Appl Phys Lett 92:162902
88. Hinkle CL, Sonnet AM, Vogel EM, McDonnell S, Hughes GJ, Milojevic M, Lee B, Aguirre-Tostado FS, Choi KJ, Kim J, Wallace RM (2007) Appl Phys Lett 91:163512
89. Hinkle CL, Sonnet AM, Vogel EM, McDonnell S, Hughes GJ, Milojevic M, Lee B, Aguirre-Tostado FS, Choi KJ, Kim HC, Kim J, Wallace RM (2008) Appl Phys Lett 92:071901
90. Aguirre-Tostado FS, Milojevic M, Hinkle CL, Vogel EM, Wallace RM, McDonnell S, Hughes GJ (2008) Appl Phys Lett 92:171906
91. Shin B, Choi D, Harris JS, McIntyre PC (2008) Appl Phys Lett 93:052911
92. Chang CC, Citrin PH, Schwartz B (1977) J Vac Sci Technol 14 (4):943–952
93. Ishikawa Y, Ishii H, Hasegawa H, Fukui T (1994) New Paltz, New York (unpublished)
94. Abrahams MS, Buiocchi CJ (1965) J Appl Phys 36(9):2855–2863
95. Adachi S, Kikuchi D (2000) J Electrochem Soc 147(12):4618–4624
96. Liu Z, Sun Y, Machuca F, Pianetta P, Spicer WE, Pease RFW (2003) J Vac Sci Technol, A 21(1):212–218

97. DeSalvo GC, Bozada CA, Ebel JL, Look DC, Barrette JP, Cerny CLA, Dettmer RW, Gillespie JK, Havasy CK, Jenkins TJ, Nakano K, Pettiford CI, Quach TK, Sewell JS, Via GD (1996) J Electrochem Soc 143(11):3652–3656
98. Yablonovitch E, Cox HM, Gmitter TJ (1988) Appl Phys Lett 52(12):1002–1004
99. Lebedev MV, Ensling D, Hunger R, Mayer T, Jaegermann W (2004) Appl Sur Sci 229(1–4):226–232
100. Arthur JR (1974) Surf Sci 43(2):449–461
101. Kawai NJ, Nakagawa T, Kojima T, Ohta K, Kawashima M (1984) Electron Lett 20(1):47–48
102. Chang RPH, Darack S (1981) Appl Phys Lett 38(11):898–899
103. Callegari A, Hoh PD, Buchanan DA, Lacey D (1989) Appl Phys Lett 54(4):332–334
104. Sugata S, Takamori A, Takado N, Asakawa K, Miyauchi E, Hashimoto H (1988) J Vac Sci Technol, B 6(4):1087–1091
105. Sugaya T, Kawabe M (1991) Jpn J Appl Phys 30(3A):L402–L404
106. Grant RW, Waldrop JR (1987) J Vac Sci Tech B 5(4):1015–1019
107. Fountain GG, Hattangady SV, Vitkavage DJ, Rudder RA, Markunas RJ (1988) Electron Lett 24(18):1134–1135
108. Tiwari S (1988) IEEE Electr Device L 9 (3):142–144
109. Callegari A, Sadana DK, Buchanan DA, Paccagnella A, Marshall ED, Tischler MA, Norcott M (1991) Appl Phys Lett 58(22):2540–2542
110. Freeouf JL, Silberman JA, Wright SL, Tiwari S, Batey J (1989) Bozeman, Montana (unpublished)
111. Ivanco J, Kubota T, Kobayashi H (2005) J Appl Phys 97(7):073712–073717
112. Brennan B, Milojevic M, Kim HC, Hurley PK, Kim J, Hughes G, Wallace RM (2009) Electrochem Solid State Lett 12:H205
113. Byun Y-C, Mahata C, An C-H, Oh J, Choi R, Kim H (2012) J Phys D Appl Phys 45:435305
114. Madan H, Veksler D, Chen YT, Huang J, Goel N, Bersuker G, Datta S (2011) IEEE 117
115. Chung KJ, Park TJ, Sivasubramani P, Kim J, Ahn J (2012) Microelectron Eng 89:80
116. Ogawa A, Iwamoto K, Ota H, Morita Y, Ikeda M, Nabatame T, Toriumi A (2007) Microelectron Eng 84(9–10):1861
117. Delabie A, Caymax M, Brijs B, Brunco DP, Conard T, Sleeckx E, Van Elshocht S, Ragnarsson L-A, De Gendt S, Heyns MM (2006) J Electrochem Soc 153:F180
118. Lee T, Rhee SJ, Kang CY, Zhu F, Kim H-S, Choi C, Ok I, Zhang M, Krishnan S, Thareja G, Lee JC (2012) Electron Device Lett 27:640
119. Ando T, Frank MM, Choi K, Choi C, Bruley J, Hopstaken M, Copel M, Cartier E, Kerber A, Callegari A, Lacey D, Brown S, Yang Q, Narayanan V (2009) Technical digest—International electron devices meeting, p 423
120. Ragnarsson L-Å, Li Z, Tseng J, Schram T, Rohr E, Cho MJ, Kauerauf T, Conard T, Okuno Y, Parvais B, Absil P, Biesemans S, Hoffmann TY (2009) Technical digest—International electron devices meeting, p 663
121. Cho M, Lee J-D, Aoulaiche M, Kaczer B, Roussel P, Kauerauf T, Degraeve R, Franco J, Ragnarsson L-A, Groeseneken G (2012) IEEE Trans Electron Devices 59(8):2042
122. International Technology Roadmap for Semiconductors 2003 (or 2009) Edition, downloaded on 18 July 2011. http://www.itrs.net/Links/2009ITRS/2009Chapters_2009Tables/2009_PIDS.pdf
123. Jeppson KO, Svensson CM (1977) J Appl Phys 48(5):2004
124. Ogawa S, Shiono N (1995) Phys Rev B 51(7):4218
125. Alam MA, Mahapatra S (2005) Microelectron Reliab 45(1):71
126. Houssa M, Aoulaiche M, De Gendt S, Groeseneken G, Heyns MM, Stesmans A (2005) Appl Phys Lett 86(9):093506-1
127. Cartier E, Pantisano L, Kerber A, Groeseneken G (2003) Presented at the IEEE Insulating Films on Semiconductors Conference (INFOS), Leuven, Belgium

128. Aoulaiche M, Kaczer B, Cho M, Houssa M, Degraeve R, Kauerauf T, Akheyar A, Schram T, Roussel Ph, Maes HE, Hoffmann T, Biesemans S, Groeseneken G (2009) In: Proceedings of the international reliability physics symposium, p 1014
129. Zafar S, Callegari A, Gusev E, Fischetti MV (2003) J Appl Phys 93(11):9298
130. Cho M, Aoulaiche M, Degraeve R, Kaczer B, Franco J, Kauerauf T, Roussel P, Ragnarsson LÅ, Tseng J, Hoffmann TY, Groeseneken G (2010) In: Proceedings of the international reliability physics symposium, p 1095
131. Sahhaf S, Degraeve R, Srividya V, Cho M, Kauerauf T, Groeseneken G (2010) In: Proceedings of the international reliability physics symposium, p 1078
132. Cho M, Aoulaiche M, Degraeve R, Kaczer B, Kauerauf T, Ragnarsson L-Å, Adelmann C, Van Elshocht S, Hoffmann TY, Groeseneken G (2011) Solid State Electron 63(1):5
133. Kauerauf T, Degraeve R, Ragnarsson L-Å, Roussel P, Sahhaf S, Groeseneken G (2011) In: Proceedings of the international reliability physics symposium, p 7
134. Sahhaf S, Degraeve R, Roussel PJ, Kaczer B, Kauerauf T, Groeseneken G (2009) IEEE Trans Electron Devices 56(7):1424
135. Park J, Park TJ, Cho M, Kim SK, Hong SH, Kim JH, Seo M, Hwang CS (2006) J Appl Phys 99:094501
136. Lee T, Park I-S, Ko H-K, Lee S, Kim K-R, Ahn J (2005) International microprocesses and nanotechnology conference, p 94
137. Mitard J, De Jaeger B, Leys FE, Hellings G, Martens K, Eneman G, Brunco DP, Loo R, Lin JC, Shamiryan D, Vandeweyer T, Winderickx G, Vrancken E, Yu CH, De Meyer K, Caymax M, Pantisano L, Meuris M, Heyns M (2008) Technical digest—International electron devices meeting, p 873
138. Witters L, Takeoka S, Yamaguchi S, Hikavyy A, Shamiryan D, Cho MJ, Chiarella T, Ragnarsson L-Å, Loo R, Kerner C, Crabbe Y, Franco J, Tseng J, Wang W-E, Rohr E, Schram T, Richard O, Bender H, Biesemans S, Absil P, Hoffmann T (2010) Symposium on VLSI Technology, p 181
139. Sze SM (1981) Physics of semiconductor devices, 2nd edn. Wiley, New York
140. De Jaeger B, Bonzom R, Leys F, Richard O, Van Steenbergen J, Winderickx G, Van Moorhem E, Raskin G, Letertre F, Billon T, Meuris M, Heyns M (2005) Microelectron Eng 80(17):26
141. Leys FE, Bonzom R, Kaczer B, Janssens T, Vandervorst W, De Jaeger B, Van Steenbergen J, Martens K, Hellin D, Rip J, Dilliway G, Delabie A, Zimmerman P, Houssa M, Theuwis A, Loo R, Meuris M, Caymax M, Heyns MM (2006) Mater Sci Semicond Process 9:679
142. Yeo CC, Cho BJ, Gao F, Lee SJ, Lee MH, Yu C-Y, Liu CW, Tang LJ, Lee TW (2005) Electron Device Lett 26(10):761
143. Mitard J, Martens K, De Jaeger B, Franco J, Shea C, Plourde C, Leys FE, Loo R, Hellings G, Eneman G, Wang W-E, Lin JC, Kaczer B, De Meyer K, Hoffmann T, De Gendt S, Caymax M, Meuris M, Heyns MM (2009) European Solid-State Device Research Conference, p 411
144. Zhang R, Huang PC, Taoka N, Takenaka M, Takagi S (2012) Symposium on VLSI Technoogy, p 161
145. Mitard J, Eneman G, Hellings G, Witters L, Hikavyy A, Vincent B, Loo R, Bender H, Horiguchi N, Collaert N, Thean A (2012) SiGe, Ge, and related compuncs 5: materials, processing, and devices, p 131
146. Aoulaiche M, Kaczer B, De Jaeger B, Houssa M, Martens K, Degraeve R, Roussel P, Mitard J, De Gendt S, Maes HE, Groeseneken G, Meuris M, Heyns MM (2008) In: Proceedings of the international reliability physics symposium, p 358
147. Franco J, Kaczer B, Eneman G, Roussel PhJ, Grasser T, Mitard J, Ragnarsson L-Å, Cho M, Witters L, Chiarella T, Togo M, Wang W-E, Hikavyy A, Loo R, Horiguchi N, Groeseneken G (2011) Technical digest—International electron devices meeting, p 445
148. Loh W-Y, Majhi P, Lee S-H, Oh J-W, Sassman B, Young C, Bersuker G, Cho B-J, Park C-S, Kang C-Y, Kirsch P, Lee B-H, Harris HR, Tseng H–H, Jammy R (2008) Symposium on VLSI Technology, p 56

149. Franco J, Kaczer B, Eneman G, Roussel PhJ, Cho M, Mitard J, Witters L, Hoffmann TY, Groeseneken G, Crupi F, Grasser T (2011) In: Proceedings of the international reliability physics symposium, p 624
150. Sioncke S, Lin D, Nyns L, Delabie A, Thean A, Horiguchi N, Struyf H, De-Gendt S, Caymax M (2012) ECS Trans 45(4):97
151. Brammertz G, Lin HC, Martens K, Alian A, Merckling C, Penaud J, Kohen D, Wang W-E, Sioncke S, Delabie A, Meuris M, Caymax M, Heyns M (2009) ECS Trans 19(5):375
152. Heyns M, Bellenger F, Brammertz G, Caymax M, Cantoro M, De Gendt S, De Jaeger B, Delabie A, Eneman G, Groeseneken G, Hellings G, Houssa M, Iacopi F, Leonelli D, Lin D, Magnus W, Martens K, Merckling C, Meuris M, Mitard J, Penaude J, Pourtois G, Scarrozza M, Simoen E, Soree B, Van Elshocht S, Vandenbergheb W, Vandooren A, Vereecke P, Verhulst A, Wang W-E (2010) Optical Microlithogr XXIII:764003
153. Lin D, Alian A, Gupta S, Yang B, Bury E, Sioncke S, Degraeve R, Toledano ML, Krom R, Favia P, Bender H, Caymax M, Saraswat KC, Collaert N, Thean A (2012) Technical digest—International electron devices meeting, p 645
154. Degraeve R, Cho M, Govoreanu B, Kaczer B, Zahid MB, Van Houdt J, Jurczak M, Groeseneken G (2008) Technical digest—International electron devices meeting, p 1
155. Cho M, Degraeve R, Roussel P, Govoreanu B, Kaczer B, Zahid M, Simoen E, Arreghini A, Jurczak M, Houdt JV, Groeseneken G (2010) Solid State Electron 54(11):1384
156. Afanas'ev VV, Chou H-Y, Thoan NH, Adelmann C, Lin HC, Houssa M, Stesmans A (2012) Appl Phys Lett 100:202104
157. Colinge J-P (2004) Solid-State Electron 48:897
158. Auth C, Allen C, Blattner A, Bergstrom D, Brazier M, Bost M, Buehler M, Chikarmane V, Ghani T, Glassman T, Grover R, Han W, Hanken D, Hattendorf M, Hentges P, Heussner R, Hicks J, Ingerly D, Jain P, Jaloviar S, James R, Jones D, Jopling J, Joshi S, Kenyon C, Liu H, McFadden R, McIntyre B, Neirynck J, Parker C, Pipes L, Post I, Pradhan S, Prince M, Ramey S, Reynolds T, Roesler J, Sandford J, Seiple J, Smith P, Thomas C, Towner D, Troeger T, Weber C, Yashar P, Zawadzki K, Mistry K (2012) Symposium on VLSI Techology, p 131
159. Zschätzsch G, Sasaki Y, Hayashi S, Togo M, Chiarella T, Kambham AK, Mody J, Douhard B, Horiguchi N, Mizuno B, Ogura M, Vandervorst W (2011) Technical digest—International electron devices meeting, p 841
160. Frank MM (2011) IEEE European solid-state device research conference, p 25
161. Witters L, Mitard J, Veloso A, Hikavyy A, Franco J, Kauerauf T, Cho M, Schram T, Sebai F, Yamaguchi S, Takeoka S, Fukuda M, Wang W-E, Duriez B, Eneman G, Loo R, Kellens K, Tielens H, Favia P, Rohr E, Hellings G, Bender H, Roussel P, Crabbe Y, Brus S, Mannaert G, Kubicek S, Devriendt K, De Meyer K, Ragnarsson L-Å, Steegen A, Horiguchi N (2011) Technical digest—International electron devices meeting, p 654
162. Boccardi G, Ritzenthaler R, Togo M, Chiarella T, Kim MS, Yuichiro S, Veloso A, Chew SA, Vecchio E, Locorotondo S, Devriendt K, Ong P, Brus S, Horiguchi N, Thean A (2012) International conference on solid state devices and materials, p 723
163. Kapila G, Kaczer B, Nackaerts A, Collaert N, Groeseneken GV (2007) Elecron Device Lett 28(3):232
164. Hu VP-H, Fan M-L, Hsieh C-Y, Su P, Chuang C-T (2011) IEEE Trans Electron Devices 58(3):805
165. Park H, Song SC, Woo SH, Abu-Rahma MH, Ge L, Kang MG, Han BM, Wang J, Choi R, Yang JW, Jung SO, Yeap G (2010) In: Proceedings of the international reliability physics symposium, p 1008
166. Maeda S, Choi J-A, Yang J-H, Jin Y-S, Bae S-K, Kim Y-W, Suh K-P (2004) In: Proceedings of the international reliability physics symposium, p 8
167. Feijoo PC, Cho M, Togo M, San Andrés E, Groeseneken G (2011) Microelectron Reliab 51:1521
168. Cho ES, Lee CH, Fayrushin A, Park HB, Park D (2006) In: Proceedings of the international reliability physics symposium, p 663

169. Kim JJ, Cho M, Pantisano L, Jung U, Lee YG, Chiarella T, Togo M, Horiguchi N, Groeseneken G, Lee BH (2012) Electron Device Lett 33(7):937
170. Shickova A, Collaert N, Rooyackers R, De Keersgieter A, Kauerauf T, Jurczak M, Kaczer B, Groeseneken G (2006) In: Proceedings of the ULIS conference, p 141
171. Crupi F, Kaczer B, Degraeve R, Subramanian V, Srinivasan P, Simoen E, Dixit A, Jurczak M, Groeseneken G (2006) IEEE Trans Electron Devices 53(9):2351

Chapter 8
Back End of the Line

Hyungjun Kim, Soo-Hyun Kim and H. -B. -R. Lee

8.1 Introduction

For the fabrication of high performance logic device, reduction of RC delay in back end of the line (BEOL) process as well as parasitic resistance are essential tasks [1–5]. For the reduction of RC delay, the utilization of low-k dielectrics and low resistivity metal should be realized [3–9]. Cu metallization has been implemented after the introduction of damascene process using chemical mechanical planarization (CMP) and electroplating of Cu films since late 1990's. In contrast to the conventional Al-based metallization, Cu metallization has several difficulties to be solved as a metallization material such as easy diffusivity, nonetchability, etc. Thus, effective diffusion barrier is needed prior to Cu deposition [10, 11]. Also, Cu electroplating process requires thin conducting layer. Besides metallization, with device scaling, the low resistance contact becomes more and more important. Moreover, the plug between contact and first-level metallization becomes smaller in size and the aspect ratio becomes higher. Currently, sputtering or chemical vapor deposition is used for the deposition of these layers. But with device down scaling into sub-20 nm regime, very high conformality is essential more than ever.

Although the currently used deposition processes have been improved in terms of conformality, it is almost impossible to meet the requirements for the deposition

H. Kim (✉)
School of Electrical and Electronic Engineering, Yonsei University, Seoul 120-749, Korea
e-mail: hyungjun@yonsei.ac.kr

S.-H. Kim
School of Materials Science and Engineering, Yeungnam University, Gyeongsangbuk-do 712-749, Korea
e-mail: soohyun@ynu.ac.kr

H.-B. -R.Lee
Department of Materials Science and Engineering, Incheon National University, Incheon 406-772, Korea
e-mail: hbrlee@incheon.ac.kr

C. S. Hwang (ed.), *Atomic Layer Deposition for Semiconductors*,
DOI: 10.1007/978-1-4614-8054-9_8,

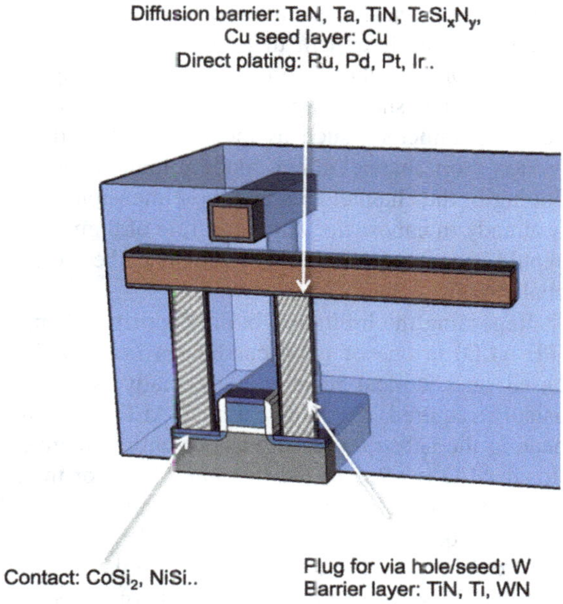

Fig. 8.1 The application of ALD for BEOL process in modern semiconductor devices

of very thin films with high conformality. Regarding this, atomic layer deposition (ALD) is receiving a great attention for its ability to deposit highly conformal film with nanoscale thickness control. ALD is a layer-by-layer thin film deposition technique by alternating exposure of chemical species, enabling highly conformal thin films on nanoscale trenches or vias, which makes it one of the most promising thin film deposition techniques for BEOL process [12–14]. Figure 8.1 shows the potential applications of ALD for BEOL of nanoscale logic devices. As can be seen, metals and nitrides are the specifically required materials in BEOL processes. While the ALD of high k oxides has been already implemented in the front end of the line process such as gate dielectrics and capacitor dielectrics, ALD is not yet a standard process in any of the BEOL process. However, at sub-20 nm technology regime, ALD is expected to play an important role in BEOL process for logic device. In this chapter, we will review the efforts on the application of ALD at BEOL process; including metallization and contact/plug formation.

8.2 Overview on the ALD for BEOL Process

8.2.1 The Characteristics of ALD Processes for BEOL

Although ALD of metals and nitrides are actively studied these days for many different applications, there are specific requirements specifically important for BEOL process. Since the BEOL processes are carried out after the completion of transistor fabrication, the allowed thermal budget of BEOL process is usually quite small. This limitation in thermal budget is becoming more serious with device scaling. One of the severe problems accompanying with the extreme device scaling is the short channel effect, which results in the unwanted deviation of device parameters, such as the threshold voltage, from the designed value. To reduce short channel effect, modern logic device is fabricated using very delicate doping in the channel area and the dimension of each component of the transistor is already in nanoscale. Thus, as a rule of thumb, the process temperature in BEOL should not exceed 400 °C, which makes the ALD for BEOL even more challenging.

Regarding the limitation in the deposition temperature, plasma-enhanced ALD (PE-ALD) is one of good candidates in the BEOL process [13]. Since ALD is based on chemical reaction, (especially, surface reaction) activation energy is usually required. For conventional ALD, this energy is provided thermally by heating the substrate (cold wall) or entire chamber (hot wall). Thus, the conventional ALD is often called thermal ALD. For thermal ALD of metals or nitrides, the deposition temperature is often higher than 400 °C, especially for halide precursor based process [14]. Meanwhile, the required growth temperature for PE-ALD can be significantly lowered, since the required activation energy is provided by plasma source in addition to the thermal energy [13]. For PE-ALD, plasma of only the reactant, not the metal precursor, is used, since excitation of metal precursors causes cracking of gas molecules, possibly resulting in the disturbance of the self-saturation of adsorption [14]. The PE-ALD chamber configuration is usually more complicated compared to thermal ALD, since plasma generation unit should be included. Plasma can be generated by several ways, including rf, microwave, and electron cyclotron resonance (ECR).

Usually, PE-ALD produces better quality film at lower deposition temperatures than thermal ALD. For example, the film density of PE-ALD thin films is higher than that of thermal ALD, which could be a great benefit in the deposition of barrier materials [14]. Also, PE-ALD often produces higher purity metal/nitride thin films than thermal ALD, resulting in low resistivity. However, there is a concern about recombination of the active species (such as atomic hydrogen) produced by plasma [13]. This could result in spatial nonuniformity across the chamber or perhaps less deposition in deep vias or trenches due to a reduction in the effective flux of the reactive species after multiple surface or wall collisions. Thus, the conformality of PE-ALD film could be limited in very high AR structure [15].

8.2.2 ALD Processes for Nitrides

Nitrides are used as a diffusion barrier in Al and Cu metallization as well as barrier for plug application in BEOL. For ALD of nitrides, NH_3, a good reducing agent as well as nitrogen source, has been the most typical reactant [16–20]. Halide precursor is a usually good choice for ALD of nitride with NH_3. For example, the combination of $TiCl_4$ and NH_3 produces good quality TiN film. However, undesirable phase is often deposited by ALD of nitrides which can form several different phases. The ALD of tantalum nitride is a typical example. To obtain desirable, low resistivity cubic TaN rather than the insulating Ta_3N_5, an additional reducing source in addition to NH_3 was used such as Zn or TMA (trimethyl Aluminum). Instead of NH_3, other nitrogen containing reducing agents have been tried such as hydrazine (Me_2NNH_2) [21] and amines [22, 23]. However, significant improvements in the ALD process or film properties were not obtained compared to NH_3. Meanwhile, PE-ALD process gives more wide range of process margin and the choice of precursors. There are numerous reports on PE-ALD of diffusion barriers such as TaN and TiN as will described in Sect. 8.5.

Instead of the use of NH_3, metal precursors containing nitrogen have been studied for ALD of nitride. For this case, metal organic (MO, or organometallic) compounds are used as precursor with the use of hydrogen or hydrogen plasma as a reactant. For example, alkylamides are the most widely studied precursor for nitride ALD [24–26]. Since alkylamides such as $Ti[NMe_2]_4$ (tetrakis(dimethylamino)titanium, TDMAT) contain nitrogen in the ligands, ALD using only with reducing agent without additional nitrogen source is possible [27–29]. The potential advantages of MO precursors are the high volatility and the low growth temperature. There is a potential contamination issue of carbon and hydrogen for MO precursor, but the carbon contamination is generally considered not a critical problem. Meanwhile, the low density of the MO ALD films is more significant problems, which can be partially improved by densification effect of plasma [30].

8.2.3 ALD Processes for Metals

Usually, a metal ALD process is more difficult to achieve compared to other ALD processes such as oxides or nitrides. Thus, compared to ALD of metal oxides, a relatively less research on ALD of metals has been carried out. Metal ALD is typically composed of adsorption of metal precursors on the substrates and the reduction of adsorbed precursors to metal films. Thus, selection of proper reducing agent is very important. The most readily available and simplest reducing agent is molecular hydrogen. And, regarding metal precursors, halides were most widely studied in early studies. For example, WF_6 is a very good precursor which has very high vapor pressure and easily reduced by H_2 produces metallic W [31]. However, for many metals, the reduction of halide is difficult by molecular hydrogen. For example, ALD Ti thin films with good film properties have been rarely reported.

Moreover, the thermal budget is limited for BEOL process as mentioned earlier. Thus, high reactivity of reactant and precursors is essential. For example, the cracking of molecular hydrogen to produce hydrogen radical is a typical method for achieving high reactivity. Ta, Ti, and Sn films have been deposited by reaction between halides and atomic hydrogen [32, 33]. In fact, the reducing capability of atomic hydrogen is such that if the proper amount of halides vapors could be generated and fed into the chamber, most of the metals are deposited by atomic hydrogen reduction at moderate sample temperatures. However, for highly reactive metal, although the ALD of the film is possible by hydrogen radical, metallic film is not stable in the open environment. Ti ALD using $TiCl_4$ and atomic hydrogen was reported, but the use of ALD for Ti barrier is still yet to be accomplished in real application [32].

Other reducing agent has been reported for metal ALD. For example, metallic Zn vapor, which is generated by heating of a Zn containing boat to a high temperature inside the chamber, is another powerful reducing agent for producing metal films for ALD [34]. However, Zn contamination could be problematic to the device performance in real semiconductor device process. Si_2H_6, SiH_4, and B_2H_6 have been employed for tungsten ALD with WF_6 [35]. In this case, the reactant molecules reduce adsorbed tungsten fluoride to tungsten metal film while the Si atoms are not incorporated during deposition. For $TaCl_5$, alternate exposure of $TaCl_5$ and SiH_4 produced $TaSi_2$, rather than metallic Ta [36].

For some special cases, molecular oxygen or air was used as a nonmetal precursor to produce metal films (Ru or Pt) [37–39]. Additionally, O_3 or H_2O was used to produce metal oxide films by ALD and the produced oxide has been reduced to metal films for some metal films such as Ni or Cu [40–42]. For some transition metals, NH_3 plasma or N_2/H_2 plasma were used to produce metallic film by ALD [43–46]. For example, Co and Ni ALD have been reported by the use of NH_3 plasma. Also, NH_3 molecules could be used to produce metallic film, which will be further described in Sect. 8.3. Recently, many novel precursors are being reported for ALD of metals. However, still many research and development efforts should be invested in the precursor development.

8.3 ALD for Contact Application

8.3.1 Contact in Nanoscale Si Devices

Contacts in Si devices are very important because high contact resistance between source and drain (S/D) and metal plugs induces problems on overall device performance, such as high power consumption and long RC delay [47]. Metal silicides are silicon compounds with metals, and they have low resistivity like metals as well as good compatibility with Si and metals [48]. So, metal silicides have been used for contact materials in Si devices. $TiSi_2$ has been widely used for contacts; however, NiSi and $CoSi_2$ are applied to sub-100 nm Si devices since $TiSi_2$ shows narrow line width effect that is increase in $TiSi_2$ resistivity with decreasing line width [47]. Generally, metal silicides are fabricated by depositing metal thin films on metal oxide semiconductor field effect transistor (MOSFET) devices followed by annealing. In addition, Ti or TiN is used for capping layer to prevent oxygen contamination of metal films. Physical vapor deposition (PVD) has been typically employed for metal film deposition. In current dynamic random

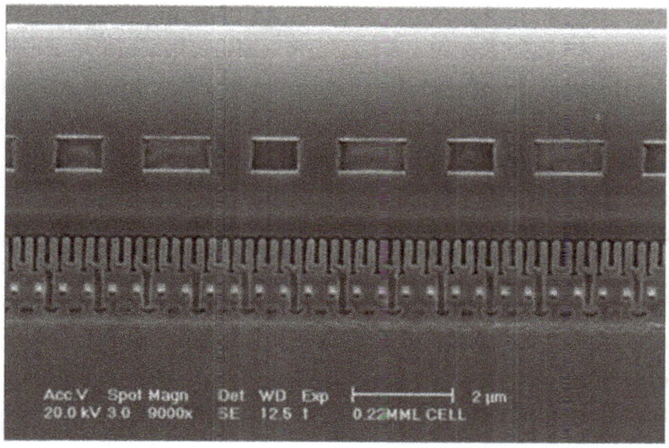

Fig. 8.2 A cross-sectional SEM image of DRAM structure [50] (Reprinted with permission from ref. [50], Copyright 2004 Elsevier B.V)

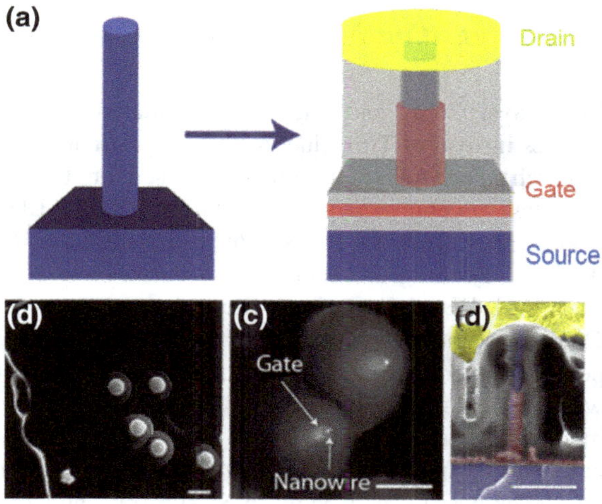

Fig. 8.3 The schematic drawings and SEM images of vertical transistor [52] (Reprinted with permission from ref. [52], Copyright 2006 American Chemical Society)

access memory (DRAM) technology, stacked capacitor structure has benefits over trench capacitor for abiding by the scaling down of the memory devices [49]. For stacked capacitor structure, the COB (capacitor-over-bitline) requires that the contact material should be formed in deep contact holes with high aspect ratio as shown in Fig. 8.2 [50]. Thus, inherent poor step coverage of PVD is becoming more problematic for sub-50 nm technology node high-density DRAM fabrication [50, 51].

In order to overcome limitations of current Si devices in downscaling, emerging nanodevices have been intensively studied, and their structures are moved from planar 2D to 3D structure. So, the schemes of contact fabrication used for 2D devices are changed in 3D emerging nanodevices, and conformality of metal thin films is the most important requirement to be achieved. For instance, the vertical transistors using nanowires shown in Fig. 8.3 require conformal deposition of metal films to form contacts [52]. Therefore, excellent conformality of metal film deposition techniques is essential for emerging nanodevices as well as conventional Si devices, so ALD is a good alternative to conventional PVD techniques.

Besides conformality, using ALD has another potential advantage in terms of direct deposition of metal silicides. Because metal silicides are formed through solid-state reaction between Si substrate and metals, the silicidation consumes certain amount of Si. For example, 3.6 nm of $CoSi_2$ is formed in the side of Si substrate from 1 nm of Co thin film since Co atoms diffuse into Si side and react with Si. This consumption of Si is significant problem in the shallow junction where the junction depth of S/D is below 5 nm [47]. One of the solutions is direct deposition of metal silicides without annealing. Since ALD film is formed through chemical reactions between two precursors, ALD has possibility to directly deposit metal silicide films with high conformality.

8.3.2 Transition Metal ALD for Contact

Nowadays, $CoSi_2$ and NiSi are considered for contact materials of nanoscale devices instead of $TiSi_2$ due to immunity of narrow line width effects. So, most of researches on ALD for contact focused on these two metal silicide systems. Although several papers on transition metals ALD were published in the early studies, they were not focused on the contact fabrication. Lee et al. reported PE-ALD Co by using metal organic precursors and NH_3 plasma for contact applications [45]. PE-ALD Co films showed very low resistivity down to 10 μΩcm and $CoSi_2$ was obtained from annealing of PE-ALD Co films. In addition, bis-cyclopentadienyl cobalt ($CoCp_2$) and dicarbonyl cyclopentadienyl cobalt ($CoCp(CO)_2$) were comparatively studied for Co PE-ALD, and it was found that $CoCp_2$ was much more suitable than $CoCp(CO)_2$ in terms of ALD growth mode and film purity [45]. In their following report, it was found that thin amorphous SiN_x interlayer was formed between Si substrate and PE-ALD Co film due to direct exposure of Si surface to NH_3 plasma during initial growth, resulting in the formation of $CoSi_2$ epitaxial growth through the nitride-mediated epitaxy mechanism [53]. PE-ALD Co using $CoCp(CO)_2$ and H_2 plasma was reported for fabrication of $CoSi_2$ [54]. Although PE-ALD Co films were relatively pure, 10 % of carbon impurity was detected which was consistent with Lee et al.'s report [45, 54]. PE-ALD Co films were in situ annealed with and without Ti capping layer, and the effects of Ti on silicidation were investigated. The $CoSi_2$ annealed from PE-ALD Co film with Ti capping layer showed flatter and sharper interface than that without Ti capping layer [54]. Same group published other Co PE-ALD process using different Co precursor, dicobalt octacarbonyl ($Co_2(CO)_2$) [55, 56]. PE-ALD Co films were contaminated by carbon and oxygen impurities due to thermal instability of $Co_2(CO)_2$ precursor, so various plasma reactants, H_2 plasma, $H_2 + Ar$ mixed plasma, NH_3 plasma, and N_2 plasma, were used to reduce contaminants [55, 56].

Do et al. investigated Ni ALD using Ni metal organic precursor, bis(dimethylamino-2-methyl-2-butoxo)nickel (Ni(dmamb)$_2$), and H_2 as a counter reactant [57]. Although large amount of carbon impurity was detected in ALD Ni films, NiSi and $NiSi_2$ were formed after annealing at 600 °C and 900 °C, respectively, without capping layer. They attributed the formation of silicides without capping layer to the present of carbon layer because carbon impurity diffused out into surface during annealing [57]. In the following paper, they exploited the carbon contamination in the Ni film and compared silicidation of ALD Ni–C film with ALD Ni capped by Ti, and they applied Ni ALD process to fabricate fully silicided gate [58, 59].

Except for contact application, various transition metal ALD processes have been studied for general purposes. Although silicidations were not investigated by using the ALD transition metal films, potentially these ALD can be adopted to fabrication of metal silicides for contact applications. In early study, IBM researchers investigated Ti and Ta PE-ALD using chloride precursors, $TiCl_4$ and $TaCl_5$, and H_2 plasma; however, the films were easily contaminated by oxygen [32, 33]. In order to deposit metallic Ni film, two consecutive processes were employed. NiO_x was deposited by ALD from nickel bisacetylacetonate (Ni(acac)$_2$) and O_3 as a precursor and reactant, respectively, then NiO was reduced into metallic Ni film during annealing under H_2 ambient [60]. Similarly, $NiCp_2$ and H_2O were reacted with each other by ALD to form NiO, and NiO film was exposed to H_2 plasma for the transformation into Ni film [41]. This reduction method was used for fabricating Co and Ni nanotubes. CoO_x and NiO_x films were deposited inside anodic aluminum oxide (AAO) templates by ALD from metallocene precursors and oxidants, and converted into Co and Ni nanotubes by H_2 annealing [61]. Lim et al. synthesized several metal organic precursors properly designed for ALD process and tested ALD processes for Co, Ni, and Fe [62, 63]. Kim and coworkers dedicated their efforts to investigation of transition metal ALD processes. [43, 44, 64, 65] Based on their early report on Co PE-ALD using NH_3 plasma, Ni PE-ALD was investigated using H_2 plasma as well as NH_3 plasma. PE-ALD Ni using NH_3 plasma was more sensitive to substrate temperature than that using H_2 plasma [43]. Low temperature PE-ALD Co was studied using cyclopentadienyl isopropyl acetamidinato-cobalt (Co(CpAMD)) precursor and NH_3 plasma down to 100 °C of substrate temperature, and feasibility of liftoff process was proved [64]. They developed Co ALD using diisopropyl acetamidinato cobalt (Co(iPr-AMD)$_2$) precursor and NH_3, and obtained low resistivity Co film with excellent conformality as shown in Fig. 8.4 [65]. By using surface modification by self-assembled monolayers (SAMs), Co was selectively deposited through area selective ALD (AS-ALD). They proposed that this Co AS-ALD can be utilized for future contact fabrication because of etching-free process [65]. In the following paper, Ni(dmamb)$_2$ precursor was used for Ni AS-ALD using same scheme as Co AS-ALD [44]. Co PE-ALD was tried to apply for AS-ALD, however, no selectivity was observed due to degradation of SAMs layer by NH_3 plasma exposure [66].

As mentioned in the previous section, direct deposition of metal silicide is important in narrow junction regime. Lemonds et al. reported $TaSi_x$ ALD using

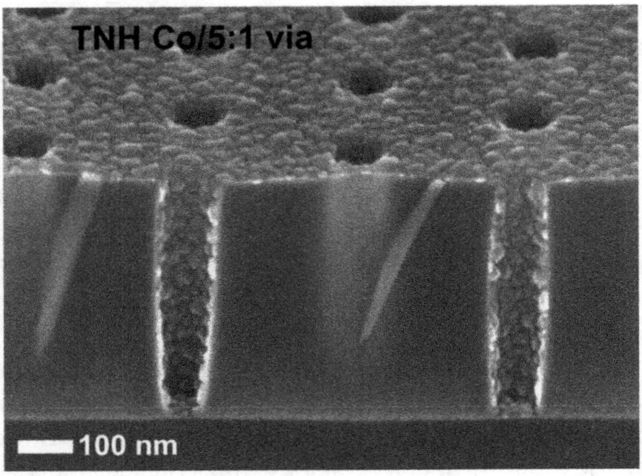

Fig. 8.4 The FE-SEM images of ALD Co on nanosized hole pattern (450-nm depth and 90-nm diameter) [65] (Reprinted with permission from ref. [65], Copyright 2010 The Electrochemical Society)

TaF_5 as a Ta precursor and Si_2H_6 as a counter reactant and Si source [67]. $TaSi_x$ alloy-like film rather than exact stoichiometric tantalum silicide was achieved [67]. Similarly, Lee et al. added SiH_4 as a Si source to reactant exposure of Co PE-ALD using $CoCp_2$ and NH_3 plasma [68]. $CoSi_2$ nanocrystals were observed instead of film. In addition, thick SiN_x was formed when flow of SiH_4 was high over certain value due to direct reaction of NH_3 plasma and SiH_4. [68].

8.4 ALD for Plug

The ''contact'' is the electrical structure that makes connection between the Si-level device and the first level of the interconnect circuitry. Since the contacts are short enough that the net resistance is minimal compared to the silicide and junction resistance, the more important material properties for contact application are chemical and thermal stability and good electromigration property [69]. In these points of view, as a representative refractory metal, tungsten has been typically used as a contact plugging material due to its relatively low resistivity ($\sim$5.6 μΩcm) and high melting point, and its hard and inert nature. Thus, chemical vapor deposited (CVD)-W films have been used as a plugging material for contact of high aspect ratio due its excellent step coverage [70, 71] as well as metal lines such as word line and bit line of memory devices such as DRAM and NAND Flash to decrease their resistance and capacitive (RC)-delay due to its low resistivity than WSi_2 [72, 73].

CVD-W thin film has been typically formed in two steps; the first step is the W nucleation layer deposition which is basically the reduction of WF_6 using SiH_4 due to its fast nucleation on typical glue/barrier layer (TiN film) in spite of its marginal step coverage, and the second is the W-plug fill deposition accomplished by H_2 reduction of WF_6 without any supply of SiH_4 because of its excellent step coverage [70]. Though CVD-W nucleation layer has been successfully used in the semiconductor industry as a nucleation layer, its limited conformality at ultrahigh aspect-ratio (UHAR) contact can induce potential problems such as the seam or void in final W-plug, leading to the degradation in contact resistance [74]. The above problem can be solved by using ALD-based W nucleation layer and will be reviewed below.

ALD-W is perhaps the most successful case among elemental metal ALD processes. This is fundamentally due to the fact that W ALD can be carried out using well-known good metal-containing precursor, WF_6, and good reducing agents, disilane (Si_2H_6), silane (SiH_4), and diborane (B_2H_6). The successful deposition of ALD-W was first reported by Klaus et al. [75]. They used a sequential supply of WF_6 and Si_2H_6 at the temperatures lower than 325 °C, which is low enough for semiconductor device fabrication. The smooth ALD-W films were deposited with growth rates close to the ideal 1 ML/cycle, ~2.5 Å/cycle and the resistivity of 122 μΩcm. Neither Si nor F was found in the film, suggesting a complete surface reaction between these two precursors. Later, it was reported that the ALD-W films could be deposited by using B_2H_6 as a reducing agent of WF_6 at 300 °C [76]. The smooth ALD-W films with low F content below 0.1 at. % were obtained with the resistivity of 150 μΩcm by using B_2H_6 as a reducing agent. The ALD-W process using very similar reducing agent with Si_2H_6, SiH_4 was also extensively investigated [77–79]. Figure 8.5 shows the step coverage of ALD-W films with deposition chemistry. All the ALD-W films showed excellent step coverage irrespective of B_2H_6 pretreatment and reducing agents of WF_6 at UHAR contact holes (AR: ~15). Here, prior to ALD-W depositions, $TiCl_4$-based CVD-TiN films were deposited as barrier layers.

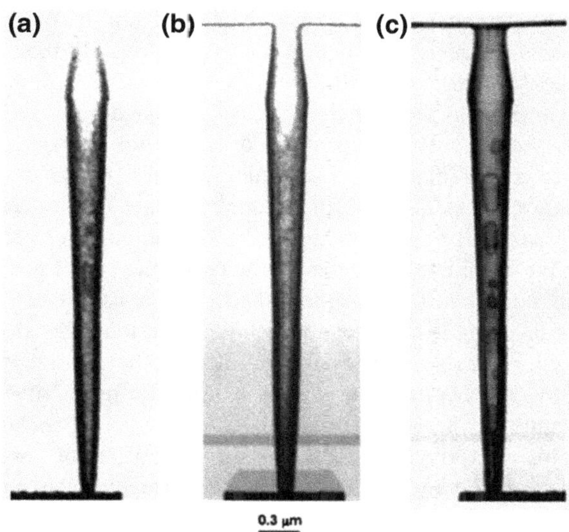

Fig. 8.5 The step coverages of ALD-W films at UHAR contact hole the contact hole (height ~3.52 μm and aspect ratio ~15); **a** ALD-W (A), **b** ALD-W(B), and **c** ALD-W(C). 5 nm-thick $TiCl_4$-based CVD-TiN as a barrier layer was deposited prior to ALD-W deposition. Two ALD-W films, ALD-W(A) and ALD-W(B) were deposited using SiH_4 as a reactant and ALD-W(C) using B_2H_6. The ALD-W(B) was pretreated with B_2H_6 for 5 s before ALD [79] (Reprinted with permission from ref. [79] of Chap. 5, Copyright 2006 The Electrochemical Society)

One of the drawbacks of W ALD is a poor nucleation on SiO_2. There has been a report that for W ALD on SiO_2 at 300 °C, the nucleation delay for the film growth is significant, up to ~10 cycles [80]. Furthermore, the thin surface oxide layer impedes W film nucleation and growth during WF_6 exposure on Si at 300 ~550 °C [81]. This in turn, causes the degradation in the conformality of ALD-W film to occur, particularly when the step coverage of TiN under-layer is not good enough and SiO_2 layer is exposed in the contact and the ALD-W film should grow on it. The nucleation of ALD-W could be accelerated by using a larger initial Si_2H_6 exposure and by repeatedly probing with an electron beam [80]. It was also reported that a SiH_4 pre-exposure on TiN [82, 83] enhances W nucleation during WF_6 exposure. These results indicate that W nucleation is critically dependent on the condition of the surface on which W is grown. In these respects, the effect of B_2H_6 pretreatment on the nucleation of ALD-W film deposited using alternating supply of WF_6 and SiH_4 was reported [84]. The results showed that the B_2H_6 pretreatment could enhance the nucleation of ALD-W on SiO_2 and also enhance the step coverage of ALD-W films at UHAR contact. The application of ALD-W as a nucleation layers for W-plug process was reported [79]. Here, W-plug process was used at the contact connecting the metal line and W bit line of DRAM with a stacked capacitor. The results showed that B_2H_6 pretreatment during the deposition of ALD-W nucleation layer provides some flexibility in terms of choosing the method for depositing TiN under-layer for W-plug process.

One of the issues in W-plug technology is to obtain the low resistivity W film as possible with continuous device shrinkage. Generally, the technology roadmap for contact or via plug and metal line of semiconductor devices shows ever-shrinking lateral dimensions as well as vertical dimensions. Eventually, their dimensions start to approach the mean free path for electron scattering of the material used in the interconnect structure, such as ~41 nm for W [85]. This leads to the increased electrical resistivity due to the proportionately increased electron scattering from the surfaces as its line width, thickness, and contact diameter decrease, often described as the "size effect". As discussed before, CVD-W films are deposited by 2-step process; a deposition of very thin W nucleation layer and a subsequent growth of relatively thick bulk-W film. This gives the possibility that the properties of the CVD-W film, including its resistivity can be controlled and improved by controlling the process of W nucleation layer because the underlying film can have considerable effects on the properties of the metal film growing on it [86–89]. In fact, it has been reported that the resistivity of bulk CVD-W film could be reduced by using B_2H_6-based W nucleation layer as compared to SiH_4-based one, and, even more, the resistivity increase by the size effect could be mitigated [90–95]. The results clearly showed that the resistivity of CVD-W film was significantly lower on B_2H_6-based ALD-W nucleation layer than on SiH_4-based one. This was obviously due to the formation of larger-sized grains, which was determined by the nucleation behavior on each nucleation layer with different phase and microstructure [91]. It was also reported that the structural properties of the B_2H_6-based ALD-W nucleation layer, such as its phase (amorphous, β, and α-phase), crystallinity, and grain size, could be controlled by varying B_2H_6 flow rate and its pulsing time during deposition, which had a significant effect on the final grain size and resistivity of the CVD-W film subsequently grown on it [92]. Thus, all the results showed that a B_2H_6 used for preparing an ALD-W nucleation layer seemed to play an important role in reducing the resistivity of CVD-W film. However, generally, the adhesion performances of CVD-W films growing on the B_2H_6-based ALD-W nucleation layer were poor compared to those on the SiH_4-based ALD-W nucleation layer [95]. The boron penetration into the interface between underlying TiN and SiO_2 during the deposition of W nucleation layer is a possible reason to degrade the adhesion performances of CVD-W films with B_2H_6-based ALD-W nucleation layers. Noticeably, the application of 5 nm-thick sputter-deposited WN_x film as a glue layer was found to present an excellent adhesions performance, which was due to its excellent diffusion barrier performance with amorphous structure.

Low resistivity W (LRW) technology mentioned above was applied into W-plug process to fill the contact connecting between the metal and W bit line or bit line in DRAM [91]. As expected, the integration scheme of CVD-W/B_2H_6-based ALD-W nucleation layer had lower resistances at the contact connecting the metal line and bit line contact. The average value was reduced by ~29 % as compared with CVD-W/SiH_4-based ALD-W nucleation layer, which is due to much lower resistivity of the total film stack of W-plug consist of CVD-W/B_2H_6-based ALD-W nucleation layer/TiN. It was also shown that with the integration scheme of CVD-W/B_2H_6-based ALD-W nucleation layer, a significantly lower bit line resistance was obtained. Thus, it could be concluded that the W deposition scheme enlarging the grain size by modifying the nucleation layer investigated in this study is a very suitable approach for ever-shrinking devices and potentially could be used as low resistivity metal gate or local interconnect for logic device.

8.5 ALD for Cu Metallization

Current copper (Cu) metallization technology involves deposition of a diffusion barrier layer and a seed layer prior to the electroplating (EP) of Cu. Here, the diffusion barrier (generally, Ta/TaN bilayer) prevents Cu diffusion into the dielectric and the seed layer (generally, Cu) facilitates the EP process of Cu [96]. However, as the feature sizes of trenches and vias continue to shrink, a deposition of defect-free EP-Cu line on these underlying thick layers becomes increasingly difficult. Corresponding with this, an unwanted and drastic increase in Cu line resistance occurs due to the size effect on the resistivity of metal films [97]. The increase in the resistivity of the interconnect can be alleviated by increasing the grain size of EP-Cu or by modifying the Cu surface [97]. Another possible solution is to maximize the portion of the EP-Cu volume in the vias or damascene structures with the conformal diffusion barrier and seed layer. Other strategy for direct deposition of Cu on the barrier (skipping the Cu seed layer) is being explored, so-called Cu direct plating technology [98–100]. In the Cu direct plating technology, the selection of barrier materials is very important and challenging because the EP of Cu on them should be possible with a high nuclei density of Cu as well as they prevent Cu diffusion effectively. Furthermore, with ever decreasing feature sizes, the Cu direct-plateable diffusion barriers become a significant fraction of the metallization, and need to be conformal and as thin as possible. With these respects, ALD can be a viable solution for preparing them because ALD enables atomic-scale control of the film thickness and composition with excellent step coverage in nanoscale structures with high aspect ratios [101, 102]. In this chapter, three categories of ALD technologies for advanced Cu metallization are reviewed; the first one is for the deposition of diffusion barrier against Cu, the second one is for the seed layer for EP-Cu, the third one is for the Cu direct-plateable diffusion barrier.

8.5.1 ALD of Diffusion Barriers Against Cu

Refractory metal nitrides such as TiN_x, TaN_x, and WN_x have been used in many different applications because of their desirable material properties such as high melting temperatures, relatively low resistivities, and chemical inertness, etc. However, the most interesting and fastest developing area is microelectronics, where these materials are being extensively studied as diffusion barrier or glue layer at contact and via holes with UHAR at ultra-large-scale-integrated (ULSI) devices [103].

8.5.1.1 Ti-Based Nitrides

TiN has been employed in Al-based metallization for long time and the possibility of extending its application to Cu-based interconnects seems quite promising, both from technical and economic point of view. Therefore, many investigations have been focused on developing the TiN diffusion barrier for Cu metallization. ALD-TiN film with low resistivity ~200 μΩ cm, extremely good step coverage, ~75 % at the trench with an aspect ratio of 85 (~55 μm deep and 475 nm wide), and low Cl content of ~2 at. % could be deposited at 400 °C using $TiCl_4$ and NH_3 [104]. In addition, the introduction of additional reducing agents such as Zn [105] and trimethylaluminum [TMA, $(CH_3)_3Al$] [106] and more effective nitrogen sources such as 1,1-dimethylhydrazine [DMHy, $(CH_3)_2NNH_2$] [107], *tert*-butylamine [^{t}BuNH$_2$, $(CH_3)_3CNH_2$,], and allylamine [allylNH$_2$, $(CH_2)_5(CHCH_2)NH_2$,] [108] were used to deposit a better quality TiN film. However, the use of $TiCl_4$ and NH_3 is known to bring about Cu pitting when the TiN film is deposited directly on the Cu surface as a result of Cl-containing by-products [104]. It requires deposition temperatures higher than 350–400 °C to reduce the Cl content in the film in order to prevent corrosion, which is often not suitable with integration of Cu/Low-k metallization. In addition, the barrier properties of this TiN film are not good enough due to the formation of columnar grain structure, which provides an easy pathway for Cu migration [109–111]. It was reported that 20 nm-thick ALD-TiN deposited using $TiCl_4$ and NH_3 with a columnar grain structure was only stable up to the annealing temperature of 550 °C for 30 min while 12 nm-thick ALD-WCN with a nanocrystalline equiaxed grain structure was stable up to the annealing temperature of 650 °C [111].

Apart from ALD-TiN deposited by inorganic precursor, $TiCl_4$, ALD-TiN processes using metal organic (MO) precursors, such as tetrakis-dimethyl-amino-titanium [TDMAT, $Ti(NMe_2)_4$] [112, 113] and tetrakis-diethyl-amino-titanium [TDEAT, $Ti(NEe_2)_4$] [113] and tetrakis-ethylmethyl-amino-titanium [TEMAT, $Ti(NEtMe)_4$] [114], were reported. Using them, amorphous TiN films were deposited at growth temperatures around 200 °C. For both precursors, the self-limiting surface reaction was not observed at a growth temperature above 220–230 °C due to the thermal decomposition of the metal precursors. Accordingly, a high amount of carbon (>~20 at. %) was incorporated in the film, resulting in a high film resistivity above ~1000 μΩ cm at the deposition temperature of 250 °C. To lower the resistivity of ALD-TiN film deposited using MO precursors, PE-ALD process using H_2 plasma as a reactant was reported [115]. Low resistivity values from 210 to 275 μΩ cm were achieved at the deposition temperature of 150 °C. PE-ALD TiN processes using $TiCl_4$ and H_2/N_2 plasma were also reported [116, 117]. Even at 100 °C, TiN film was deposited with the growth rate of 0.26 Å/cycle and the film resistivity was 209 μΩ cm with the Cl content of 6.7 at. %. With increasing the deposition temperature, the resistivity was decreased and the film resistivity deposited at 400 °C was 79 μΩ cm with a negligible Cl content (0.07 at. %).

To address a poor diffusion barrier performance of ALD-TiN film against Cu, ternary diffusion barriers such as ALD-TiSiN [112, 118–120], or ALD-TiAlN [121] were prepared by adding Si source pursing or Al source pulsing. This concept was based on the extensive work using sputtering by Nicolet and his coworkers [122] that as the addition of the third element into the transition metal nitride disrupts the crystal lattice and leads to the formation of a stable ternary amorphous material and expected to improve the diffusion barrier performance against Cu. All the results showed that the diffusion barrier performance of TiSiN and TiAlN was superior to that of TiN counterpart due to the amorphous microstructure.

8.5.1.2 Ta-Based Nitrides

Compared with Ti-based materials, Ta-based materials shows the advantage as diffusion barriers against Cu. Tantalum and tantalum-nitrides (Ta, Ta_2N, and TaN) have high melting temperatures (2996, 2050, 3087 °C, respectively) and are thermodynamically stable with respect to Cu, because of the absence of Cu–Ta or

Cu–N compound from the binary phase diagrams [123]. The barrier properties of Ta might be further improved by using its nitride compounds such as Ta_2N and TaN [124–127]. Thus, extensive studies to make TaN_x film by ALD were performed.

The most critical issue in the TaN_x-ALD process is to obtain highly conductive TaN films. Ta_3N_5, which is nitrogen-rich phase in the Ta–N binary system and a dielectric material, was deposited using $TaCl_5$ and NH_3 and its resistivity was as high as ~200 Ω cm [128]. Highly conductive TaN film can be deposited using an additional reducing agent, such as Zn [128] or TMA [129] but its resistivity was still high, from ~900 μΩ cm to ~1,300 μΩ cm. Moreover, small amounts of Zn impurities (~4 at. %) can cause severe problems if they are diffused into the Si substrate. In addition, the use of inorganic precursors, such as $TaCl_5$ and $TaBr_5$, has the potential problem of the incorporation of corrosive halogen elements in the film, particularly when the deposition temperature is as low as 400 °C [128, 129]. The successful deposition of low resistivity TaN (ρ: 350–610 μΩ cm) with inorganic precursors of $TaCl_5$ [130] and TaF_5 [131] is possible using N_2/H_2 plasma as a reducing agent of the precursors, in contrast to the highly resistive Ta_3N_5 growth by the reaction of $TaCl_5$ with molecular NH_3.

Ta nitride ALD was also reported using a range of metal–organic precursors, such as Ta-based alkylamides and alkylimides such as pentakis (dimethylamido) tantalum [PDMAT, $Ta(NMe_2)_5$] [132–135], pentakis (diethylamido) tantalum [PDEAT, $Ta(NEt_2)_5$] [136], *tert*-butylimidotris (diethylamido) tantalum [TBTDET, $Ta(=N^tBu)(NEt_2)_3$] [137–142], and tertiary-amylimido-tris(dimethylamido) tantalum [TAIMATA, $Ta[NC(CH_3)_2C_2H_5][N(CH_3)_2]_3$] [143, 144]. Basically, the use of metal–organic precursors have benefits in producing halogen-free films. Similar to the results with inorganic precursors, thermal ALD using these precursors and molecular NH_3 or molecular H_2 produce highly resistive Ta nitride films, possibly Ta_3N_5 with a considerable carbon impurity in the film [133, 136, 137, 139, 140]. Although N_2 plasma [132] or NH_3 plasma [133] was used to reduce PDMAT, the resulting film was Ta_3N_5, whose resistivity was too high to be measured. Only when the H_2-based plasma such as H_2 or methane/H_2 or Ar/H_2 plasma [132, 137, 140, 142–144] was used as a reactant, highly conductive Ta-based nitride film, actually TaC_xN_y could be deposited, where the C or N content in the film could be controlled by the deposition conditions, such as growth temperature, plasma condition, and type of reactant. It was shown that the diffusion barrier properties of deposited PE-ALD-TaN films for Cu interconnects were excellent. The PE-ALD TaN films were good diffusion barriers even at a small thickness as 0.6 nm. Better diffusion barrier properties were obtained for higher nitrogen content. It was also suggested the nanocrystalline microstructure of the films deposited by MO precursor was responsible for the better diffusion barrier properties compared to polycrystalline PE-ALD TaN films deposited from $TaCl_5$ [132] as shown by Fig. 8.6.

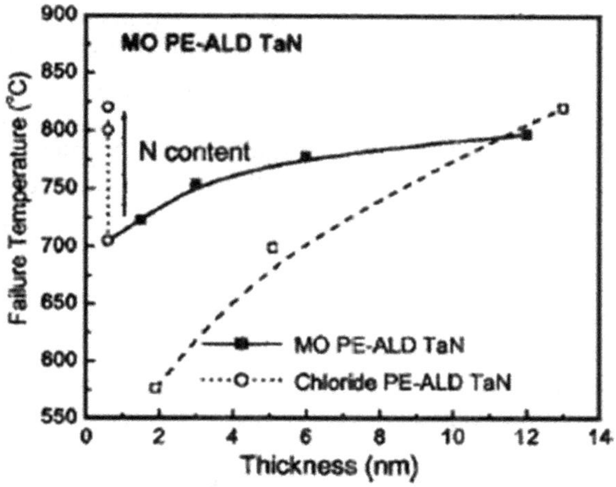

Fig. 8.6 Barrier failure temperatures determined using synchrotron x-ray diffraction analysis for MO PE-ALD TaN (*black square*) and chloride PE-ALD TaN (*open circle*). Also shown are the failure temperatures of 0.5 nm MO PE-ALD TaN_x deposited at different nitrogen plasma conditions open square.132 (Reprinted with permission from reference [132] of Chap. 6, Copyright 2005 American Institute of Physics)

8.5.1.3 W-Based Nitrides

W and its nitride such as W_2N and WN are also thermodynamically stable to Cu. The Cu-W phase diagram shows that the absence of any reaction between the two elements and there are negligible mutual solubility between them. These WN_x is particularly noteworthy because WN_x films have been frequently reported to be formed in amorphous or amorphous-like phase, which gives a desirable form as a diffusion barrier or contact glue layer. This is contrary to the case of TiN_x and TaN_x systems [103]. Besides, in terms of their material properties, WN_x has remarkable advantages from the viewpoint of the ALD process. It has a well-known, metal-containing precursor, WF_6, with good properties such as its being in the gas-state at room temperature, and this can ensure the stability of the reactant transport and process. In the case of the other transition metal nitrides. on the other hand, $TiCl_4$ used for TiN_x is liquid and $TaCl_5$ used for TaN_x is solid. In addition, the typical deposition temperature of WN_x films is 100–200 °C lower than those of the ALD-TiN_x or TaN_x processes [105, 128], resulting in its having little adverse effect on the device reliability especially when ALD-WN_x is used at the metal contact of DRAM. ALD-WN_x film was first deposited using a sequential supply of WF_6 and NH_3 at 350 °C [145] but it suffered from too high a resistivity (~4,500 μΩ cm) [145, 146] to be used in the interconnect stack of semiconductor device. Such a high resistivity of ALD-WN_x film from WF_6 and NH_3 indicates that the direct reduction of WF_6 with NH_3 is highly difficult. This is because W is at a higher oxidation state in the W precursor, WF_6 than in the desired tungsten nitrides such as W_2N or WN. ALD-WN_x film deposited at 250 and 350 °C using MO precursor, bis(tert-butylimido)-bis-(dimethylamido) tungsten(IV) [TBIDMW, $(^tBuN)_2(Me_2N)_2W$] and NH_3 also demonstrated high resistivity ranging from 1,500 to 4,000 μΩ cm though the deposition was self-limiting in both ammonia and $(^{t}BuN)_2(Me_2N)_2W$ doses [147, 148]. Due to the formation of amorphous structure, ALD-WN_x showed excellent diffusion barrier performance against Cu. Films as thin as 1.5 nm proved to be good barriers up to 600 °C between Cu and Si [147]. More recently, very low temperature growth of ALD-WN_x film ranging from 150 to 250 °C was reported using low valent MO precursor of tungsten(III) dimer, $W_2(NMe_2)_6$. [149] X-Ray diffraction (XRD) and transmission electron microscopy indicated that the as-deposited films were amorphous but its resistivity was still high as ~810 μΩ cm possibly due to the formation of nitrogen-rich WN_x films.

As mentioned above, additional reducing agents such as Zn [105] TMA [106] were used to deposit a low resistivity ALD-TiN film at a relatively lower temperature with a sequential supply of $TiCl_4$ and NH_3. The same concept was also used for the growth of low resistivity ALD-WN_x film. A triethylboron [TEB, $(CH_3)_3B$] [111, 150–152] or C_2H_4 and SiH_4 [153] were used as additional reducing agents for ALD-WN_x deposition with WF_6 and NH_3. Even though the deposited film using TEB contained ~35 % of carbon ($WN_{0.3}C_{0.7}$), its resistivity was reported to be 600–900 μΩ cm for 7 nm–thick films and 350–400 μΩ cm for 25 nm-thick films. The failure temperature of ALD-WNC diffusion barrier (12 nm) determined by XRD and sheet resistance measurement was as high as 700 °C at 30 min annealing due to both high density and noncolumnar grain structure while the sputter-deposited Ta (12 nm) and ALD-TiN film (20 nm) failed after annealing at 650 and 600 °C, respectively [151]. The thermal stability of ALD-WCN was also excellent [111].

In view of adding another reducing agent for the formation of low resistivity ALD-WN_x film, the use of B_2H_6 was quite interesting [154–156]. The B_2H_6 contains no carbon and is a well-known gas chemical that has been used for many years in semiconductor industry for doping the Si or SiO_2. By allowing the B_2H_6 exposure to take place before the WF_6 exposure, an ALD-W layer can be formed and this layer can be successfully converted to an ALD-WN_x layer by the final exposure to NH_3. Using the proposed reaction scheme, a low resistivity (~350 μΩ cm), high-purity (no B, F, and O impurities below the level of the XPS detection limit), high-density (~15 g/cm^3), and highly conformal (~90 % of step coverage at 16:1 contact, contact height ~2.23 μm)ALD-WN_x film at a low-deposition temperature (300 °C) could be deposited. As-deposited ALD WN films showed the formation of nanocrystalline structure and successfully integrated as a contact barrier layer for W-plug process [155]. Recently, it was also demonstrated that this ALD-WN_x can be a viable candidate as a diffusion barrier between Al and Cu for the implementation of Cu interconnects in memory devices. ALD-WN_x effectively prevented the inter-diffusion of Al and Cu until 550 °C annealing while TiN failed after annealing at 450 °C and TaN failed after annealing at 500 °C [156].

8.5.2 ALD of Seed Layers for Cu Electroplating

So far, the Cu seed layer has been mainly deposited by PVD techniques, primarily by various versions of ionized PVD (IPVD). However, with any PVD deposition technology, there is a slow build-out at the upper corners of trenches or vias, forming a slight overhang and as the interconnect nodes approach the 22 nm node or beyond. The overhang, which does not scale down as rapidly as technology node, will significantly close off the top opening of the via or trench, inhibiting the fill by electroplating. Therefore, a very conformal seed layer deposition technique will be required by that generation, and this may only be achievable by ALD. In view of this, we will focus on the ALD of Cu and Ru, which have been extensively studied as seed layers for Cu-EP.

8.5.2.1 ALD of Cu

Cu thin films made from ALD techniques have been reported using several different copper sources. They include copper(I) monochloride [$(CuCl)_3$] [157, 158], β-diketonate derivatives, such as copper (II) bis(2,4-pentanedionate) [Cu(acac)2] [159], copper bis(2,2,6,6-tetramethyl-3,5-heptanedionate) [$Cu(thd)_2$] [160], copper(II) hexafluoroacetylacetonate hydrate [$Cu(hfac)_2 \cdot xH_2O$] [161], and copper (I) amidinate of N,N'-di-iso-propylacetamidinato copper(I) [Cu (iPr-amd$)_2$] [162]. Cu ALD was first reported by Mårtensson et al. where H_2 molecules were used to reduce copper(I) monochloride. However, in this case, significantly high deposition temperatures (360–410 °C) were required. Subsequently, the use of Zn as a reducing agent for copper(I) monochloride was reported [158]. In this report, 3 at. % of Zn was incorporated in the film deposited at 500 °C. Moreover, due to a reversible dissolution of Zn into the Cu film, no self-limiting growth characteristics could be achieved. Also, the volatility of copper(I) monochloride is so low that very high temperatures (>400 °C) must be used to evaporate and transport the precursor.

By using metal–organic precursors, in contrast, lower sublimation temperatures as well as reduced processing temperatures were possible. Mårtensson and Carlsson [160] have produced self-limited Cu deposition in the 190–260 °C range using Cu $(thd)_2$ as a Cu source that was reduced by H_2. These processes turned out to be highly substrate dependent, as film growth relied on catalytic under-layers such as Pt or Pd. With $Cu(hfac)_2$, ALD was reported from 230 °C. However, the use of $Cu(hfac)_2$ caused fluorine contamination of the films, reduced adhesion, and etching of substrates by hydrofluoric acid (HF) which is the reaction by-product. The use of Cu amidinate precursor enhanced the reactivity with molecular hydrogen and could deposit pure Cu film at relatively low temperature (200 °C) and excellent conformality (AR: 35–40). As an alternative route for Cu ALD by direct precursor reduction, the formations of a Cu compounds such as oxides and nitrides followed by reduction at each cycle was also reported [161, 163, 164].

In view of preparing metal film, PE-ALD has addition advantages compared with thermal ALD, including high film density and low processing temperature, and, thus PE-ALD of Cu using H_2 plasma or H radical as reactants were reported [165, 166]. The results showed that the low resistivity Cu film (5.3 μΩ cm at 30 nm thickness) with an excellent conformality at the trench with the 400 nm feature size and ~5:1 aspect ratio. However, in general, obtaining a continuous ultrathin ALD Cu film on several surfaces at elevated temperatures has been an ongoing challenge, due to the propensity for Cu to nucleate in an island-type fashion, thus leading to noncontinuous films with increasing resistivity as the Cu thickness decreases to such levels [162, 166]. So, other materials have been evaluated as seed layers for Cu electroplating.

8.5.2.2 ALD of Ru

New materials for Cu electroplating were investigated in order to increase the volume of EP-Cu filled into the trench and via. Recent investigations have been focused on the noble materials such as Ru, Pt, Os, Rh, Ir, and many reports suggested Ru to be successfully used as a seed layer/diffusion barrier layer for direct plating of Cu [167, 168] leading a considerable research attention to ALD-Ru.

As a metal precursor, cyclopentadienyl (Cp)-based Ru precursors, such as bis(cyclopentadienyl) ruthenium(II) [$RuCp_2$, $Ru(C_5H_5)_2$] [169–172] and bis(ethylcyclopentadienyl) ruthenium(II) [$Ru(EtCp)_2$, $Ru(C_2H_5C_5H_4)_2$] [172–176] have been used widely. Here, the metal valence in the precursor is 2 and O_2 molecules were used to reduce the Ru film from the precursors. Altonen et al. [170] suggested that the subsurface oxygen produced by a molecular O_2 pulse reacted with the ligands of the Ru precursor provided by the subsequent precursor pulse, producing Ru metal based on quartz crystal microbalance (QCM) study (Fig. 8.7). However, ALD-Ru with these two precursors and molecular O_2 reactant shows a relatively long incubation time with a lower film growth rate at the initial stages of film deposition, which is related to the delayed nucleation of the film material. The use of a β-diketonate Ru precursor of tris(tetramethyl-heptane-dionate) ruthenium(III) [$Ru(thd)_3$, $Ru(C_{11}H_{19}O_2)_3$], where the metal valence is 3, showed similar results in terms of Ru nucleation [177]. This results in a deposited film with a rough morphology that is unsuitable for conformal film deposition on high aspect-ratio structures [178]. To reduce the incubation time shown in ALD using a Cp-based Ru precursor, PE-ALD, which employs NH_3 plasma as the reactant instead of O_2 gas has been studied [172, 176, 179, 180]. Ru films deposited by PE-ALD showed a negligible incubation cycle [179], and the nuclei density is as high as 10^{12} cm^{-2}, which is ~50 times higher than that of ALD using molecular O_2 [180].

Ru ALD with molecular O_2 showed a different incubation time according to the substrates or surface conditions. Park et al. [171] indicated that Ru films deposited on HfO_2 surface showed a shorter incubation time than on the SiO_2 and H-terminated Si substrate. In addition, the results showed that thermally grown SiO_2

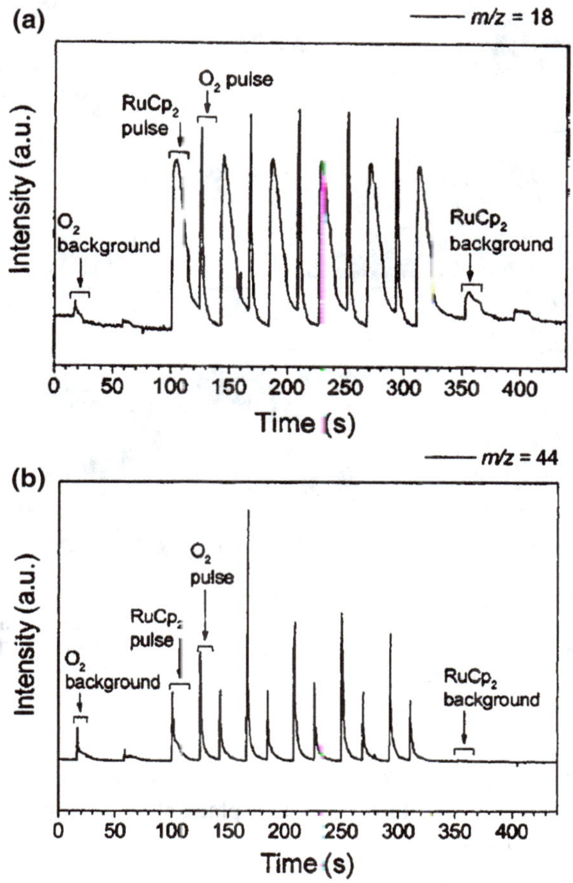

Fig. 8.7 QMS signal obtained **a** from H_2O (m/z 5 18) and **b** from CO_2 (m/z 5 44) during ALD of Ru. The growth cycles consisted of 12 s $RuCp_2$ pulse, 10 s purge period, 6 s oxygen pulse, and 10 s purge period. Two successive background pulses for oxygen and for $RuCp_2$ were introduced before and after the growth cycles, respectively. 170 (Reprinted with permission from ref. [170] of Chap. 6, Copyright 2003 The Electrochemical Society)

and chemical SiO_2 provides different film growth conditions. The nucleation of ALD-Ru was improved greatly on thin ALD-Al_2O_3 according to TEM analysis [174, 181]. These aspects suggest that the initial nucleation behavior of an ALD-Ru film differs according to the surface conditions, and the nucleation of Ru films can be enhanced by making the surface more active to the adsorption of the Ru precursor. Indeed, it was suggested that the initial growth of Ru thin films is determined by the adsorption of the metal precursor rather than by the dissociative adsorption of molecular oxygen on the substrate or Ru film itself [182]. This means that one can improve the nucleation of ALD-Ru using Ru precursors to have active adsorption properties on the surface of interest. ALD-Ru films deposited using another Ru(II) precursor, (dimethylpentadienyl)(ethylcyclopentadienyl) ruthenium (II) (called as DER) were reported to have a small incubation cycle on Si or SiO_2 [183]. However, other studies using the same chemistry reported that the nucleation of Ru had long incubation cycles on SiO_2, TiO_2 and

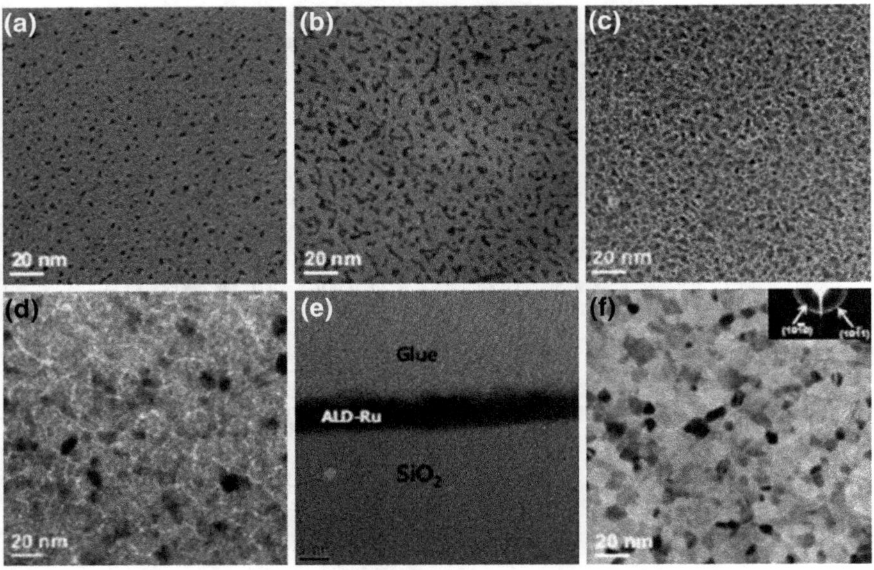

Fig. 8.8 TEM analysis of ALD-Ru deposited on a thermally grown SiO_2 substrate using IMBCHRu as a precursor. Plan-view TEM image after **a** 7, **b** 10, **c** 30, and **d** 50 ALD cycles, **e** XTEM image of ALD after 50 ALD cycles, and **f** Plan-view TEM image after 100 ALD cycles. [185] (Reprinted with permission from ref. [185] of Chap. 6, Copyright 2009 The Electrochemical Society)

TiN surfaces [182], which are an important contacting surface for Ru layers in an integrated structure as a direct-plateable diffusion barrier, capacitor electrode, and a seed layer for Cu-EP. Among them, the result on a SiO_2 surface which has covalent bonding was the worst, and that on TiN and TiO_2 surfaces with ionic bonding was slightly better in terms of nucleation. Nucleation was only good on surfaces with metallic bonding, such as Au and Pt, which showed ~10 incubation cycles irrespective of the deposition temperature. Though the use of other Ru(II) precursor, bis(*N*,*N′*-di-*tert*-butylacetamidinato) Ru(II)dicarbonyl [181, 184] could produce ~2 nm-thick continuous ALD-Ru film on WN, the films deposited on SiO_2 were not continuous even after 80 ALD cycles. Accordingly, the above consideration indicates that more study is needed to enhance the ALD-Ru process and successfully integrate it into microelectronic or emerging nanodevices. A recent investigation reported that Ru film nucleation was enhanced considerably using a zero metal valence precursor, isopropyl-methylbenzene-cyclohexadiene Ru(0) (IMBCHRu, $C_{16}H_{22}Ru$) and molecular oxygen (O_2), compared to the utilization of precursors with higher metal valences [185, 186]. The Ru nuclei were clearly visible after only three ALD cycles [185]. The maximum density of the Ru nuclei was 2.1×10^{12} cm^{-2} on SiO_2 at seven ALD cycles (Fig. 8.8a). After 50 ALD cycles (Fig. 8.8d), a continuous Ru film with a thickness of ~3.5 nm (Fig. 8.8e) was formed. There was negligible and no incubation time for Ru film growth on the covalent SiO_2 and ionic TiN substrate, respectively. Due to the enhanced nucleation, ALD-Ru deposited with IMBCHRu showed excellent conformality with a smooth film surface. The electroplating of Cu on ALD-Ru film was also reported [186]. The potentiostatic deposition method in the Cu plating bath was used to grow Cu films directly on the surface of the 10 nm-thick ALD-Ru film. The cross-section view SEM image showed that a thin (~60 nm) Cu film was deposited directly on the 10 nm-thick Ru A continuous Cu film was formed even at only ~60 nm in thickness, indicating that very high area density of extremely small Cu nuclei was produced in the initial stages of nucleation and growth. This means that the ALD-Ru film can be used successfully as a seed layer for Cu interconnects.

When it comes to the application of ALD-Ru as a seed layer for electroplating of Cu, the use of O_2 molecules as reactants might cause diffusion barrier layer to oxidize, forming interfacial metal oxide layer even though the effect of O_2-based Ru process on device integration has not been reported yet. To avoid this potential problems, ALD process in nonoxidizing ambient was introduced, since the resistivity value needs to be maintained as low as possible for the application into Cu metallization. PE-ALD with NH_3 plasma instead of O_2 molecules as reactants was investigated [172, 176, 179, 180], since the use of NH_3 plasma in ALD process resulted in pure Ru film without any detectable impurities, having smooth and dense film. It was also reported that the use of N_2/H_2 plasma mixture or H_2 plasma can effectively reduce the Ru MO precursor to form the Ru and the lowest resistivity (~20 μΩ cm at 7 nm) was obtained at approximately 0.3 of the ratio of H_2 flow rate versus total gas flow rate [187]. The step coverage of this PE-ALD-Ru was also good at 50 nm via pattern.

8.5.3 ALD of Cu Direct-Plateable Diffusion Barriers

Although Ru has been also suggested as a diffusion barrier that is compatible with the direct plating of Cu, previous studies have shown that the Ru by itself is not a suitable diffusion barrier for Cu metallization [188–191]. It has been reported that a PVD Ru film (between Cu and Si) with a thickness of 20 nm could prevent Cu diffusion up to 450 °C [188], but a similar Ru film with a thickness of 5 nm lost its barrier property at only just above 300 °C [189]. Thus, the diffusion barrier performance of the Ru film should be improved in order for it to be successfully incorporated as a Cu direct-plateable diffusion barrier. It was suggested that a poor diffusion barrier performance of Ru is mainly due to its microstructure with polycrystalline columnar grains which provides a fast diffusion pathway for Cu [191]. Two types of approaches were reported to address the poor diffusion barrier performance and obtain reliable seedless Cu interconnects with Ru. The first is to use Ru as a seed layer and combine it with superior materials to function as a diffusion barrier against Cu. Sputtered–deposited Ru/TaN [192], Ru/Ta [193], Ru/WN_x [191], and ALD-Ru/TaCN bilayers [194] have been suggested and their

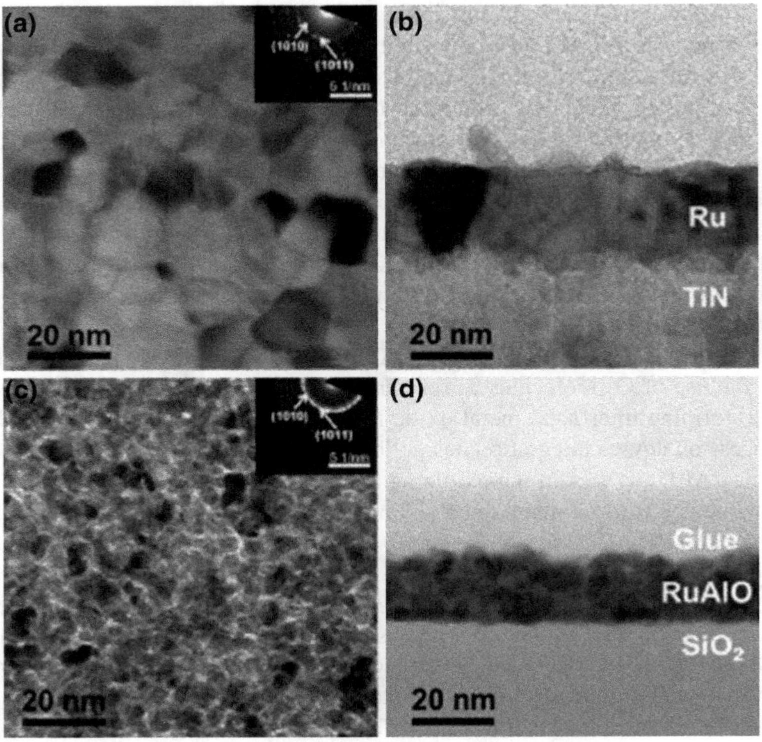

Fig. 8.9 **a** Plan-view, **b** cross-sectional view TEM (XTEM) image of ALD-Ru film, **c** plan-view, and **d** XTEM image of ALD-RuAlO film [199] (Reprinted with permission from ref. [199] of Chap. 6, Copyright 2011 The Electrochemical Society)

diffusion barrier performance against Cu were superior to a Ru single layer with the same thickness. Nevertheless, bilayer diffusion barriers require additional process that can increase the processing cost. Moreover, the volume of the trench for Cu to be filled becomes narrow due to the additional materials prior to the electroplating of Cu. The second approach is to modify the microstructure of Ru to an amorphous or nanocrystalline structure. A basic idea is to incorporate materials into Ru during deposition to produce a Ru-based ternary thin film. Ru-TiN, [195] Ru-TaN [196], Ru-WCN [197], RuSiN [198], RuAlO [199] films prepared by ALD were suggested and the successful direct plating of Cu on them was also reported. Here, the results on ALD-Ru-based ternary thin film [199] are briefly summarized. In this process, ALD-Al_2O_3 was incorporated into ALD-Ru to prevent its columnar growth, and a ternary thin film RuAlO with a nanocrystalline and noncolumnar grain structure, was fabricated at 225 °C. TEM analysis showed that a RuAlO film formed with noncolumnar grains and a nanocrystalline microstructure consisting of Ru nanocrystals separated by amorphous Al_2O_3, whereas the Ru films were polycrystalline with columnar grains as shown by Fig. 8.9. The step coverage of ALD-RuAlO was excellent, ~93 % at contact holes with an aspect ratio of ~29. The sheet resistance measurements and XRD showed that the structure of Cu (100 nm)/RuAlO(15 nm)/Si was stable after annealing at 650 °C for 30 min, whereas the Cu (100 nm)/Ru(15 nm)/Si structure failed after annealing at 450 °C [191].

8.6 Conclusions

Among various potential applications in logic device fabrication, the ability of ALD to deposit various metal and nitride films with excellent conformality and high quality at low temperature makes it a viable deposition process for BEOL process of sub-20 nm technology node logic device fabrication. In this chapter, the efforts on the application of ALD at BEOL process including contact/plug formation and metallization have been reviewed. ALD processes of transition metal thin films have been studied recently, showing great potential in contact applications. Additionally, a wide range of ALD processes have been carried out showing great advance in the deposition of nanoscale thickness diffusion barrier and seed layer. However, even further extensive understanding on the processes as well as the deposited film properties would be required for the implementation of ALD technique in BEOL processes.

References

1. Sakurai T (1983) Solid-state circuits. IEEE J 18(4):418–426
2. Luan S, Neudeck GW (1992) J Appl Phys 72(2):766–772
3. Bohr MT (1995) Presented at the electron devices meeting
4. Stamper AK, Fuselier MB, Tian X (1998) Presented at the interconnect technology conference, 1998. In: Proceedings of the IEEE 1998 international, 1998
5. Fayolle M, Passemard G, Louveau O, Fusalba F, Cluzel J (2003) Microelectron Eng 70(2–4):255–266
6. Jan CH, Bielefeld J, Buehler M, Chikamane V, Fischer K, Hepburn T, Jain A, Jeong J, Kielty T, Kook S, Marieb T, Miner B, Nguyen P, Schmitz A, Nashner M, Scherban T, Schroeder B, Wang PH, Wu R, Xu J, Zawadzki K, Thompson S Bohr M (2003) Presented at the interconnect technology conference, 2003. In: Proceedings of the IEEE 2003 international, 2003
7. Hoofman RJOM, Verheijden GJAM, Michelon J, Iacopi F, Travaly Y, Baklanov MR, Tökei Z, Beyer GP (2005) Microelectron Eng 80:337–344
8. Torres J (1995) Appl Surf Sci 91(1–4):112–123
9. Venkatesan S, Gelatos AV, Hisra S, Smith B, Islam R, Cope J, Wilson B, Tuttle D, Cardwell R, Anderson S, Angyal M, Bajaj R, Capasso C, Crabtree P, Das S, Farkas J, Filipiak S, Fiordalice B, Freeman M, Gilbert PV, Herrick M, Jain A, Kawasaki H, King C, Klein J, Lii T, Reid K, Saaranen T, Simpson C, Sparks T, Tsui P, Venkatraman R, Watts D, Weitzman EJ, Woodruff R, Yang I, Bhat N, Hamilton G, Yu Y (1997) Presented at the electron devices meeting, IEDM '97. Technical Digest, Internationa

10. Marc-A N (1995) Appl Surf Sci 91(1–4):269–276
11. Istratov AA, Flink C, Hieslmair H, Weber ER, Heiser T (1998) Phys Rev Lett 81(6):1243–1246
12. Leskelä M, Ritala M (2003) Angew Chem Int Ed 42(45):5548–5554
13. Hyungjun K (2011) Thin Solid Films 519(20):6639–6644
14. Kim H (2003) J Vac Sci Technol B: Microelectron Nanometer Struct 21(6):2231–2261
15. Kim H, Cabral JC, Lavoie C, Rossnagel SM (2002) J Vac Sci Technol B: Microelectron Nanometer Struct 20(4):1321–1326
16. Min JS, Son YW, Kang WG, Chun SS, Kang SW (1998) Japan J Appl Phys 37 (part 1): 4999–5004
17. Fang Z, Aspinall HC, Odedra R, Potter RJ (2011) J Cryst Growth 331(1):33–39
18. Dezelah CL, El-Kadri OM, Kukli K, Arstila K, Baird RJ, Lu J, Niinistö L, Winter CH (2007) J Mater Chem 17(11):1109–1116
19. Alén P, Ritala M, Arstila K, Keinonen J, Leskelä M (2005) Thin Solid Films 491(1–2):235–241
20. Miikkulainen V, Suvanto M, Pakkanen TA (2006) Chem Mater 19(2):263–269
21. Huang S-H, Pilvi T, Wang X, Leskelä M, Richmond MG (2010) Polyhedron 29(7):1754–1759
22. Alen P, Aaltonen T, Ritala M, Leskela M, Sajavaara T, Keinonen J, Hooker JC, Maes JW (2004) J Electrochem Soc 151(8):G523–G527
23. Alén P (2005) *Atomic Layer Deposition of TaN, NbN, and MoN Films for Cu Metallizations*
24. Kim D-J, Jung Y-B, Lee M-B, Lee Y-H, Lee J-H, Lee J-H (2000) Thin Solid Films 372(1–2):276–283
25. Lim J-W, Park J-S, Kang S-W (2000) J Appl Phys 87(9):4632–4634
26. Min J-S, Park J-S, Park H-S, Kanga S-W (2000) J Electrochem Soc 147(10):3868–3872
27. Furuya A, Tsuda H, Ogawa S (2005) J Vac Sci Technol B: Microelectron Nanometer Struct 23(3):979–983
28. Kim H, Detavenier C, van der Straten O, Rossnagel SM, Kellock AJ, Park DG (2005) J Appl Phys 98(1):014308–014308
29. Rayner JGB, George SM (2009) J Vac Sci Technol A: Vac Surf Films 27(4):716–724
30. Park J-S, Lee M-J, Lee C-S, Kang S-W (2001) Electrochem Solid-State Lett 4(4):C17–C19
31. Elam JW, Nelson CE, Grubbs RK, George SM (2001) Thin Solid Films 386(1):41–52
32. Kim H, Rossnagel SM (2002) J Vac Sci Technol A: Vac Surf Films 20(3):802–808
33. Rossnagel SM, Sherman A, Turner F (2000) J Vac Sci Technol B: Microelectron Nanometer Struct 18(4):2016–2020
34. Leskelä M, Ritala M (2002) Thin Solid Films 409(1):138–146
35. Chu SS (2009) Advanced nanoscale ULSI interconnects: fundamentals and applications. 207–220
36. Vaartstra BA (2006) (Google Patents, 2006)
37. Aaltonen T, Alén P, Ritala M, Leskelä M (2003) Chem Vap Deposition 9(1):45–49
38. Kessels WMM, Knoops HCM, Dielissen SAF, Mackus AJM, van de Sanden MCM (2009) Appl Phys Lett 95(1):013114-013114-013113
39. Min YS, Bae EJ, Jeong KS, Cho YJ, Lee JH, Choi WB, Park GS (2003) Adv Mater 15(12):1019–1022
40. Lindahl E, Ottosson M, Carlsson J-O (2009) Chem Vap Deposition 15(7–9):186–191
41. Chae J, Park H-S, Kang S-W (2002) Electrochem Solid-State Lett 5(6): C64–C66
42. Huo J, Solanki R, McAndrew J (2002) J Mater Res 17(09):2394–2398
43. Lee H-B-R, Bang S-H, Kim W-H, Gu G, Lee Y, Chung T-M, Kim C, Park C, Kim H (2010) Japan J Appl Phys 49:05FA11
44. Kim WH, Lee H-B-R, Heo K, Lee YK, Chung TM, Kim CG, Hong S, Heo J, Kim H (2011) J Electrochem Soc 158:D1
45. Lee H-B-R, Kim H (2006) Electrochem Solid-State Lett 9(11):G323–G325
46. Yoon J, Lee H-B-R, Kim D, Cheon T, Kim S-H, Kim H (2011) J Electrochem Soc 158(11):H1179–H1182

47. Zhang S-L, Östling M (2004) Crit Rev Solid State Sci 28(1):1–129
48. d'Heurle FM, Gas P (1986) J Mater Res 1(1):205–221
49. Ha D, Shin D, Koh G-H, Lee J, Lee S, Ahn Y-S, Jeong H, Chung T, Kim K (2000) IEEE Trans Electron Devices 47(7):1499–1506
50. Lee J, Cha J (2004) Microelectron Eng 71(3–4):321–328
51. Fitz C, Goldbach M, Dupont A, Schmidbauer S (2005) Microelectron Eng 82(3–4):460–466
52. Goldberger J, Hochbaum AI, Rong Fan A, Yang P (2006) Nano Lett 6(5):973–977
53. Lee H-B-R, Son JY, Kim H (2007) Appl Phys Lett
54. Lee K, Kim K, Park T, Jeon H, Lee Y, Kim J, Yeom S (2007) J Electrochem Soc 154(10):H899
55. Kim K, Lee K, Han S, Jeong W, Jeon H (2007) J Electrochem Soc 154(3):H177
56. Kim K, Lee K, Han S, Park T, Lee Y, Kim J, Yeom S, Jeon H (2007) Jpn J Appl Phys 46(8):L173–L176
57. Do K-W, Yang C-M, Kang I-S, Kim K-M, Back K-H, Cho H-I, Lee H-B, Kong S-H, Hahm S-H, Kwon D-H, Lee J-H, Lee J-H (2006) Jpn J Appl Phys 45(4):2975–2979
58. Yang C-M, Yun S-W, Ha J-B, Na K-I, Cho H-I, Lee H-B, Jeong J-H, Kong S-H, Hahm S-H, Lee J-H (2007) Jpn J Appl Phys 46(4):1981–1983
59. Ha J-B, Yun S-W, Lee J-H (2010) Curr Appl Phys 10(1):41–45
60. Utriainen M, Kröger-Laukkanen M, Johansson L-S, Niinistö L (2000) Appl Surf Sci 157(3):151–158
61. Daub M, Knez M, Goesele U, Nielsch K (2007) J Appl Phys 101(9):09J111-109J111-113
62. Lim BS, Rahtu A, Gordon RG (2003) Nat Mater 2(11):749–754
63. Lim BS, Rahtu A, Park JS, Gordon RG (2003) Inorg Chem 42(24):7951–7958
64. Kim J-M, Lee H-B-R, Lansalot C, Dussarrat C, Gatineau J, Kim H (2010) Japan J Appl Phys 49(5):6928
65. Lee H-B-R, Kim W-H, Lee J, Kim J-M, Heo K, Hwang I, Park Y, Hong S, Kim H (2010) J Electrochem Soc 157(1):D10–D15
66. Lee H-B-R, Kim W-H, Lee JW, Kim J, Hwang IC, Kim H (2010) J Korean Phys Soc 56(1):104–107
67. Lemonds A, Bolom T, Ahearn W, Gay D, White J, Ekerdt J (2005) Thin Solid Films 488(1–2):9–14
68. Lee H-B-R, Kim H (2010) J Cryst Growth 312(15):2215–2219
69. Kim H (2003) J Vac Sci Technol, B 21:2231
70. Schmitz JEJ (1991) Chemical vapor deposition of tungsten and Tungsten silicides Noyes. Park Ridge, New Jersey
71. Kim S-H, Kwak N, Kim J, Sohn H (2006) J Electrochem Soc 153:G887
72. Suh DH, Lee J (2004) J Electrochem Soc G151:G618
73. Shang H, White MH, Guarini KW, Solomon P, Cartier E, McFeely FR, Yurkas JJ, Lee W-C (2001) Appl Phys Lett 78:3139
74. Omstead T, Chris D'Couto G, Lee S-H, Wongsenkaum P, Collins J, Levy K (2002) Solid State Technol 51:45
75. Klaus JW, Ferro SJ, George SM (2000) Thin Solid Films 360:145
76. Yang M, Chung H, Yoon A, Fang H, Zhang A, Knepfler C, Jackson R, Byun JS, Mak A, Eizenberg M, Xi M, Kori M, Sinha AK (2002) In: Conference proceedings ULSI **XVII**, 655 materials research society
77. Lee S-H, Gonzalez L, Collins J, Ashitani K, Levy K (2002) Conference proceedings ULSI **XVII**, 649 materials research society
78. Mizoguchi Y, Suzuki K, Tachibana M, Abe D (2003) Conference proceedings ULSI XVIII, 451 Materials Research Society
79. Kim S-H, Kwak N, Kim J, Sohn H (2006) J Electrochem Soc 153:G887
80. Elam JW, Nelson CE, Grubbs RK, George SM (2001) Thin Solid Films 386:41
81. Lantz SL, Ford WK, Bell AE, Danielson D (1993) J Vac Sci Tech A 11:911
82. Lantz SL, Bell AE, Ford WK, Danielson D (1994) J Vac Sci Tech A 12:1032
83. Herner SB, Desai SA, Nak A, Ghanayem SG (1999) Electrochem Solid-State Lett 2:398

84. Kim S-H, Hwang E-S, Kim B-M, Lee J-W, Sun H-J, Hong TE, Kim J-K, Sohn H, Kim J, Yoon T-S (2005) Electrochem Solid-State Lett 8:C155
85. Fawcett E, Griffithsm D (1962) J Phys Chem Solids 23:1631
86. Rossnagel SM (2002) J Vac Sci Technol, B 20:2328
87. Steinhögl W, Schindler G, Steinlesberger G, Engelhardt M (2002) Phys Rev B 66:075414
88. Schindler G, Steinhögl W, Steinlesberger G, Traving M, Engelhardt M (2003) Advanced metallization conference 2002 (AMC 2002), 13 Materials Research Society
89. Wu W, Brongersma SH, Hove MV, Maex K (2004) Appl Phys Lett 84:2838
90. Ivanov IP, Sen I, Keswick P (2006) J Vac Sci Technol, B 24:523
91. Kim S-H, Kim J-T, Kwak N, Kim J, Yoon T-S, Sohn H (2007) J Vac Sci Technol, B 25:1574
92. Kim S-H, Yeom S, Kwak N, Sohn H (2008) J Electrochem Soc 155:D148
93. Buerke A, Shmidauer S (2006) Conference proceedings., AMC XXI, p 239, Material Research Society
94. Smith S, Aouadi K, Collins J, Van der Vegt E, Basso M-T, Juhel M, Pokrant S (2005) Microelect Eng 82:261
95. Kim C-H, Rho I-C, Kim S-H, Han I-K, Kang H-S, Ryu S-W, Kim H-J (2009) J Electrochem Soc 156:H685
96. Edelstein D, Uzoh C, Cabral Jr C, DeHaven P, Buchwalter P, Simon A, Cooney III E, Malhotra S, Klaus D, Rathore H, Agarwala B, Bguyen D (2002) In: McKerrow AJ, Shacham-Diamand Y, Zaima S, Ohba T advanced metallization conference in 2001 (Mater. Res. Soc. Proc. Warrendale, PA, 2002), pp. 541–547
97. Rossnagel SM, Kuan TS (2004) J Vac Sci Technol, B 22:240
98. Josell D, Wheeler D, Witt C, Moffat TP (2003) Electrochem Solid-State Lett 6:C143
99. Lane MW, Murray CE, McFeely FR, Vereecken PM, Rosenberg R (2003) Appl Phys Lett 83:2330
100. Chyan O, Arunagiri TN, Ponnuswamy T (2003) J Electrochem Soc 150:C347
101. Kim H (2003) J Vac Sci Technol, B 21:2231
102. Kim H, Lee H-B-R, Maeng W-J (2009) Thin Solid Films 517:2563
103. Kaloyeros AE, Eisenbraun E (2000) Annu Rev Mater Sci 30:363
104. Elers K-E, Saanila W, Soininen PJ, Li W-M, Kostamo JT, Haukka S, Juhanoja J, Besling WFA (2002) Chem Vap Deposition 8:149
105. Ritala M, Leskelä M, Rauhala E, Haussalo P (1995) J Elcecotrchem Soc 142:2731
106. Juppo M, Alén P, Ritala M, Leskelä M (2001) Chem Vap Deposition 7:211
107. Juppo M, Ritala M, Leskelä Markku (2000) J Electrochem Soc 147:3377
108. Juppo M, Alén P, Ritala M, Sajavaara T, Keinonen J, Leskelä M (2002) Electrochem Solid-State Lett 5:C4
109. Jeon H, Lee J-W, Kim Y-D, Kim D-S, Yi K-S (2000) J Vac Sci Technol, A 18:1595
110. Satta A, Beyer G, Maex K, Elers K-E, Haukka S, Vantomme A (2000) Presented at 2000 materials research society spring meeting
111. Kim S-H, Oh SS, Kim H-M, Kang D-H, Kim K-B, Li W-M, Haukka S, Tuominen M (2004) J Electrochem Soc 151(4):C272–C282
112. Min J-S, Park J-S, Park H-S, Kang S-W (2000) J Electrochem Soc 147:3868
113. Kim JY, Choi GH, Kim YD, Kim Y, Jeon H (2003) Jpn J Appl Phys 42:4245
114. Kim D-J, Jung Y-B, Lee M-B, Lee Y-H, Lee J-H (2000) Thin Solid Films 372:276
115. Caubet P, Blomberg T, Benaboud R, Wyon C, Blanquet E, Gonchond J-P, Juhel M, Bouvet P, Gros-Jean M, Michailos J, Richard C, Iteprat B (2008) J Electrochem Soc 155:H625
116. Elers K-E, Winkler J, Weeks K, Marcus S (2005) J Electrochem Soc 152:G589
117. Heil SBS, Langereis E, Roozeboom F, van de Sanden MCM, Kessels WMM (2006) J Electrochem Soc 153:G956
118. Min JS, Park HS, Kang SW (1999) Appl Phys Lett 75:1521
119. Park J-S, Kang S-W (2004) Electrochem Solid-State Lett 7:C87
120. Park J-S, Kang S-W, Kim H (2006) J Vac Sci Technol, B 24:1327
121. Kim JY, Kim HK, Kim Y, Kim YD, Kim WM, Jeon H (2002) J Korean Phys Soc 40:176

122. Nicolet M-A (1995) Appl Surf Sci 91:269
123. Massalski TB (1990) Binary Alloy Phase Diagrams 2:1486
124. Holloway K, Fryer PM, Cabral C Jr, Harper JME, Bailey PJ, Kelleher KH (1992) J Appl Phys 71:5433
125. Oku T, Kawakami E, Uekubo M, Takahiro K, Yamaguchi S, Murakami M (1996) Appl Surf Sci 99:265
126. Min K-H, Jun G-C, Kim K-B (1996) J Vac Sci Tech B 14:3263
127. Stavrev M, Fischer D, Praessler F, Wenzel C, Drescher K (1999) J Vac Sci Tech A 17:933
128. Ritala M, Kalsi P, Riihelä D, Kukli K, Leskelä M, Jokinen J (1999) Chem Mater 11:1712
129. Juppo M, Ritala M, Leskelä M (2000) J Electrochem Soc 147:3377
130. Kim H, Kellock AJ, Rossnagel SM (2002) J Appl Phys 92:7080
131. Chung H-S, Kwon J-D, Kang S-W (2006) J Electrochem Soc 153:C751
132. Kim H, Detavenier C, van der Straten O, Rossnagel SM, Kellock AJ, Park D-G (2005) J Appl Phys 98:014308
133. Maeng WJ, Park S-J, Kim H (2006) J Vac Sci Technol *B* 24:2276
134. Knoops HCM, Baggetto L, Langereis E, van de Sanden MCM, Klootwijk JH, Roozeboom F, Niessen RAH, Notten PHL, Kessels WMM (2008) J Electrochem Soc 155:G287
135. Langereis E, Knoops HCM, Mackus AJM, Roozeboom F, van de Sanden MCM, Kessels WMM (2007) J Appl Phys 102:083517
136. Park S-J, Na K-I, Jeong W-C, Kim S-H, Boo S-E, Bae N-J, Lee J-H (2002) American vacuum society topical conference on atomic layer deposition, Seoul
137. Park J-S, Park H-S, Kang S-W (2002) J Electrochem Soc 149:C28
138. Burton BB, Lavoie AR, George SM (2008) J Electrochem Soc 155:D508
139. Song M-K, Rhee S-W (2008) J Electrochem Soc 155:H823
140. Hossbach C, Teichert S, Thomas J, Wilde L, Wojcik H, Schmidt D, Adolphi B, Bertram M, Mühle U, Albert M, Menzel S, Hintze B, Bartha JW (2009) J Electrochem Soc 156:H852
141. Cho G-H, Rhee S-W (2010) Electrochem Solid-State Lett 13:H426
142. Kim S-H, Song M-K, Rhee S-W (2010) J Electrochem Soc 157:H652
143. Adelmann C, Meersschaut J, Ragnarsson L-Å, Conard T, Franquet A, Sengoku N, Okuno Y, Favia P, Bender H, Zhao C, O'Sullivan BJ, Rothschild A, Schram T, Kittl JA, Van Elshocht S, De Gendt S, Lehnen P, Boissière O, Lohe C (2009) J Appl Phys 105:053516
144. Park TJ, Kim JH, Jang JH, Na KD, Hwang CS, Kim JH, Kim G-M, Choi JH, Choi KJ, Jeong JH (2007) Appl Phys Lett 91:252106
145. Klaus JW, Ferro SJ, George SM (2000) J Electrochem Soc 147:1175
146. Elers K-E, Saanila W, Soininen PJ, Li W-M, Kostamo JT, Haukka S, Juhanoja J, Besling WFA (2002) Chem Vap Deposition 8:149
147. Becker JS, Gordon RG (2003) Appl Phys Lett 82:2239
148. Becker JS, Suh S, Wang S, Gordon RG (2003) Chem Mater 15:2969
149. Dezelah IV CL, El-Kadri OM, Kukli K, Arstila K, Baird RJ, Lu J, Niinistö L, Winter CH (2007) J Mater Chem 17:1109
150. Li W-M, Elers K-E, Kostamo J, Kaipio S, Huotari H, Soininen M, Soininen PJ, Tuominen M, Haukka S, Smith S, Besling W (2002) Proceedings of the IEEE 2002 international interconnect technology conference, June 3–5, 2002, pp 191
151. Smith S, Li W-M, Elers K-E, Pfeifer K (2002) Microelectron Eng 64:247
152. Kim S-H, Oh SS, Kang D-H, Kim K-B, Li W-M, Haukka S, Tuominen M (2003) Appl Phys Lett 82:4486
153. Bystrova S, Aarnink AAI, Holleman J, Wolters RAM, Elcecotrchem J (2005) Soc. 152:G522
154. Kim S-H, Kim J-K, Kwak N, Sohn H, Kim J, Jung S-H, Hong M-R, Lee SH, Collins J (2006) Electrochem Solid-State Lett 9:C54
155. Kim S-H, Kim J-K, Lee JH, Kwak N, Kim J, Jung S-H, Hong M-R, Lee SH, Collins J, l Sohn H (2007) J Elcecotrchem Soc 154:D435
156. Kim S-H, Yim S–S, Lee D-J, Kim K-S, Kim H-M, Kim K-B, Sohn H (2008) Electrochem Solid-State Lett 11:H127

157. Juppo M, Ritala M, Leskelä M (1997) J Vac Sci Technol A A15:2330
158. Mårtensson P, Carlsson JO (1997) Chem Vap Deposition 3:45
159. Utriainen M, Kröger-Laukkanen M, Johansson L-S, Niinistö L (2000) Appl Surf Sci 157:151
160. Mårtensson P, Carlsson JO (1998) J Electrochem Soc 145:2926
161. Solanki R, Pathangey B (2000) Electrochem Solid-State Lett 3:479
162. Li Z, Rahtu A, Gordon RG (2006) J Elcecotrchem Soc 153:C787
163. Huo J, Solanski R, McAndrew J (2002) J Mater Res 17:9
164. Waechtler T, Oswald S, Roth N, Jakob A, Lang H, Ecke R, Schulz SE, Gessner T, Moskvinova A, Schulz S, Hietschold M, Elcecotrchem J (2009) Soc. 156:H453
165. Niskanen A, Rahtu A, Sajavaara T, Arstila K, Ritala M, Leskelä M (2005) J Elcecotrchem Soc 152:G25
166. Wu L, Eisenbraun E (2007) J Vac Sci Technol B 25:2581
167. Josell D, Wheeler D, Witt C, Moffat TP (2003) Electrochem Solid-State Lett 6:C143
168. Lane MW, Murray CE, McFeely FR, Vereecken PM, Rosenberg R (2003) Appl Phys Lett 83:2330
169. Aaltonen T, Alën P, Ritala M, Leskelä M (2003) Chem Vap Dep 9:45
170. Aaltonen T, Rahtu A, Ritala M, Leskelä M (2003) Electrochem Solid-State Lett 6:C130
171. Park KJ, Terry DB, Stewart M, Parsons GN (2007) Langmuir 23:6106
172. Park S-J, Kim W-H, Lee H-B-R, Maeng WJ, Kim H (2008) Microelectron Eng 85:39
173. Lee D-J, Yim S–S, Kim K-S, Kim S-H, Kim K-B (2010) J Appl Phys 107:013707
174. Lee D-J, Yim S–S, Kim K-S, Kim S-H, Kim K-B (2008) Electrochem Solid-State Lett 11:K61
175. Kwon O-K, Kim J-H, Park H-S, Kang S-W (2004) J Electrochem Soc 151:G109
176. Yim S–S, Lee D-J, Kim K-S, Kim S-H, Yoon T-S, Kim K-B (2008) J Appl Phys 103:113509
177. Aaltonen T, Ritala M, Arstila K, Keinonen J, Leskelä M (2004) Chem Vap Dep 10:217
178. Kim J-H, Kim Y-D, Kil D-S, Yeom S-J, Roh J-S, Kwak N-J, Kim J-W, Park S-K (2008) 8th international conference on atomic layer deposition held at Bruges, Belgium
179. Kwon O-K, Kwon S-H, Park H-S, Kang S-W (2004) J Electrochem Soc 151:C753
180. Yim S–S, Lee D-J, Kim K-S, Lee M-S, Kim S-H, Kim K-B (2008) Electrochem Solid-State Lett 11:K89
181. Farmer DB, Gordon RG (2007) J Appl Phys 101:124503
182. Kim SK, Han JH, Kim GH, Hwang CS (2010) Chem Mater 22:2850
183. Kim W-H, Park S-J, Son J-Y, Kim H (2008) Nanotechnology 19:045302
184. Li H, Farmer DB, Gordon RG, Lin Y, Vlassak J (2007) J Electrochem Soc 154:D642
185. Eom T-K, Sari W, Choi K-J, Shin W-C, Kim J-H, Lee D-J, Kim K-B, Sohn H, Kim S-H (2009) Electrochem Solid-State Lett 12:D85
186. Choi S-H, Cheon T, Kim S-H, Kang D-H, Park G-S, Kim S (2011) J Electrochem Soc 158:D351
187. Jeong D, Inoue H, Ohno Y, Namba K, Shinriki H (2008) Proceedings of the IEEE 2008 international interconnect technology conference, pp 95–96
188. Chan R, Arunagiri TN, Zhang Y, Chyan O, Wallace RM, Kim MJ, Hurd TQ (2004) Electrochem Solid-State Lett 7:G154
189. Arunagiri TN, Zhang Y, Chyan O, El-Bouanani M, Kim MJ, Chen KH, Wu CT, Chen LC (2005) Appl Phys Lett 86:083104
190. Damayanti M, Sritharan T, Gan ZH, Mhaisalkar SG, Jiang N, Chan L (2006) J Electrochem Soc 153:J41
191. Sari W, Eom T-K, Jeon C-W, Sohn H, Kim S-H (2009) Electrochem Solid-State Lett 12:H248
192. Qu X-P, Tan J–J, Zhou M, Chen T, Xie Q, Ru G-P, Li B-Z (2006) Appl Phys Lett 88:151912
193. Tan J–J, Qu X-P, Xie Q, Zhou Y, Ru G-P (2006) Thin Solid Films 504:231

194. Kim S-H, Kim H-T, Yim S–S, Lee D-J, Kim K-S, Kim H-M, Kim K-B, Sohn H (2008) J Electrochem Soc 155:H589
195. Kim S-W, Kwon S-H, Jeong S-J, Kang S-W (2008) J Electrochem Soc 155:H885
196. Kumar S, Greenslit D, Chakraborty T, Eisenbraun ET (2009) J Vac Sci Technol A 27:572
197. Greenslit D, Kumar S, Chakraborty T, Eisenbraun ET (2008) ECS Trans 13(8):63–70
198. Eom T-K, Kim S-H, Park K-S, Kim S, Kim H (2011) Electrochem Solid-State Lett 14:D10
199. Cheon T, Choi S-H, Kim S-H, Kang D-H (2011) Electrochem Solid-State Lett 14:D57

Part V
ALD Machines

Chapter 9
ALD Machines

Schbuert Chu

State-of-the-art wafer fabrication equipment for semiconductor manufacturing is represented by the systems used to process 300 mm wafers. Equipment used for leading edge manufacturing is complex with stringent requirements on within-wafer and wafer-to-wafer nonuniformity of thickness, composition, and resistivity. In addition, particle defects, step coverage, and wafer throughput, as measured by number of wafers processed per hour, are also key criteria for selecting the optimum equipment. The ALD machines used for semiconductor manufacturing are also highly sophisticated in terms of wafer handling, FOUP transfer, factory automation, host communication, ambient contamination control, and safety interlocks. Unlike the ALD reactors used in research, ALD equipment used for semiconductor manufacturing also needs to be installed with proper abatement systems for effluent treatment such that the exhaust and waste water streams are in compliance with the industrial regulations of various local jurisdictions. The following sections will describe major ALD reactor types used in semiconductor manufacturing as well as some of the necessary support equipment.

9.1 Reactor Design Concepts

The self-limiting nature of ALD makes the design of an experimental reactor relatively simple. In fact, research ALD reactors are very simple and affordable as gas distribution and temperature do not need to be strictly uniform. Furthermore, in a research environment, pulse cycle time, particle defects, and automated wafer handling are also not necessities. As a result, there are a myriad of commercial and home-grown research ALD reactors available. Most of these reactors include either a hot wall reaction cavity or a hot plate. The gases are typically injected

S. Chu (✉)
Applied Materials Inc, Santa Clara, CA 95054, USA
e-mail: schubert_chu@amat.com

C. S. Hwang (ed.), *Atomic Layer Deposition for Semiconductors*,
DOI: 10.1007/978-1-4614-8054-9_9,

through a single opening; although showerheads with multiple gas flow paths are also available in some research reactors. However, as ALD gains increased adoption in semiconductor manufacturing, the numbers of ALD process steps and the types and complexities of ALD reactor designs have also drastically increased. In general, an industrial ALD reactor design is driven by four key process requirements: (1) deposition uniformity, (2) deposition conformality (step-coverage), (3) particle defects, and (4) wafer throughput or the number of wafers processed each hour. A typical one-sigma film thickness uniformity requirement for ALD is <1 %. Deposition step coverage, as defined by the thickness inside a feature divided by the thickness on the wafer surface, needs to be >95 % in order to be considered conformal. Partical defects are also a very important consideration as chip manufacturing yields are directly impacted by the presence of particle defects as small as the minimum feature size. At press time, particle defects as small as <50 nm in diameter are routinely monitored by leading semiconductor manufacturers. Finally, ALD equipment must be able to process as many wafers as possible in the highly cost-competitive semiconductor market. Typically, each piece of equipment must be able to process at least 25 wafers per hour regardless of target thickness in order to be considered for manufacturing. Wafer throughput as high as >100 wafers per hour have also been demonstrated for some applications.

There is usually a preferred ALD reactor design architecture for each application based on technical and/or economic considerations. The leading reactor architectures can be categorized either by the primary source providing the energy required for reaction or by the number and arrangements of the substrates within the reactor. The various types of reactor designs are shown in Fig. 9.1. These various types of reactor design will be discussed in the sections that follow; starting with single-wafer reactors, but not according to the classification shown in Fig. 9.1, since they are often intermixed.

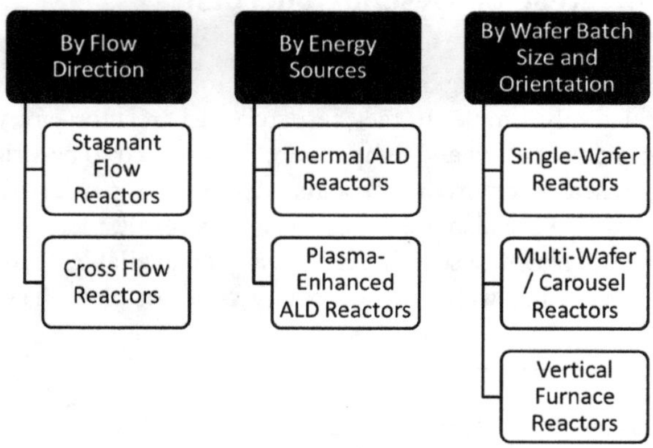

Fig. 9.1 Major ALD reactor categories

9.1.1 Single-Wafer Reactors

Amongst its many advantages, single-wafer ALD reactors can be the most easily integrated with other process chambers on the same vacuum cluster for sequential processing without air-break. This capability is particularly important for depositing air-sensitive stacks, such as the copper barrier and seed for interconnects. Figure 9.2 shows a copper barrier-seed system used to deposit the barrier metal TaN and Cu seed material for subsequent electroplating of Cu interconnects.

In this system, single-wafer ALD TaN reactors are integrated with chambers for degassing moisture off the incoming wafer, chambers for pre-cleaning the substrate, and chambers for physical vapor deposition (PVD) of Ta and Cu. In this configuration, ALD TaN is followed by an optional PVD Ta layer and capped by a PVD Cu seed layer without air-break.

Besides being able to be integrated with other process chambers on the same vacuum cluster, there are other advantages of single-wafer ALD reactors as well. Single-wafer ALD reactors typically have very small reaction volume. This allows very fast switching of gases which leads to short processing times. It is also much easier to conduct experiments and to develop unit processes on a single-wafer reactor instead of a batch-wafer reactor. An additional benefit of single-wafer reactors is that since each wafer is processed under the same conditions with the same environment, wafer-to-wafer repeatability is usually better compared to batch-wafer reactors.

9.1.2 Stagnant-Flow Single-Wafer Reactor

Stagnant-flow single-wafer ALD reactors are a logical extension of single-wafer CVD reactors.

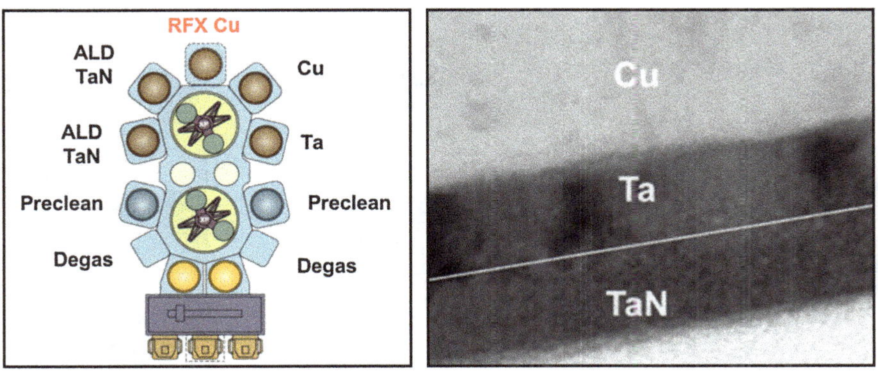

Fig. 9.2 Integrated barrier-seed system with *ALD TaN* chambers (Courtesy of applied materials)

In the most basic form, a stagnant-flow single-wafer ALD reactor retains a traditional CVD showerhead whose primary purpose is to uniformly distribute reactant gases over the substrate. The uniform gas distribution is achieved by the relatively low conductance of the holes in the showerhead. Instead of simultaneously flowing both the precursor and reactant gases, only one gas is flowing at a time in ALD mode. The two halves of the ALD reaction are separated by purge and/or pump steps which are critical and need to be sufficiently long in order to have effective separation of the precursors. If the precursors are not separated, the resultant film will likely exhibit CVD behavior such as inferior uniformity and conformality. Furthermore, ineffective precursor separation can also result in gas phase reaction and particle defects on the substrate. However, since the holes of the showerhead have low gas flow conductance, the purge/pump steps can take up to minutes to complete. The long purge/pump times inevitably make showerheads impractical for use in semiconductor manufacturing.

One approach to reducing the long pump/purge times created by the showerhead is to drastically reduce the flow restriction. In the example shown in Fig. 9.3, the low conductance showerhead is replaced by a high flow conductance inverted vortex funnel. Reactant gases sequentially enter the top or narrow part of vortex funnel. The expanding conical shape of the funnel directs the gas flow in a spiral manner and distributes the gases evenly across the surface of the wafer substrate.

It is important to note that although an ideal ALD reaction is supposed to be self-limiting, very few, if any, ALD processes used in semiconductor manufacturing exhibit true self-limiting behavior. Most ALD processes used in semiconductor manufacturing contain significant CVD characteristics. These CVD characteristics are the results of several factors including decomposition of metal organic precursors, virtual sources within the reactor such as reactor walls, and compromised pumping and purging of the reactor space. The last factor is worth noting. Costs and wafer outputs are very important considerations in semiconductor manufacturing. Real life industrial ALD processes routinely make

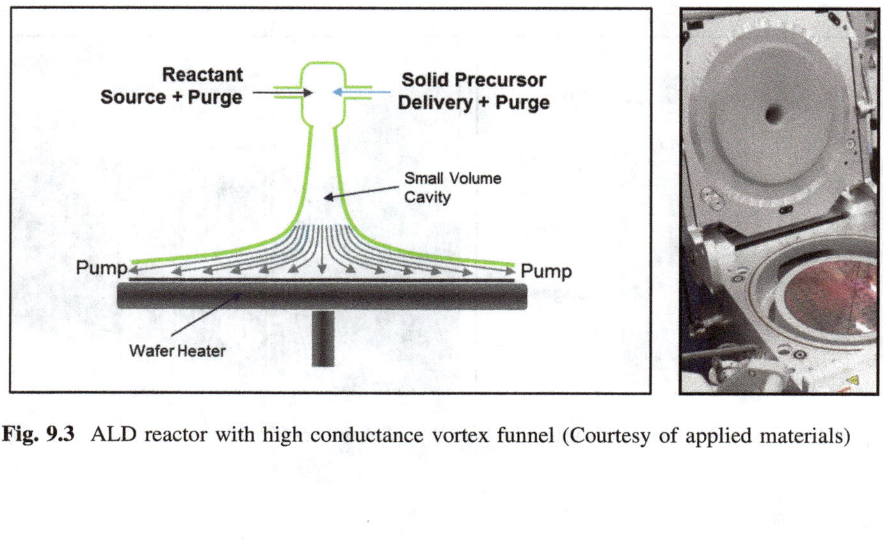

Fig. 9.3 ALD reactor with high conductance vortex funnel (Courtesy of applied materials)

compromises between wafer output and ideal ALD characteristics; i.e., perfect uniformity or conformality is often sacrificed by shortening the pumping and purging duration in order to gain additional wafer output. The vortex funnel is a design aimed at maximizing wafer output while making acceptable trade-offs on academically pure ALD behavior. While the uniformity of the gas distribution of a vortex funnel is certainly not as even when compared to that of a CVD showerhead, it is capable of depositing a film with uniformity of around or <1 %, 1–sigma with pump and purge times of just a few seconds instead of minutes. Additionally, the vortex funnel design provides the added benefit of reduced precursor decomposition when combined with a cold-wall reactor configuration where the walls of the reaction space are kept at a temperature that is substantially below that of substrate so as to minimize the amount of deposition on the reactor walls.

9.1.3 Plasma-Enhanced Single-Wafer ALD Reactor

Many process requirements are continually pushing the thermal budget limit for film deposition lower and lower. Some of the limitations include nickel silicide (NiSi) stability, void formation in Cu lines, dopant diffusion in the transistor junction area, and thermal stability of photoresist materials. In some other cases, the thermal instability of the precursor precludes it from being used at temperatures required for the second reactant to be active. Plasma-enhanced ALD activates the reactant(s) and offers the possibility to deposit films at lower temperatures. In many cases, plasma-enhanced ALD provides the possibility to deposit films that are otherwise inaccessible.

The stagnant-flow single-wafer ALD reactor is the most common type of reactor to be adapted for plasma-enhanced ALD. Both direct plasma and remote plasma configurations are possible to be adapted to a stagnant-flow single-wafer ALD reactor. In the case of a direct reactor, the showerhead or a modified version of the vortex funnel is connected to a radio-frequency (RF) power supply via a matching network and is electrically isolated from the rest of the reactor. The grounding path is typically through the wafer pedestal or through the chamber body. The capacitively coupled plasma is ignited and sustained between the showerhead and the wafer in this configuration. Typical RF frequency used in plasma-enhanced ALD can range from hundreds of kHz to 13.56 MHz. Higher frequency is possible but is generally less popular. Typical power delivered during a plasma-enhanced ALD process is generally less than a few hundred watts. An example of a direct plasma reactor can be seen in Fig. 9.4 This plasma configuration is quite similar to a typical PECVD reactor and is the most common design for plasma-enhanced ALD. Another direct plasma configuration including inductively coupled plasma is also possible. However, they are far more complex and generally not used in manufacturing.

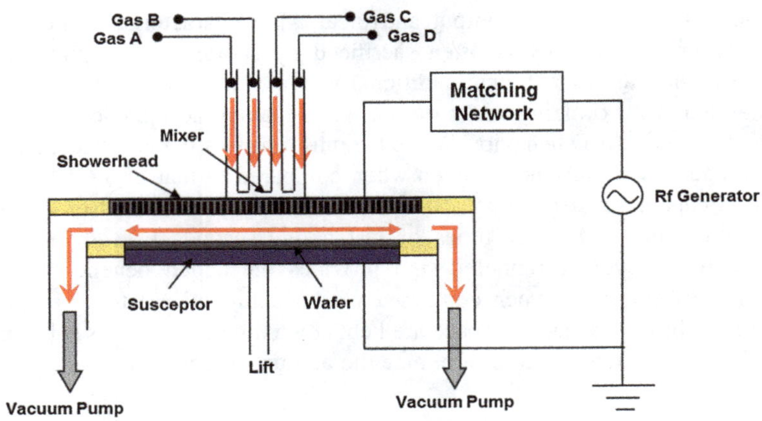

Fig. 9.4 Direct plasma reactor (Courtesy of ASM international)

Although direct plasma reactors are relatively straight forward to design, direct plasma may cause damage to the substrate in some applications. Furthermore, direct plasma tends to be anisotropic in nature. Films deposited with direct plasma may have inconsistent properties on the sidewalls of high aspect ratio structures. In such cases, remote plasma configuration is preferable. In a remote plasma configuration, few to none of the more energetic particles can reach the wafer thus reducing the probability of plasma damage to the substrate. Furthermore, remote plasma generally produces a higher radical-to-ion ratio thus creating a more isotropic deposition. There are two main types of remote plasma configurations. The more common approach is to install a remote plasma system (RPS) on the reactant line just before it enters the reactor. In this configuration, the reactant (e.g., O_2 or NH_3) is ionized by the RPS unit. The walls of the gas lines and feed-throughs to the reactor quickly neutralize most of the ionized species leaving primarily energized radicals to enter into the reactor. This approach works well with some gases such as nitrogen and ammonia which produce radicals with long lifetimes.

However, for some other gases such as hydrogen, the radical lifetimes are generally not long enough to survive through the showerhead to reach the wafer. Chemical compatibility to the wall of the remote plasma cavity is also another important consideration with this remote plasma configuration (Fig. 9.5).

Another remote plasma configuration that is gaining increased popularity is one where the plasma is generated capacitively between an RF hot plate and a grounded showerhead. The grounded showerhead in this case serves to neutralize ionized species and also to uniformly distribute the energized radical reactants into the reactor space. The proximity to the wafer substrate and the flexibility of process conditions are the primary advantages of this second type of remote plasma configuration. It is particularly beneficial for radical species that have a very short lifetime such as hydrogen.

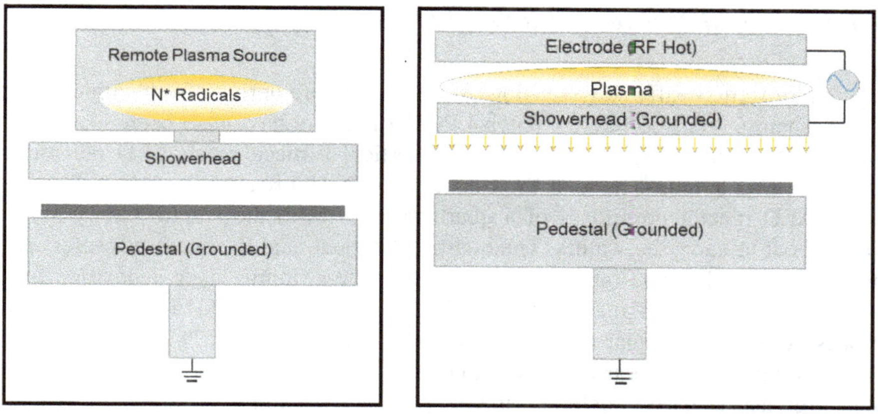

Fig. 9.5 Two remote plasma reactor configurations (Courtesy Applied Materials)

9.1.4 *Cross-Flow Single-Wafer ALD Reactor*

In cases where the precursor chemistry exhibits strong ALD behavior, a cross-flow reactor can be a very good alternative to the stagnant-flow style reactors. Cross-flow reactors are also known as traveling-wave reactors, because the precursor and reactants spread across the width of the wafer and are swept from one side of the wafer to the other sequentially in a traveling wave. Cross-flow reactors are often made to be hot-walled reactors. The cross-flow style ALD reactors were pioneered by Microchemistry in Finland in the 1980s and have been adopted for high volume semiconductor manufacturing in recent years. Cross-flow style reactors have the advantages of simplicity in design and fast gas switching capability. An example of a cross-flow ALD reactor is shown in Fig. 9.6.

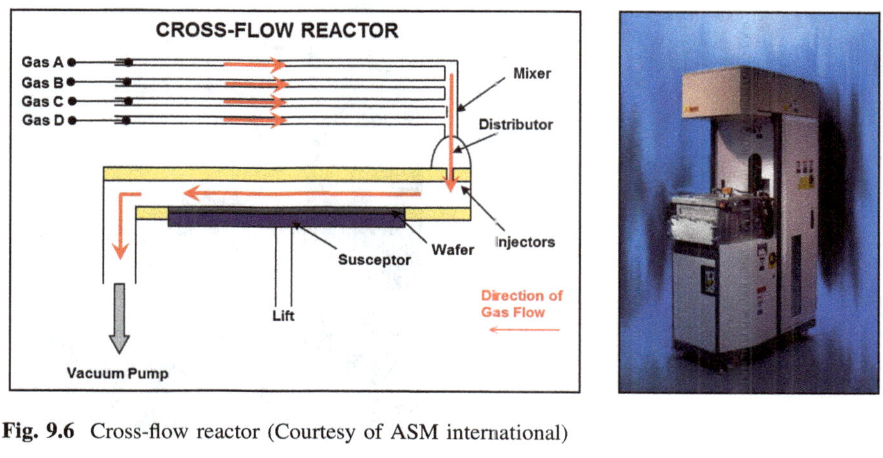

Fig. 9.6 Cross-flow reactor (Courtesy of ASM international)

9.1.5 Vertical Furnace Batch ALD Reactor

Much like single-wafer stagnant-flow ALD reactors are derived from single-wafer CVD reactors, vertical furnace batch LPCVD reactors have also been adapted to become ALD reactors. An example of a vertical furnace batch ALD reactor is shown in Fig. 9.7. Just as with the case of an LPCVD reactor, a vertical furnace batch ALD reactor comprises of a quartz tube to form the reactor cavity and a quartz boat to carry the wafers. The quartz tube, boat, and the enclosed wafers are heated by the elements outside of the quartz tube. Atomic layer deposition in a vertical furnace batch reactor enables a large batch of 100–150 wafers to be processed in a single quartz tube reactor simultaneously, which leads to very high wafer outputs. This is particularly valuable for thicker layers or for deposition in very high aspect ratio structures, such as DRAM capacitor structures, where long pulse and purge times are necessary to ensure true ALD behavior such as conformality and uniformity inside the high aspect ratio structures.

In a typical vertical furnace batch LPCVD reactor, the precursor and reactant gases are delivered through a single injection point. However, since precursors used in ALD often decompose at elevated temperatures, gases in modern vertical furnace batch ALD reactors are delivered through a series of injectors on the side of the wafers. The distributed injection of gases ensures even precursor flow across the entire batch of wafers and serves to minimize precursor decomposition.

One of the primary applications of vertical furnace batch ALD reactors is the deposition of the high-k dielectric material (e.g., Al_2O_3, HfO_2, and ZrO_2) in

Fig. 9.7 Example of vertical furnace batch ALD reactor (Courtesy of ASM International)

DRAM metal–insulator-metal (MIM) capacitors. Metalorganic precursors are used for this application, and ozone (O_3) is typically used as a reactant. Ozone is a very important and frequently used reactant in ALD. Details of ozone generators used in semiconductor manufacturing are discussed later in this chapter. Other energized reactants are also of great interest in order to obtain films otherwise inaccessible while preserving the throughput advantage of the large batch reactor. Leading equipment makers have made modifications in order to introduce plasma-enhanced ALD to vertical furnace reactors. Just as it is the case with single-wafer reactors, both direct and remote plasmas are possible. Direct plasma is typically obtained by modifying the reactor and the insertion of a plasma rod at the outlet of the precursor injectors. Remote plasma generators can also be installed at the gas inlet of the vertical furnace reactors to provide energized radical reactants to the reactors.

9.1.6 Carousel ALD Reactors

The time required to purge the reactants from the reactor is the primary limitation to the time required to complete an ALD cycle for both single-wafer and vertical furnace batch ALD reactors. In a single-wafer reactor, time required per cycle is typically a few seconds to <20 s, whereas in a vertical furnace ALD reactor the time required per cycle can be a few minutes. One solution is to separate the reactants spatially rather than temporally. In a spatially separated ALD reactor, the precursor and reactants are continuously flowing instead of intermittently pulsed into the reactor. The difference in gas introduction schemes between spatially separated and temporally separated ALD reactors is illustrated in Fig. 9.8.

The gases in a spatially separated ALD reactor are separated by space and isolated from each other with effective inert gas purge as shown in Fig. 9.9. The effectiveness of gas isolation is strongly influenced by the gap between the gas injector and the wafer substrate. Spatially separated ALD in fact is not a new idea;

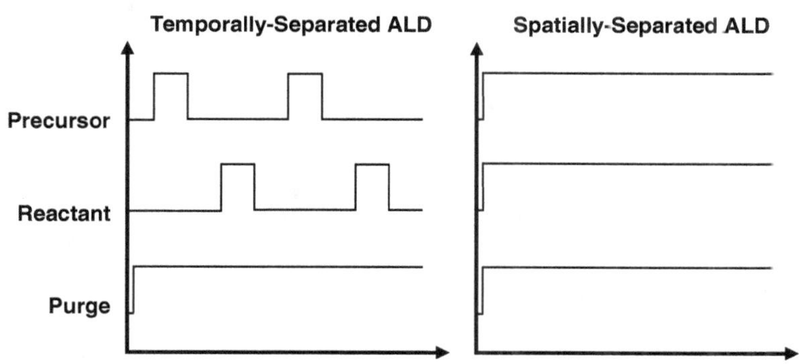

Fig. 9.8 Differences in flow schemes between temporally-separated and spatially separated ALD

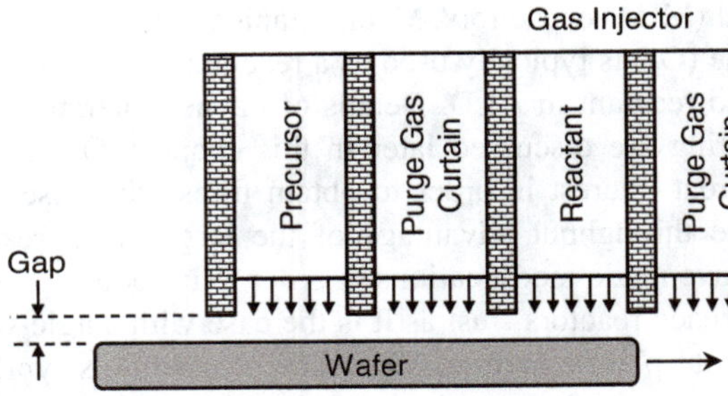

Fig. 9.9 Representative physical layout of spatially separate ALD reactor

Suntola and colleagues made carousel-style spatially separated ALD reactors back in the 1970s [1]. However, spatially separated ALD reactors came to the forefront once again in the late 2000s due to increased adoption of ALD in semiconductor manufacturing and due to an increased availability of precision motion components capable of maintaining the critical gap distance and thus achieving superior gas separation.

Several spatially separated ALD reactors have been commercialized for semiconductor manufacturing in pursuit of ever greater wafer output per hour. They typically take the form of a mini-batch carousel whereby several wafers are loaded onto a rotational carrier. There are different zones with precursor, purge, and reactant flow. As the carrier rotates, each wafer is alternatingly exposed to precursor, purge, and then reactant in order to complete an ALD cycle. The rotation continues until the desired thickness is reached. Typical rotation speeds of the carrier ranges from a few RPM to nearly 100 RPM. The carousel reactors are generally cold-walled, and the rotational carrier provides heat to the substrates. Carousel reactors are also capable of PEALD. Both direct and remote plasma configurations are available on carousel reactors much in the same manner as they are on single-wafer reactors.

An example of a carousel ALD tool is shown in Fig. 9.10.

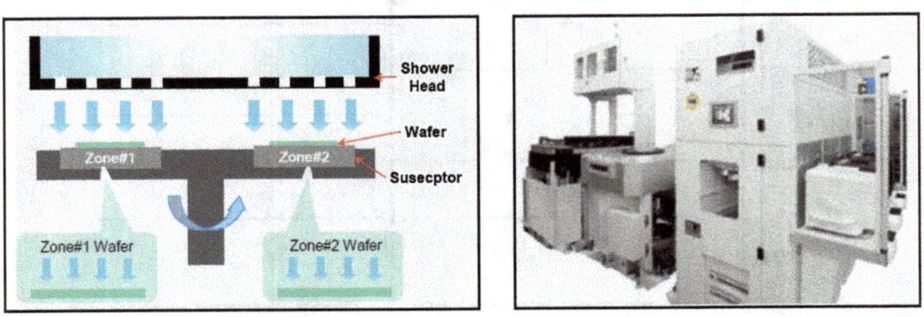

Fig. 9.10 Example of carousel spatially separated ALD Reactor (Courtesy of Wonik-IPS)

9.2 Support Equipment

In order to reach acceptable levels of yield, process variability and defectivity have to be very tightly controlled for ALD equipment used in semiconductor manufacturing. Additionally, semiconductor fabrication lines that house these reactors must treat and abate any unreacted chemicals as well as reaction by-products in order to meet the strict emission guidelines that are standard for industrial production. Consequently, nearly as much consideration is taken in selecting support equipment as in selecting the ALD reactor itself. The sections below will describe some of the major support equipment used in semiconductor manufacturing: Vacuum pumps, abatement units, and ozone generators.

9.2.1 Vacuum Pumps

ALD processes used in microelectronics require low pressures and hence require vacuum pumps. These ALD processes provide a number of challenges to a vacuum pumping system that requires individual consideration when choosing appropriate equipment and optimizing its set-up to achieve reliable operation 24 h a day and 7 days a week. The challenges center around the highly reactive nature of many of the precursors used in ALD.

Dry vacuum pumps (i.e., pumps without any oil or fluid in the swept volume) using a claw, roots, or screw type mechanism are used for ALD equipment in semiconductor fabrication lines. The size of vacuum pump (i.e., the pumping speed) required depends on the gas flow rate and operating pressure. Fast evacuation of each precursor can reduce cycle time and hence increase production rates. This can lead to increased pumping speeds. The actual flow rates of precursor and carrier gas can vary from a few standard cubic centimeters per minute (sccm) to tens of standard liters per minute (slm) depending on the application. Required chamber pressures can vary from 10^{-3} mbar to tens of mbar. Pumps with peak pumping capabilities in the range of 100–6,000 m^3/h are typically required across a in place of the wide range of ALD applications. Figure 9.11 shows the inner workings of a typical dry vacuum pump with a roots/claw mechanism.

In addition to the basic function of removing gas from a chamber, there are a number of challenges presented by ALD applications that could lead to early pump failure if not addressed properly. Many of the precursors used in ALD applications have relatively low vapor pressure and are either liquid or solid at atmospheric pressure and room temperature. This is also true for some of the potential ALD by-products, for example, ammonium chloride can be produced from the ALD of TiN films using $TiCl_4$ and NH_3 precursors. In the process chamber, the low pressure and elevated temperature keeps the precursors and by-products in the vapor phase, but as a vacuum pump compresses the vapor it will tend to condense as the pressure rises unless properly controlled. It is also sometimes necessary to insulate

Fig. 9.11 Cut-away diagram of a typical roots/claw dry pump system (Courtesy of Edwards Ltd.)

and heat the foreline (the pipe between process chamber and vacuum pump) to prevent condensation in this pipe which can lead to pump failure. Since many precursors are highly reactive and some are pyrophoric, condensation in the pump or exhaust could lead to safety issues if subsequently exposed to an oxidant (e.g., if a fluorine based chamber clean is used) or when opening the system for maintenance.

Condensation is prevented by controlling the partial pressure of the vapor as it passes through the pump and by controlling the pump temperature. The control of partial pressure is achieved by admitting controlled flows of purge gas (usually nitrogen) progressively to different stages of the pump mechanism. The required purge flow can be determined from the knowledge of each precursor flow, and carrier gas flow, the precursor vapor pressure curve, and the pump stage speeds. If purge flows are too great they can affect the pump performance and add to operating costs, and if too low then condensation may not be prevented.

Many ALD precursors are large organometallic molecules that readily decompose at elevated temperatures. Since vacuum pumps compress gas they have a tendency to get hot if not thermally controlled. This is particularly the case towards the exhaust end of a pump where the pressures are higher. If allowed to get too hot the precursor will decompose leading to a form of CVD causing material to build-up on surfaces. This material build-up would eventually fill clearances and lead to early pump failure. Good thermal control is required to ensure the pump runs hot enough to prevent condensation but cool enough to prevent thermal decomposition of precursors.

There are further mechanisms that may lead to unwanted solid films building up on the internal surfaces of a vacuum pump. One is through the ALD mechanism itself where a thin film deposits in the same way as it does in the process chamber. This can be minimized through thermal control by operating the pump in a region where it is either cool enough to discourage precursor reactivity, or hot enough to encourage desorption before the following precursor pulse. Poor film adhesion or poor film density on pump surfaces can be dealt with by designing features into the pump mechanism to aid the transport of powders through and out of the mechanism to the exhaust.

9.2.2 *Effluent Abatement*

ALD processes use organometallic or inorganic precursors that are often toxic and create equally dangerous by-products which can be abated efficiently using thermal techniques. Moreover, ALD chamber cleaning exhaust streams are usually corrosive, and laden with particles and condensable compounds which compel semiconductor manufacturers to treat these effluents as close as possible to the process tool. Thus, a number of "Point of Use" (PoU) abatement techniques have been developed to meet the needs of the industry. The systems most commonly used for the treatment of ALD effluents combine combustion with water scrubbing as shown in Fig. 9.12. Typically, such systems can be divided in five distinct sections:

(1) Process gas inlets;
(2) Combustion chamber, designed to oxidize or reduce process gases into a water-reactive form;
(3) Water-cooling stage, designed to reduce the temperature and capture solids;
(4) Collection tank, where the contaminated water and solids are collected; and
(5) Counter-current packed tower, designed to react the combustion by-products with water.

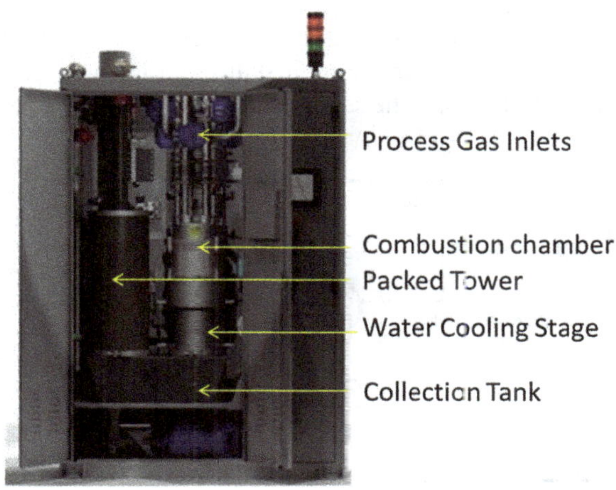

Fig. 9.12 Thermal abatement system combining combustion with wet scrubbing (Courtesy of Edwards Ltd.)

9.2.3 Ozone Generator

Ozone has three bonded oxygen molecules and it is easily broken down into active oxygen atoms by heat and light energy. Because of its reactive nature, it is used as a strong oxidation agent in the various fields of semiconductor manufacturing process. Ozone is a particularly important reactant in the ALD of oxide thin films since it can be easily purged from the reactor space or from deep inside the structures on the wafer when compared to water. Since any residual oxidizer during precursor pulse will lead to CVD characteristics, ozone is often the oxidizer of choice; especially, in high aspect ratio structures. One exception where water is preferred is in the case of HfO_2 gate oxide deposition where ozone would lead to an unacceptable amount of silicon substrate oxidation.

A typical ozone generation scheme is shown in Fig. 9.13. When AC voltage is applied between two facing electrodes, silent discharge generates in the reaction field. During discharge, diatomic oxygen is broken down to monatomic oxygen atoms, and ozone is generated by three body collision reaction. Nitrogen is often added as a catalyst to increase the rate of reaction and subsequently create higher concentration ozone. In recent years, ozone generators that use non-nitrogen catalytic material are starting to gain attention.

9.3 Summary

Semiconductor technology is advancing at a furious pace. Trends such as continued scaling, adoption of new materials, and migration from planar to FinFET transistors will continue to drive an increasing number of ALD process steps used in semiconductor manufacturing. Similarly, equipment for ALD in semiconductor manufacturing will continue to evolve to include new capabilities, tighter controls, fewer variations and defects, and ever increasing wafer output per hour. This chapter reviewed the current status and features required for ALD equipment for mass production. The salient features that are required for an appropriate ALD tool

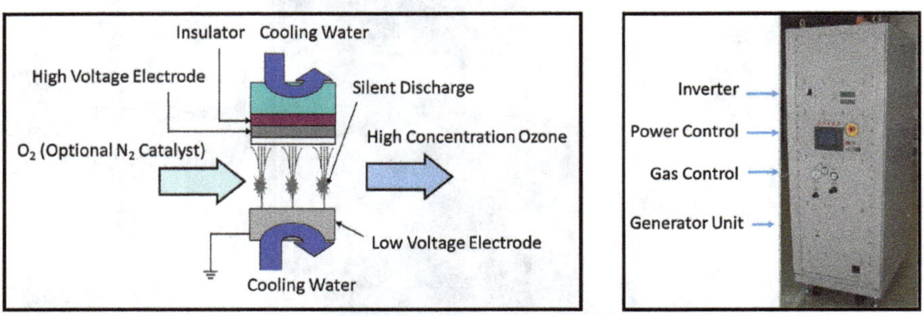

Fig. 9.13 Example of ozone generator (Courtesy of TMEIC)

are described in detail based on the reactor geometry, energy source, and gas flow pattern. There have been several innovations that have made the adoption of ALD tools in semiconductor fabrication lines successful, such as the vortex funnel gas distribution system. There has been a certain level of compromise between the genuine ALD-type reaction and CVD-like deposition for achieving a in place of the sufficient level of productivity. In addition to the reactor fabrication, which is largely dependent on the target film materials and adopted precursors, design and adoption of support equipment, such as vacuum pump and abatement tools, are another crucial part of the mass-production compatible ALD equipment. Ozone, which is of specific importance as the oxygen source for oxide ALD production equipment was also discussed.

Acknowledgments The author would like to thank Applied Materials, ASM International, Wonik-IPS, Edwards Ltd., TMEIC, and MRK International for their assistance in providing graphics used in this chapter to illustrate the key features of various equipment.

Reference

1. Suntola T, Hyvarinen J (1985, Aug) Atomic layer epitaxy. Annu Rev Mater Sci 15:177–195

Chapter 10
Erratum to: Atomic Layer Deposition for Semiconductors

Cheol Seong Hwang

Erratum to:
C. S. Hwang (ed.), *Atomic Layer Deposition for Semiconductors*, DOI 10.1007/978-1-4614-8054-9

The volume was incorrectly published with an additional editor (Cha Young Yoo) but there is just one editor for this volume: Cheol Seong Hwang.

The online version of the original chapter can be found under DOI 10.1007/978-1-4614-8054-9

C. S. Hwang (✉)
Applied Materials Inc, Santa Clara, CA 95054, USA
e-mail: cheolsh@snu.ac.kr

C. S. Hwang (ed.), *Atomic Layer Deposition for Semiconductors*,
DOI: 10.1007/978-1-4614-8054-9_10,

Index

C. S. Hwang (ed.), *Atomic Layer Deposition for Semiconductors*,
DOI: 10.1007/978-1-4614-8054-9,

R

GPSR Compliance
The European Union's (EU) General Product Safety Regulation (GPSR) is a set of rules that requires consumer products to be safe and our obligations to ensure this.

If you have any concerns about our products, you can contact us on

ProductSafety@springernature.com

In case Publisher is established outside the EU, the EU authorized representative is:

Springer Nature Customer Service Center GmbH
Europaplatz 3
69115 Heidelberg, Germany

www.ingramcontent.com/pod-product-compliance
Lightning Source LLC
Chambersburg PA
CBHW070920090925
32286CB00003B/77
* 9 7 8 1 4 6 1 4 8 0 5 3 2 *